D1127145

A Global Corporation

E. P. NEUFELD

A Global Corporation

A HISTORY OF THE INTERNATIONAL DEVELOPMENT OF MASSEY-FERGUSON LIMITED

UNIVERSITY OF TORONTO PRESS, 1969

SBN 8020 3204 4

I dedicate this book to my wife

Preface

The history of the development of a corporation may be written in more than one way. An author may, for example, have in mind a eulogy to management or he may decide to emphasize incidents of human interest and value. Perhaps in recent years the attempt to use company histories for creating or sharpening predetermined public images has tended to predominate. An author who is prepared and permitted to be completely objective must still decide what he is being objective about. Some delineation of the field of enquiry is therefore necessary, and, the conditions of freedom under which that enquiry is pursued must be understood.

This study of the growth and development of Massey-Ferguson Limited is intended to be both descriptive and analytical. It records unfolding events that helped to shape the company's character and structure, but it also analyses the company's successes and failures in adjusting to a changing national and international business environment. Reasons for successes scored and failures suffered are recognized. The study is not obsessed with personalities, but neither does it studiously ignore them; sources of major decisions are identified whenever this is possible. Developments of interest to industrial corporations in general are in several important instances outlined in greater detail than is usual in company histories. The study concentrates on the years after the Second World War, a period in which the company's international operations became increasingly complex, and for this reason it is not inaccurate to regard this volume as a history of the international development of Massey-Ferguson.

What were the forces behind the company's growth and why, at times, did it falter? How was it that the company developed into a global corporation, one that considers the whole of the international environment when making decisions relating to the allocation of its marketing, purchasing, manufacturing, and engineering activities? What have been the experiences of the company in the various countries in which it has become heavily committed? What adaptations became necessary for it to operate successfully in the international environment? What dominant characteristics has it acquired in the process? These are the kinds of questions that will concern us along the way, and the answers to them provide us with a theme – the development of a global corporation.

That development has not been simple. It has been much more than the growth of a single company in an expanding market. Because mergers also played a vital role, the history of other companies must be recounted, and the experiences of the local Massey-Ferguson companies abroad must be traced. Management reorganization and the introduction of new management planning and control techniques contributed a good deal to the company's development. My decision to combine description with analysis has made the exposition more complex, and the danger of becoming lost in a mass of detail was always present. Footnote references to internal documentation have largely been eliminated, for they seemed to be distracting.

The introductory chapter should help minimize the danger of becoming lost. It sketches briefly several major long-term trends in the development of the company and relates these to the development of the industrial corporation as such. It also outlines the fundamental economic forces that have formed the basis of the company's existence. The second chapter provides some historical detail relating to the Massey-Harris Company prior to 1939, and chapter 3 outlines the effect on Massey-Harris of its wartime activities. Then we enter the period that interests us most.

A discussion of the experiences of Massey-Harris up to 1953 is followed by a brief history of the Harry Ferguson Company, an order of discussion that facilitates a more complete understanding of the significance of the 1953 merger between Massey-Harris and Ferguson. The years from the merger to 1956, covered in chapter 9, were exceedingly difficult ones for the company, and after 1956 chronology has been sacrificed in favour of the separate and detailed treatment of the many important and concurrent developments that took place at that time. These developments included management reorganization; reorganization of operations in North America; additional capital commitments on the European continent; the acquisition of tractor manufacturing facilities at Coventry from the Standard Motor Company and diesel engine manufacturing facilities at Peterborough from F. Perkins Limited; the quite novel development of the Special Operations Division and the fascinating activities of that division in many countries; acquisition of manufacturing facilities in Italy; and reorganization of operations in South Africa and Australia.

The sacrifice of chronology in those chapters is compensated for by chapter 18. This chapter presents an over-all view of the corporation's development and of its financial achievements. In the concluding chapter we yield to the temptation of making some generalized comments on the character of the global corporation as seen from the experiences of Massey-Ferguson.

The conditions under which this research was undertaken were clarified before the project was begun. They were simply that I would have access to all company documents, that I would be given necessary assistance to accumulate required material, and that my name would be associated with published material only if that material represented accurately my own analysis and judgment. It must be said that Massey-Ferguson Limited on no occasion suggested any deviation from these terms. My salary, while I was on leave of absence from the University of Toronto, was paid to me by the University, and Massey-Ferguson reimbursed the University for that amount. All additional expenses were paid directly by the company.

It is not possible to acknowledge by name the many people of the parent company in Canada, and of the subsidiary and associated companies abroad, who have rendered assistance. But I can express appreciation for the willingness with which information was provided and the candidness with which the company's affairs were discussed. This candidness was nowhere more evident than in my discussions with Mr. A. A. Thornbrough, President of the company from December 1956. Also much appreciated was the willingness of Mr. J. S. Duncan, President of the company from March 1941 to July 1956, to relate to me at length his recollections concerning the company's history.

Specific acknowledgment must be made of the contribution to this project of the late Lt. Col. W. E. Phillips, who was Chairman of the Board of Directors and Chief Executive Officer of the company from July 6, 1956, until his death on December 26, 1964. The project itself was his idea, and instructions that it be attempted came from him. He read a preliminary draft of this book and seemed pleased with its general approach. I benefited greatly from discussing the company's affairs with him.

Sarah Kaskovitch, always cheerful, typed all the material, R. H. Johnston helped me collect information, G. S. Freeman proof-read the manuscript and compiled the index, and H. G. Kettle helped in many ways from beginning to end. To all of them I am grateful.

The University of Toronto is not to be forgotten. By granting me a year's leave of absence, it made it possible for me to undertake this study.

E. P. NEUFELD
University of Toronto

Contents

Tables

Charts

Organization Charts

A Global Corporation

CHAPTER 1

Introduction

The history of Massey-Ferguson Limited is, in an important sense, the history of the development and the growth of the industrial corporation. There are surprisingly few large business corporations in the western world that can make a similar claim. To reflect fully the history of the development of industrial organizations, a company would have had to survive a series of evolutionary changes extending over many decades, and it would today have to have the distinguishing characteristics of a large, modern, international, industrial corporation – one that stands at the forefront of the development of such organizations.

Massey-Ferguson generally fits that description. Through its predecessor companies it can trace its history to 1847 and so has survived long periods of industrial change. Beginning as a private local family enterprise, it was transformed into the legal entity of a public corporation, although it retained for a long while many of the managerial characteristics of a family enterprise. But eventually those characteristics had to disappear, to be replaced by ones that permitted the company to survive the major revolution in business organization necessitated by the seemingly inevitable increase in size and complexity of the industrial business unit. In the farm machinery industry Massey-Ferguson today is among the largest international corporations in the western world. That a substantial portion of this expansion was achieved through amalgamations is itself an accurate reflection of much industrial business history.

When using the phrase "large, modern, international industrial organization" we have in mind a corporation with certain characteristics. Three dominant characteristics of a modern industrial enterprise are its integrated nature, the international dimension of its operations, and the highly developed character of the concepts and techniques of organization and control that it employs to ensure efficient operation. It is integrated in the sense that it controls directly a relatively high proportion of the manufacturing activity that produces the products it sells, and also in the sense that it directly controls a high proportion of the marketing activity involved in bringing its products to its customers. The process of integration is itself the mechanism by which the optimum size of the industrial enterprise is achieved and this is made possible not merely by a reduction in the number of businesses in the industry but by expansion of the size of the market being served and, most significantly, by expansion of a corporation's market into the international environment. It is in this way that the corporation begins to acquire an international dimension.

That dimension has become increasingly complex. The development of export activity alone is a relatively new venture for many industrial companies, but what is even newer is the extension of a corporation's manufacturing and product development activity into the international environment. This is prompted sometimes by local trade barriers and sometimes by the incentive of lower production costs abroad. Regardless of the nature of the incentive, when a company decides to manufacture abroad it obviously believes that it can compete with local manufacturers. The competitive advantage of the large international corporation arises not merely (and sometimes not at all) from simple manufacturing economies of scale of operation, but from superior technology, including its management principles and practices, and from the size of its research and development budget – itself a result of large-scale operation. In effect, therefore, when the industrial corporation begins to manufacture some of its products abroad, alone or in association with local companies, it begins to be an exporter of technology as well as of goods. It receives income from selling both. This process of minimizing production costs and maximizing income through deploying manufacturing and engineering facilities in a number of countries, and so through exporting technology, is one of the most intriguing developments of Massey-Ferguson in recent years.

It is quite apparent then that there are different ways in which an industrial enterprise may be viewed as being an *international* industrial enterprise. It may simply be that the company sells its products outside as well as inside its home market; and the extent to which it has "internationalized" these operations can be seen by the completeness with which it has covered potential international markets, by the value of its export sales relative to its total sales, and by the degree to which the company sells abroad through its own local branches or subsidiaries or through independent distributors. In this respect, Massey-Ferguson, through its predecessor companies, has been an international organization for a remarkably long period of time. It began to expand its operations by selling abroad in a serious way in the 1880s, well before other North American implement manufacturers did, and even before most manufacturers in other industries took that important step. By 1908, 48 per cent of its sales and at least as high a proportion of its profits were made outside its own Canadian market – and this before it had developed a market in the United States.

But selling abroad is only the first stage in the development of an international business organization. The second stage, as far as industrial corporations are concerned, is to establish local assembly and manufacturing operations, beginning typically with a high proportion of component parts imported from the home factories and gradually replacing these with locally manufactured components. Typically also, this move is prompted by the desire to protect a market that has already been developed and has suddenly been threatened by the proposed imposition of tariffs, quotas, and exchange restrictions. These restrictions appear for varying reasons, but more often than not the encouragement of local industrialization is a prime factor.

A most important bridge for moving from the "selling" to the "manufacturing" stage has been the local sales organization, whether company branch or independent distributor. It frequently can supply local personnel, local contacts for capital, and knowledge of local government and industrial conditions relevant to beginning of manufacturing operations. A company with an experienced local sales organization has substantial advantages over one without such an organization when entering the stage of local manufacturing.

Massey-Harris, a predecessor company of Massey-Ferguson, was faced with the major decision of establishing local assembly and manufacturing operations abroad. It saw some of its export markets threatened by trade restrictions many years ago. These restrictions forced the issue of local manufacturing, although the company's first foreign acquisition (its purchase in 1910 of an implement manufacturing business in Batavia, New York) was motivated by slightly different considerations, as we shall see. It was in the late 1920s that Massey-Harris established its own assembly and manufacturing operations in France and Germany, and in 1930 it acquired an interest – although a minority interest – in a manufacturing enterprise in Australia. In each case the move was prompted by impending threats to its local market arising from government policy. Quantitatively these steps were not major ones in the company's over-all operations, and the centre of its manufacturing activity up to the Second World War remained in North America; but in terms of the evolution of the corporation they were important. They were followed by the significant decision of the company to begin manufacturing and assembling operations in the United Kingdom just after the Second World War, and by the purchase of a minority interest in a small local manufacturing company in South Africa in 1947. All of these moves were essentially of a defensive character in the sense already explained.

While these important first steps were being taken to move the manufacturing activity of the company into the international environment, a rather interesting development was occurring in the degree of "integration" in its manufacturing activity. Traditionally the company had been highly integrated in this area of its activity, for it manufactured almost all of its implements. Even the initial transition into the age of power farming involved no great compromise with this approach, for the company acquired its first successful tractor in 1927, the Wallis tractor, by buying the company that manufactured it at Racine, Wisconsin. But in 1938 it decided not to use its own engine in the new tractor models, and purchased one from another company; and during the war it decided to assemble tanks rather than both manufacture and assemble them at its Racine plant. These moves reduced the importance of the company's manufacturing activity relative to its total operations, a trend that was greatly magnified when the very important merger with Harry Ferguson Limited was effected in 1953 – for Ferguson manufactured only a minute proportion of the tractors and implements he sold.

After 1956 the company successfully reverted to its historical position of manufacturing a high proportion of the value of its products, principally by acquiring a diesel engine manufacturing company, a transmission and axle plant, and a tractor manufacturing plant. Also, after 1956 the company became an international company in two additional and important respects. It developed a policy of enquiring into and establishing local assembly and manufacturing operations as a means to establishing new markets in specific countries, and not just as a method for retaining markets already well developed through exports in past years; and it initiated a policy of deploying its own manufacturing operations internationally so as to minimize costs of production generally (including transportation costs) rather than merely to protect and develop particular markets. If, for example, some of the company's export markets could best be served by expansion of plant facilities in the United Kingdom, France, or Australia rather than by expansion of North American facilities, that would now be done. It was not any threat of trade barriers that prompted the company to purchase tractor and engine manufacturing facilities in the United Kingdom in 1959. The change from former times is epitomized by the fact that as late as 1951 the Board of the company decided that not more than 10 per cent of its assets would be located outside North America – a ruling that would be in direct contradiction to present company policy guiding deployment of manufacturing facilities.

One facet of that policy is the provision or "sourcing" of whole machines for particular markets from factories able to supply them at lowest cost. Combines, for example, are at present being manufactured in five countries. Another facet is a substantial inter-company movement of component parts. A prerequisite for this approach is interchangeability of component parts produced by various operations units, and such a policy is being pursued. In this way flexibility is achieved not merely by plant location, but also by the degree of manufacturing, as distinct from assembling, in the various plants. The latter means that manufacturing activity can subtly and swiftly be shifted from plant to plant without changing the type of machines being assembled in those plants.

An integral part of economically efficient plant location in the international sphere involves deploying the company's facilities so that it can in future compete effectively in emerging trading blocks – the most familiar one being the European Economic Community. By appropriate plant location and appropriate degree of manufacturing in the various plants, and with correct analysis of future tariff patterns, the company should be able to minimize the cost of its products in the various international markets in spite of changing trading areas devised by governments.

It is now more apparent what specifically we mean when we refer to Massey-Ferguson as a global corporation. It is an international company in the sense that its sales arise in a great many countries, its marketing activity excludes no potential market on earth, the international deployment of its manufacturing and engineering facilities is aimed at protecting individual markets, at developing new markets, and at minimizing global costs without arbitrary restrictions on the amount of capital invested in any particular market, and its marketing and manufacturing strategy is expressly designed to take account of changing international trading patterns. To operate effectively as an international corporation the company has developed a system of organization and control that meets the special requirements of its world-wide operations. Finally, because it emphasizes decentralization of decision-making and the development of local executive talent, it may properly be considered a multinational company.

Some of this is very recent history. A basic change in the company's approach to organization and control came after 1956. For industry in general great strides forward in this area were made in the 1920s when the management of General Motors Corporation realized that conscientious attention to organization and control in a rigorously logical way would become a prerequisite for coping with size and complexity. The persistent drive towards greater efficiency in the form of lower costs of production, and the development of multi-product organizations had created business corporations that bore little resemblance to the family foundries and machine shops of preceding years. The concepts that General Motors began to develop were received by a wider audience when Peter F. Drucker published his book, *The Concept of the Corporation*, in 1946. Curiously, such formalized concepts of organization and control seem for many years to have escaped the serious attention of Massey-Ferguson's predecessor company, in part perhaps, because of the impressive commercial success that it enjoyed for some years after the 1930s.

As a result of the increased volume of its business and the international dispersion of its assets, the company, even by 1953, seems to have needed new management concepts and control procedures. When the merger with Harry Ferguson Limited was effected in that year, problems of size and complexity were suddenly compounded and were aggravated by the emergence of other problems, including a depressed farm sector in North

America. It was a crucial period for the company. In 1956 a change in management began that was probably about as complete as any company has ever experienced.

While we must leave the details of these developments to later discussion, the broad nature of the change in management personnel may be commented upon now, for it is an interesting and important aspect of the evolution of the industrial corporation in general. Essentially what was involved was a change from the ambitious, loyal, self-trained and sometimes somewhat anti-intellectual men of great practical experience, who move by instinct in matters of organization and management, to the formally trained and highly mobile professional managers who fully accept and understand the developing concepts of business management, who are professionally trained in specific management fields, and who are at home in an organization that depends on detailed planning to achieve defined objectives. The managerial revolution began many years ago when the hired, but self-trained, manager replaced the owner in the management of business; now the professional manager is replacing the self-trained one. In this broad sense it is a change that few, if any, large corporations in a competitive environment can escape indefinitely. But they can plan for it, and so ensure that it will be evolutionary rather than revolutionary in nature as far as their own company is concerned.

It is rather interesting that, having lagged in applying emerging concepts of organization and control, the company then proceeded to place such emphasis on them that in several years it was probably at the forefront of the development in this area. The complexity of its operations arising out of its integrated manufacturing and marketing activities and out of its full commitment to the international environment seemed to demand innovation in organization and control. Some novel approaches to management were introduced, but, as we shall see, the results were not without turmoil.

What, it may be asked, is the fundamental economic basis for the existence and growth of this company and of the industry of which it is a part? Throughout its long history, the company has been in the business of manufacturing and selling farm machinery, although from time to time it has been concerned in minor ways with other products, and in recent years it has turned seriously toward the industrial equipment business and the selling of diesel engines. Having concentrated on farm machinery for so many decades it has naturally been intimately involved in the pronounced technological changes that mark the history of farm mechanization, and it is the mechanization of farms that forms the basis for its existence. Operating internationally the company finds itself in a number of markets that, appropriately arranged, would compose a wide spectrum of degrees of farm mechanization.

The forefront of farm implement technology has for many decades been in the United States. This must be explained not merely by the chance interest of a few individuals in the development of the technological arts, but also by a persistent shortage of labour that gave powerful impetus to that interest. Australia also experienced a shortage of labour and, at a relatively early stage, mechanized its farms with locally designed implements. The Argentine, too, turned anxiously to farm mechanization and provided markets for Australian and North American machines. Canada's proximity to the United States enabled her to draw heavily on United States farm technology, but Canada made its own contribution in the form of the self-propelled combine.

The shortage of labour that spurred farm mechanization and helped form the economic basis for the farm implement industry is not as simple a phenomenon as it may seem. Certainly labour was relatively scarce because of the abundant resources of new land. As the frontier was pushed back land became even more abundant. But an additional factor affecting the labour situation, even after new arable land was virtually exhausted, was the growing demand for labour in other industries. High productivity in these areas meant high wages and a better living standard, so that an incentive was provided to "pull" labour off the farms. This in turn induced an upward pressure on farm wages, thereby making it profitable for farmers to substitute machinery, or capital, for labour. During periods of war this process was seen in extreme form with the sudden and substantial increase in the demand for non-agricultural labour. The results of the process are reflected in agricultural statistics. In Canada implements and machinery formed 8 per cent of farm capital in 1901 and 14 per cent in 1951, and the change in the Province of Saskatchewan was from 9 to 26 per cent. Rural population was 68 per cent of total population in 1891, and only 37 per cent in 1961. In the United States there were about 1,000 tractors on farms, excluding steam engines and garden tractors, in 1910, and about 24,000,000 horses and mules. In 1956 there were no less than 4,500,000 tractors and only about 3,900,000 horses and mules. Weekly earnings of production workers in United States manufacturing industry, in money terms were 7¼ times higher in 1956 than they were in 1914, whereas farm wages were 6 times higher, so the "pull" of the former was probably strong.

The movement of farm labour into industry has probably been the most significant force behind farm mechanization. But it is also possible that from time to time mechanization has come about in another way – through the appearance of a major technological development in farm machinery that suddenly reduced costs of farm production and left some farm labour redundant. A farmer, after all, may have bought the first self-propelled combine not because the cost of threshing crews had increased, but because the self-propelled combine operated more cheaply than the crew, even at former wage levels. To put it simply, farm labour may have been "pushed" as well as "pulled" off the farms.

In North America the movement of people from farm to city is far advanced, although not complete. In many other countries in which Massey-Ferguson operates the movement to the city into industry is in its early stage. Those countries are vitally concerned with raising the standards of living of their people through economic development. Fortunately the adoption of mechanized farming by them does not depend on the slow process of technological development, for known technology is readily available through importation and, for many countries, through implement companies establishing local manufacturing operations. It is perhaps not surprising, but none the less remarkable, that Massey-Ferguson could introduce the advanced "know-how" of tractor production into countries such as Brazil and India in a matter of months.

With the company being committed to investigating possibilities of manufacturing anywhere in the world (the same process is at work in other industries) there are very few "non-financial" obstacles to the international transmission of advanced farm machinery technology. Indeed, and this we shall see time and again in succeeding chapters, the international corporation has become a major vehicle for the transmission of technology or "know-how"; it is a mistake to regard it essentially as a means for transferring financial capital. Its contribution of outside capital frequently represents a small portion of funds required for the project. Many countries are concerned when "foreign" companies become established in their industries without major injections of "foreign" capital. This view unfortunately ignores the principal advantage of such developments – the steady stream of managerial and technological "know-how" that is directed into those countries.

Much more difficult for a nation to decide is to what extent farm population should be "pushed" out through importation or local production of farm machinery, and to what extent it should be "pulled" out through industrialization. Probably no countries enjoy economic characteristics that permit them to improve their lot by concentrating entirely on farm mechanization, that is by building up an export surplus of agricultural products sufficient to import the "good things of life." Farm mechanization makes it possible for a decreasing proportion of the nation's effort to be devoted to feeding its people, and frees people to produce less elemental but no less desirable goods for production and for enjoyment. The means for producing these other goods and services – the education, the export markets, the capital, the cultural reorientation – must, however, also be developed. A programme of "balanced economic development" is usually recommended, to the point of becoming a cliché; and it is not a very useful phrase as any such programme will probably have to be balanced differently for different countries. But when fitted to meet individual requirements, it becomes a vital concept to many nations. The concept is also important to Massey-Ferguson. Its opportunities for marketing products in various countries and for establishing local manufacturing operations frequently arise out of governmental agricultural and industrial planning programmes. Thus the ability to work closely and in harmony with governmental authorities is a prerequisite for the success of the international corporation.

Although there is little doubt that the economic basis for the agricultural machinery industry is farm mechanization, the process is somewhat more complex than might be thought. This complexity has its own implications for the operation of an implement company. Farm mechanization seems to involve several frequently overlapping stages. Historically it first involved the mechanization of implements and not power; oxen, donkeys, mules and horses for a period remained a dominant source of power – as they still do in many countries of the world. In North America, beginning particularly in the 1920s, the farm tractor began to be widely adopted. This posed a difficult technological challenge for implement companies; it is one thing to build horse-drawn implements, and another thing to build a tractor and implements to go with it. Many companies were not able to meet this challenge, and Massey-Harris almost failed to do so. Still, in the end, farm mechanization proceeded and the tractor became the most important product of full-line farm machinery companies.

As mechanization proceeded, a point was eventually reached where the process was relatively complete. In the case of the farm tractor, for example, demand came initially from farmers who had never owned one or who wished to buy additional tractors; but then the stage was reached where essentially a "replacement" demand, that is one arising out of obsolescence or wear, largely accounted for the demand for tractors. It is one of the interesting aspects of the international operations of Massey-Ferguson that this time-pattern for a single country also appears when countries in which it operates are arranged in cross-section fashion at a given point in time according to the degree of mechanization in each. One of the attractions for the company in establishing markets abroad is that those markets hold out the prospects of a limited period of robust "new" demand for tractors as well as an unlimited period of stable "replacement" demand. In North America, for example, where agricultural mechanization is relatively advanced, the number of tractors used for agriculture increased by only 8 per cent from 1954 to 1964, while in the "developed" countries as a whole they increased by 58 per cent, and in the "developing" countries by 106 per cent.[1]

This "replacement" demand is itself more interesting than the word implies. It may be replacement of a given machine with a like machine because the first one can no longer be operated efficiently. It may be replacement of a machine with a different type of machine because of changes in the farmer's needs; or his requirements may change, for example through an increase in the size of farms – a most significant development in Canada and the United States, but evident to some extent in

1 See U.S. Government, Department of Agriculture, Foreign Agricultural Service, *Foreign Agriculture*, vol. 5, no. 14 (April 3, 1967), p. 5.

many other countries as well. It may be replacement of a machine because implement companies are able to offer more efficient ones. Finally in the more wealthy countries particularly, it may even be that in some cases replacement arises from a company's successful appeal to the farmer's taste for styling.

From the foregoing discussion it may be thought that the gentle evolutionary processes at work in the industry were equally gentle and inevitable in the evolution of Massey-Ferguson and its predecessor companies through the application of good reason and great light by men of intelligence. Alas, such is seldom the history of any corporation in spite of what corporate histories say. While the history of Massey-Ferguson reflects the history of the farm machinery industry, the latter can never be a history of any individual corporation in it. A theory that attempts to explain the behaviour of an industry can never be a theory that adequately explains the behaviour of a corporation. An industry survives and grows because it satisfies an identifiable and expanding demand. A corporation within a competitive industry survives and grows because in terms of price, quality, service, and product development, and through the intelligent application of skills, ideas, and energy, it can satisfy demand with sufficient efficiency relative to other corporations.

Survival of a corporation within an industry hinges heavily on the decisions taken by its management. Theories of corporations that do not emphasize the personal role of management in making and implementing decisions are not likely to explain satisfactorily the behaviour and success of corporations. The history of a corporation is therefore partly the history of decisions taken, some right some wrong, some only after much internal conflict. For Massey-Ferguson this has particularly been the case and it has made for lively and, occasionally for individuals involved, somewhat painful history. There is little that is painless and smooth about a change in management, or about the sudden application of new concepts, practices, procedures, and policies. Yet all these feature prominently in the long history of Massey-Ferguson. The interesting aspects of the history of Massey-Ferguson relate to the decisions that made it a large global corporation. The circumstances that led to those decisions being taken must be examined in the light of its experience at home and abroad. And finally – since it is very much a profit-maximizing organization – the financial success that has followed from decisions taken must be assessed.

We may now sketch briefly the profile of Massey-Ferguson, for this will assist us in maintaining perspective in the detailed story to follow. The net book value of the company's fixed assets totalled $179.4 million on October 31, 1966. Of those assets 18 per cent were located in Canada, 16 in the United States, 33 in the United Kingdom, 17 in France, 4 in Germany, 4 in Australia, 3 in Brazil, 2 in Italy, 2 in the Republic of South Africa, and 1 per cent in Mexico. Its world-wide sales of $932 million were widely distributed: Canada accounted for 10 per cent of the total, the United States 31, United Kingdom 11, France 11, Germany 5, Australia 6, Brazil 3, Italy 3, Republic of South Africa 3, Scandinavia 5, and all the rest 12.

The company has always concentrated most of its attention on the market for farm machinery. Mechanized farming involves the use of power supplied by tractors, grain harvesting machinery, now principally combine harvesters, and swathers; a great assortment of tillage and seeding equipment such as plows, harrows, disc harrows, cultivators, seed drills and planters; haying equipment including mowers, rakes, and balers; and a variety of other equipment used around the farm yard. In 1966 Massey-Ferguson's sales of tractors accounted for 42 per cent of its total sales. This percentage is similar to that of other full-line farm machinery companies, and explains why there is considerable emphasis in this study on the company's experience with tractors. Grain harvesting equipment accounted for 17 per cent of total sales, and the combine harvester in fact is the second most important machine in the company's product line. Hay harvesting machinery amounted to 4 per cent of the company's sales, other farm machinery to 9 per cent, and the very important business of spare parts sales to 10 per cent. Since Massey-Ferguson acquired the diesel engine firm of F. Perkins Limited of Peterborough, England, in 1959, the company has sold such engines to other, even competing, companies. These sales in 1966 amounted to $99 million or 11 per cent of all its sales. Sales of the industrial goods were 8 per cent of the total. Finally, the company has two office furniture and equipment plants in Canada, and sales of those products amounted to about 1 per cent of total sales in 1966.

The company faces a great range of competitors in the many countries in which it manufactures and sells its products, and its share of the market varies substantially from one market to another. In Europe, for example, Massey-Ferguson tractors accounted for about 18 per cent of all tractors sold in 1965, leading all other makes, and its major competitors were Ford (U.S. owned company), International Harvester (U.S.), Fiat (Italian), Deutz (German), and Renault/Porsche (French); but many other makes of tractors were sold as well, including John Deere-Lanz (U.S.), David Brown (British), Nuffield (British), Valmet (Finnish), Steyr (Austrian), Bolinder-Munktell (Swedish). Within Europe, Massey-Ferguson's share of the tractor market varied from a high of 35 per cent in Eire and the United Kingdom to 7 per cent in Germany, 25 in Denmark, 26 in Finland, 37 in Norway, 27 in Sweden, 22 in France, 14 in the Netherlands, 17 in Belgium, 4 in Spain, 10 in Austria, and 8 in Switzerland.

In the United States market Massey-Ferguson tractors in 1965 accounted for approximately 13½ per cent of tractors sold, and were exceeded by sales of John Deere, International Harvester, and Ford tractors; while in Canada Massey-Ferguson tractors had the largest share of the market with 23 per cent.

On a world-wide basis (excluding Communist countries and India),

Massey-Ferguson tractors seem to enjoy about 20 per cent of the market, the highest proportion of any make of tractor, followed by Ford, International Harvester, Deere, Fiat, Renault/Porsche, and David Brown. More Massey-Ferguson tractors were registered than any other make, accounting for about 20 per cent of all tractors registered.

In combines Massey-Ferguson had on the same world-wide basis about 18 per cent of the market in 1965, being slightly exceeded by sales of Claas combines manufactured in Germany. Third in size was Deere (U.S.), then International Harvester (U.S.), Clayson (Belgian), Allis-Chalmers (U.S.), Bolinder-Munktell (Swedish), Case (U.S.), and Braud (French), followed by a number of other local makes. As in tractors Massey-Ferguson's share of the combine market varies greatly from country to country. In Europe it stands second in sales, being exceeded by Claas of Germany, and in North America it is third after John Deere and International Harvester. It is obvious that Massey-Ferguson faces strong competition from both local producers and from international companies in virtually every one of its many markets.

Another aspect of the international profile of Massey-Ferguson is the location of its employees and factories. In 1966 the company had 46,040 employees of which 7,810 were in Canada, 4,914 in the United States, 18,803 in the United Kingdom, 5,746 in France, 2,640 in Germany, 2,288 in Australia, 1,249 in Brazil, 1,112 in Italy, 1,454 in the Republic of South Africa, and 24 elsewhere. The company operated six factories in Canada, eight in the United States, eight in the United Kingdom, three in France, one in Germany, two in Italy, three in Australia, one in South Africa, one in Southern Rhodesia, two in Brazil, and it was associated with companies manufacturing Massey-Ferguson machinery in India and Spain. All of the company's centres of operations made profits in 1966. How all this happened is the story of Massey-Ferguson Limited.

CHAPTER 2

Heritage

By the beginning of the Second World War, Massey-Harris Company Limited enjoyed to the full those advantages that accrue to a corporation with the passing of time; through its predecessor companies, its operations by that time had extended over ninety-two years. Massey-Harris itself was incorporated under Dominion of Canada charter on July 22, 1891, but it brought together Massey Manufacturing Company of Toronto and A. Harris, Son & Company Limited of Brantford, both Canadian companies. The name of the new company remained unchanged for sixty-two years until another significant merger, that with Harry Ferguson Limited, of Coventry, England, resulted in its being changed to Massey-Harris-Ferguson Limited on October 31, 1953, and to Massey-Ferguson Limited in March 1958.[1] The parent company still operates as a Canadian company, and a majority of its stock is held in Canada, but by any other definition it has become an international organization.

Both the Massey and the Harris companies had operated for many years prior to their union in 1891, and it is through the Massey company that the present organization traces its origin to 1847. In that year Daniel Massey purchased a plant and some equipment from R. F. Vaughan of Newcastle, Province of Canada West (now Ontario), and began to manufacture simple implements. The Harris company began operations ten

1 For the sake of simplicity we shall refer to the company as "Massey-Ferguson" even though for part of the period it should more accurately be "Massey-Harris-Ferguson." For a detailed account of the early history of the Massey-Harris Company see *Massey-Harris, An Historical Sketch 1847-1920* (Toronto: Massey-Harris Company Ltd., December 1920); *Massey-Harris, An Historical Sketch, 1846-1926* (Toronto: Massey-Harris Company Ltd., March 1926); Merrill Denison, *Harvest Triumphant* (Toronto: McClelland and Stewart Ltd., 1948).

years later, in 1857, after Alanson Harris bought a small factory in Beamsville, Province of Canada, and entered the farm implement manufacturing business. Both companies changed their location in the 1870s, Harris to Brantford in 1872, and Massey – since 1870 incorporated as the Massey Manufacturing Company – to Toronto in 1879. They were by no means the only companies in the farm implements industry. In 1861 the census reported sixty-one companies and in 1871, two hundred and fifty-two. Since by the time of their union the Massey and Harris companies accounted for over half of all farm implement sales in Canada, it is obvious that they had prospered to a degree that set them apart from others in the industry.

What was the secret of that prosperity? For the industry as a whole, the powerful economic forces leading to farm mechanization, which we discussed earlier, were at work. These were present in magnified form during the period of labour shortage and high prices of the American Civil War of the 1860s; the size of the total market was greatly expanded with the settlement of the plains of western Canada; and technological advancements, particularly in harvesting machinery, were available for general adoption by farmers.

For a number of years tariff protection in Canada played almost no part in the growth of the industry and yet the Canadian industry did not encounter competition from the United States industry.[2] Rudimentary transportation facilities, relatively low wages in Canada, and competitive steel supplies from England, combined to give Canadian manufacturers natural protection from United States competition. Two years after the increase in tariff to 17½ per cent in 1874, Mr. C. A. Massey wrote the House of Commons Select Committee on Depression of Trade that: " . . . the existing *tariff* is satisfactory to us, and is sufficient protection; perhaps even a little less would also be. A still further advance in the tariff would certainly prove adverse to our interest."[3] But other parts of Canadian industry favoured higher tariffs, and tariffs on farm machinery rose to 25 per cent in 1879 and to 35 per cent in 1883; but they were reduced to 20 per cent in 1894.

The success of the Massey and Harris companies during that early period resulted perhaps more than anything else from their success in remaining near the forefront of farm technology and in adopting with enthusiasm the selling techniques that were beginning to emerge. The first of these, the way they approached the matter of farm machinery technology, is of particular interest for it shows how much they benefited from "foreign" technological advancement – somewhat in the way that lesser developed countries today benefit when Massey-Ferguson or other

2 J. D. Woods & Gordon Ltd., *The Canadian Agricultural Machinery Industry*, Royal Commission on Canada's Economic Prospects (Ottawa: Queen's Printer, 1956).

3 Denison, *Harvest Triumphant*, p. 65.

industrial organizations establish manufacturing facilities there.

Almost all of the major technological innovations that the companies had to acquire came from the United States. In that country were developed many of the mowers, reapers, binders, and tractors that eventually led to the disappearance of the sickle, the scythe, the horse, and the mule. Both the Massey and the Harris companies secured access to those machines, and frequently modified them to suit local conditions. The Masseys had originally come from New England and this may have made it easier for them to return for ideas and innovations than it was for families without such bonds.

In 1851 Hart Massey visited American plants, attended field trials, and returned with rights to manufacture the Ketchum mower in Upper Canada – an important new grass-cutting machine. Similarly, the Manny combined hand rake reaper and mower, which Massey began to build a few years later, had also been developed in the United States. And John Harris was able to begin producing the very famous Kirby mower in the 1860s because he had acquired manufacturing rights for it from an American.

The Wood self-rake reaper which Massey began to produce in 1861, and later the reel-rake reaper introduced into Canada by Harris, originated in the United States, as did the Sharp's patent self-dumping wheel rake introduced by Massey, and the patents of the Toronto Reaper and Mower Company, which Massey acquired in 1881. The Wallis tractor acquired by the Massey-Harris company in 1928 was an American tractor. Thus it appears that fundamental engineering developments were not a prerequisite for success in the industry, but their acquisition in one way or another, and their adaptation and modification were. However, just before the Second World War the Massey-Harris company itself developed the first really successful self-propelled combine harvester for North American conditions. In view of the quite different way in which the other major new machines were acquired over the years, this development was indeed unique.

The success of the Massey and Harris companies in acquiring modern machines was accompanied by considerable attention to the art of selling in a competitive market. Displays at county and city fairs, field contests, "delivery day" parades, catalogues, newspaper advertising, price-cutting, and easy credit terms were the ingredients of selling efforts that at times reached fever pitch even before the turn of the century. In the wake of that activity were left many of the weaker companies. Larger companies absorbed smaller ones and growth through merger became common in the agricultural implement industry as in most other industries around the turn of the century.

One aspect of the Massey and the Harris marketing activity of those earlier years remains of lasting significance – its extension into foreign territories. Machines made by the Massey company were exhibited at the International Exposition in Paris as early as 1867 and were awarded two grand gold medals. Not long afterward the first export orders for mowers and reapers were received from Germany, although these were not followed by additional orders for some years. In 1885 Massey machines received considerable publicity at the International Exhibition in Antwerp, but the decisive move of the company into international trade followed contacts made by the Masseys and the Harris's at the Indian and Colonial Exhibition held in London, England, in 1886. Here Hart Massey met importers from France, Australia, Argentine, Chile, and Russia, some of whom soon became distributors for Massey machines. Massey established a branch of the company in London, England, in 1887, with Fred I. Massey as its manager until 1906, and also in Adelaide, Australia, in 1887, following a visit there of W. E. H. Massey and Fred V. Massey. The Massey catalogue of 1887 points to the whole of this development:

> We have, this year for the first time completed a systematic and efficient organization in Great Britain and on the Continent for the wider distribution of our goods. In England, Scotland, Ireland, France, Germany, Belgium, Russia, Asia Minor, South Africa, South America, West Indies, Australia, our machines are at work, and during the past season, have given remarkable satisfaction.

In a 1926 publication the company claimed that the Massey Manufacturing Company had been the first firm from North America to sell its products in Europe under its own name. What sort of influence led the Masseys to take an interest in export markets? We do not know. Perhaps the intense competition in the home market and the obvious financial advantages to be gained if the venture was successful explain it all, but these incentives do not explain why the Masseys were the North American pioneers in the development of the farm machinery export business.

Harris began to export in 1883 and greatly expanded its activity after 1886. Soon that company was selling machines as widely as the Massey company. Both took full advantage of international fairs, exhibitions, and competitions for publicity purposes, just as they were doing in Canada, and the list of prizes won is long. However, the export activity did not extend into the United States. United States duties on agricultural machinery until 1897 amounted to 45 per cent; they were reduced to 20 per cent that year, to 15 per cent in 1909, and entirely abolished in 1913. As long as the duties were high Massey-Harris could not profitably sell directly in the United States, and it did not attempt to establish manufacturing facilities there. In 1910 when United States tariffs had already been much reduced, Massey-Harris for special reasons did purchase a company in the United States, a development we shall deal with shortly.

We have already observed that growth through merger has been a common characteristic of most industrial corporations. For the Massey and Harris companies this process began before the turn of the century. The transition from family to public companies, from many small manufacturing companies to a few large companies, was under way at that time in many industries. In Canada, the merger movement in general began before the turn of the century and reached its first peak before the beginning of the First World War. Each merger had its own story, but behind most of them was the invisible influence of lower unit costs and higher labour productivity through an increased scale of operation.

For the Massey company the first instance of growth through acquisition of a rival came in 1881, two years after its move to Toronto. In that year the company experienced strong competition from a mower built by the Toronto Reaper and Mower Company, a relatively new company with American management, some American capital, and important rights to patents originating in the United States. Massey discovered that the financial position of the new competitor company was weak, and succeeded in purchasing it. By doing so it acquired not only a very successful mower but, what was subsequently much more important, patterns for a light twine binder with the revolutionary and exceedingly successful Appleby knotter.

It was the phenomenal success of another machine a few years later – the open end binder of A. Harris, Son & Company Limited of Brantford, Ontario – which was an important incentive for Massey to seek a merger with that company. The two companies had for years been competing with each other at home and in export markets, and in 1891 the Harris company was second only to Massey in sales in Canada. It was in that year that they merged to form Massey-Harris. Both companies were strong in harvesting machines, and the principal advantage of the merger arose from the economies of large-scale production and distribution that it facilitated, and possibly from improved control over profit margins at the manufacturing level. Longer production runs, elimination of duplicate agencies, more shipments by carload lots, better spare parts service, were the specific forms that economies of scale took in the case of the Massey and Harris merger.

Other mergers followed almost immediately. One with the Patterson Wisner Company, which had plants at Woodstock and Brantford, Ontario, and which itself was the result of a merger of two pioneer implement companies, occurred in December 1891. With this merger the Massey-Harris Company acquired many general farm implements to balance off its strong position in harvesting machines. In the year following, Massey-Harris acquired stock in the Verity Plow Company, Exeter, Ontario, and exclusive marketing rights to its products, and particularly its famous plow. However, for reasons of consolidation, that company's manufacturing operations were moved to Brantford. The Corbin Disc Harrow business of Prescott, Ontario, was acquired in 1893.

In 1895 the Bain Company, Woodstock, became a subsidiary company of Massey-Harris, providing it with farm wagons, sleighs, and other items, and its name survives to this day as a holding company wholly owned by Massey-Ferguson Limited as does the Verity name. The Kemp Manure Spreader Company of Stratford, Ontario, was purchased by

Massey-Harris in 1904, but operations were not continued in that city. To enter the field of stationary gasoline farm engines Massey-Harris purchased the Deyo-Macey Company of Binghampton, New York, in 1910, moved the machinery to the newly constructed plant at Weston (a suburb of Toronto) in 1916, and began producing such engines in that year.

These were the mergers and purchases of companies which located the company in Canada at Woodstock, Brantford, and Toronto, and, with the exception of the Woodstock plant which was closed down in 1966, they are also the locations retained by the company in Canada to this day. The Weston plant was acquired in 1916, expanded during the Second World War for war production, but disposed of soon after the war.

In 1910 the company also acquired its first farm implement manufacturing facility outside of Canada when it purchased a controlling interest in the Johnston Harvester Company of Batavia, New York; by 1927 it owned all the stock of the company. This significant move had both welcome and unwelcome repercussions, as we shall see. In 1928, Massey-Harris purchased the J. I. Case Plow Works of Racine, Wisconsin, sold the name "Case" to another Case Company, but for the first time acquired a tractor of good quality – the Wallis three-plow tractor. No more mergers or purchases of companies occurred prior to the Second World War, and no significant ones until the merger with Harry Ferguson Limited in 1953. The over-riding significance of these mergers was that they facilitated the transition to an organization of mass production, extended the product line of the company, and permitted it to acquire rights to important technological developments in agricultural machinery.

While the company in its earlier years was able to survive the trend towards larger manufacturing units by participating actively in the merger movement, it faced other problems not so easily overcome. Of these, competition from the United States industry, the appearance of the farm tractor, and the emergence of trade restrictions in export markets were the most important; and internally Massey-Harris had also to survive the transition from family management, control, and ownership, to management by hired managers and ownership by a diffused group of public shareholders. Unfortunately the Great Depression of the 1930s, with its particularly devastating effect on farm income, came at a time when the company was already struggling with these problems.

Competition from the United States was particularly important. The opening up of the West and the consequent growth of the Canadian machinery market made that market increasingly attractive to United States machinery companies, and the 20 per cent Canadian tariff from 1894 to 1907 (after which it was lowered progressively until by 1922 it stood at 6 per cent to 7½ per cent where it remained until it was increased in 1930) ensured that foreign competition, if it were to develop at all, would involve foreign companies establishing manufacturing facilities in Canada. The International Harvester Company (I.H.C.) – the result of a five-company merger in the United States in 1902 – took this step by building a plant at Hamilton, Ontario, in 1903. This kind of competition was not to be taken lightly when it is remembered that in 1902 I.H.C. controlled 90 per cent of the United States trade in binders and 80 per cent in mowers.[4] Massey-Harris, many times smaller in size, did not counter by establishing itself in the United States market. However, it did expand its other export activities and soon after the turn of the century these accounted for over half its sales and profits.

It is not clear why the company failed to become firmly established in the United States market. True, until after the turn of the century, United States tariffs clearly discouraged exports of machinery to that country. But neither did the company attempt to establish manufacturing facilities in the United States in an effort to circumvent the tariff. Perhaps it feared retaliatory measures from the strong United States companies in the Canadian market. The decision to manufacture outside of Canada was also a much more difficult one to make than to develop export markets for Canadian-built machines. When it did obtain its first manufacturing facilities in the United States it was, parodoxically, because of the fear of free trade and not because of any attempt to circumvent the tariff.

This is how it happened. Potential competition from the United States threatened to change with the 1910 campaign for Reciprocity, that is, for free trade between Canada and the United States. Massey-Harris seemed to fear that if the campaign were successful, its Canadian market would be threatened by a flood of imports from the United States. Such fears were not entirely ill-founded. To protect itself from such competition Massey-Harris hurriedly established itself in the United States by buying into the Johnston Harvester Company Incorporated of Batavia, New York, a company that had a few years earlier barely missed becoming part of the International Harvester Company complex.[5] There were other arguments that supported the acquisition of that company. The Massey-Harris factories in Canada were over-loaded and additional facilities would be useful. Also the Johnston Harvester Company had built up an important export business with which Massey-Harris was thoroughly familiar.

But in some respects the move appears to have been as ill-conceived as it was hurried. If costs of manufacturing were lower in the United States than they were in Canada it was largely the result of longer production runs, for wages were higher; and yet the Johnston purchase, by splitting the company's production into two organizations and two product lines, did not lead naturally to such economies. Therefore it is not at all clear how the purchase would improve the company's competitive position

4 See W. S. Phillips, *The Agricultural Implement Industry in Canada* (Toronto: University of Toronto Press, 1956), p. 14.

5 *Ibid.*

in Canada. Its position in the United States market also was not obviously enhanced through the acquisition of the Johnston company, for that company's strength lay in export territories not in the United States market.

Reciprocity, of course, did not come. Or more accurately, for the agricultural implement industry it came through the complete abolition of of the United States tariff in 1913, and through a series of reductions in the Canadian tariff, ending with the removal of all remaining Canadian tariffs in 1944. Consequently, the Johnston company, apart from transportation cost savings, had no direct cost advantage in the United States market over the Canadian Massey-Harris factories, it could not compete with the Canadian factories in the Canadian market and, as already mentioned, it was weak in the United States market. All this meant that the Johnston company depended almost entirely on its export trade. An adverse turn in that area would place it in a difficult position and could weaken somewhat the Massey-Harris organization as a whole. Even the Johnston export trade was fundamentally less advantageous to Massey-Harris than it appeared, for it involved Johnston brand implements competing with Massey-Harris brand implements. Not until 1928 was the Johnston business in France merged with the Massey-Harris business there, and not even then were the two product lines merged into one.

As it turned out the export business of the Johnston company permitted it to make good profits until 1920, but then its fortunes changed as exports suffered; a staggering loss appeared in 1921, reduced somewhat by an unusual and unwarranted upward revaluation of the Johnston company's assets. Many years of losses followed. The consequences of the lack of a strong distribution system were fully revealed and frequent shortages of working capital were also partly responsible for the difficulties. There was no evidence that Massey-Harris, through the Johnston company, knew how to cope with the United States market.

What this meant was that as Massey-Harris began to see its share of the Canadian market decline as a result of competition from United States companies, it was unable to offset this with profitable operations in the United States. This disturbing trend was made infinitely more disturbing as a result of the appearance of the farm tractor after 1910 – a machine that was in time to account for a high proportion of sales of full-line farm machinery companies.

In the United States the tractor population (excluding steam and garden tractors) increased from 1,000 in 1910 to 246,000 in 1920. Canadian farmers too began to turn to power-farming in force. Massey-Harris attempted to meet the demand by importing the Big Bull tractor in 1917 from the United States (see illustrations on p. 396) but unfortunately it was not a good tractor. Then in 1919, using its Weston plant it began to manufacture tractors modelled on the Parrett tractor belonging to the Parrett Tractor Company of Chicago. It sold these as Massey-Harris #1,

#2 and #3. But the venture was a failure and production ceased in 1923. The operation had not been helped by the complete removal of Canadian tariff protection on tractors valued at less than $1,400, but there seem to have been difficulties with the project itself. With that failure there was no longer any doubt that Massey-Harris, perhaps for the first time, was seriously lagging in a major area of farm machinery technology. When it is recalled that in the 1960s tractor sales accounted for just under half the sales of full-line farm implement companies, it can be appreciated how potentially serious this lag was for the future of the company.

Industry imports of tractors into Canada meanwhile mushroomed. In 1925 they totalled $2.6 million and in 1927 $8.6 million.[6] The company's position in other implements remained strong and these – the new reaper threshers, originally developed for the Australian market but introduced into the United States market in 1917, the binders, mowers, drills, rakes, harrows, plows, cultivators and other implements – gave it the means to survive the severe depression years of 1921, 1922, and 1923 and the ability to move into very profitable operations in Canada and many export territories during the rest of the 1920s. Emphasis on sales in foreign markets was maintained. Vincent Massey travelled to the new Russia in 1924 to see what could be done to partially restore the trade the company had had with Imperial Russia before the war, and local Massey-Harris sales companies were established from time to time in Europe, Africa, and South America.

In order to close the gap in its product line left by the failure of the Weston tractor project, Massey-Harris in late 1926 began to negotiate with the J. I. Case Plow Works Company for the right to market that company's fairly well-known Wallis tractor. Case apparently made an offer for the Massey-Harris shares held by the Massey Foundation and the Massey family, which together constituted controlling ownership of the company. The matter was discussed with the management of Massey-Harris. Thomas Bradshaw, Vice President and General Manager, received the news while on a train to Fort William en route to Australia.[7] Bradshaw returned immediately. Rumours of what was happening, mostly unfounded, led to bitter public discussion of the possible "sell-out" of the Canadian company to United States interests. Very soon J. H. Gundy, of Wood, Gundy and Company, Toronto, and Bradshaw formed a syndicate, purchased about 70,000 shares for approximately $8.0 million, and so control passed from the Massey family in 1927. The Masseys were therefore not put in a position of choosing between a Canadian bid and a much higher bid from United States interests.

In this way the Massey association with the company ended after eighty years and it had ended because Vincent Massey, recently appointed

6 See discussion in *ibid.*, chap. 7.

7 *Ibid.*, p. 65.

Ambassador to Washington, wished to avoid the possibility of conflict of interest.[8] However, he had already relinquished the presidency of the company in 1925, prior to his entry into politics in that year.

The common stock of $8 million was widely distributed after the $100 par value shares were split into four with no par value. With the Massey influence gone the Company provided an opportunity for new groups to become influential in its affairs. Until the 1940s, when E. P. Taylor became interested in it, its most influential Board member was J. H. Gundy, the company's underwriter. At the time of the change in control in 1927, the company also negotiated a bond issue of $12 million which enabled it to pay off bank loans and appeared to give it ample funds for financing future business.

The problem of obtaining a tractor had not, however, been solved, and negotiations with the J. I. Case company continued. In 1927 an arrangement was made whereby Massey-Harris was given the right to sell the Wallis Certified three-plow tractor in Canada and certain parts of the United States, and in 1928 Massey-Harris purchased that company for $1.3 million cash and the guarantee of $1.1 million bonds then outstanding. It immediately sold its rights to the name "Case" to the J. I. Case Threshing Machine Company for the sum of $700,000. With the acquisition of the Racine company, Massey-Harris acquired its first satisfactory tractor. It proceeded to reorganize its United States operations by forming a new company with $8 million capital and a Maryland charter – The Massey-Harris Company, with head office at Racine. As a result the Batavia implement factory and the Racine tractor factory became one organization. Also in 1928, the United States company sold its Johnston Harvester Company operations in France to the French Massey-Harris Company – Cie Massey-Harris S.A., a subsidiary of the Toronto parent company. The "decks" seemed to be clear for a new start in the United States market. But well before this the company had begun to observe disturbing developments in its important European markets.

One consequence of the company's early venture into and substantial penetration of export markets was that its prosperity was soon seen to depend increasingly on trade policies of governments which it could not hope to influence decisively. In the 1920s the company for the first time faced the alternative of either losing a valuable export market or extending its manufacturing activities abroad. It was a problem that the company was to encounter many times, both before and after the Second World War. The extent of the company's concern with this problem is reflected in the remarks of its General Manager, Thomas Bradshaw, who stated in 1925 that:

When . . . we contemplate that some of our most important markets in Europe and in the Southern Hemisphere are even now difficult or unprofitable for us

8 See Vincent Massey, *What's Past is Prologue* (Toronto: Macmillan, 1963), p. 118.

to operate in, on account of, mainly tariff barriers, transportation and packing costs, and cheap local manufacture, we are faced with the inevitable alternative of either taking courage and initiative into our hands and grappling with the situation in the establishment of local factories or of losing those valuable markets and the equally valuable assets that we possess at present in the prestige and reputation which we hold in our name and for our goods. The subject is one which has been engaging our most earnest and sincere consideration and must continue to do so, but it is a more complicated problem than might appear at first sight, for it not only involves finance of an important character, but it also is very closely linked up with our large plants and organization in Canada which are amply sufficient for the trades of those countries, but which must unquestionably suffer just as soon as the output is taken away from them and produced in overseas plants.[9]

In the mid 1920s such an alternative faced the company in France, after it had been exporting to that country for almost forty years. The first association with France began in 1887 when James S. Duncan, Senior, who had met Hart Massey at the Indian and Colonial Exhibition in London the year before, was appointed agent for Massey-Harris in Paris. In 1903 this agent was facing insolvency because of a trade depression and high inventories; as a result a partnership organization operating under the name "Société en nom Collectif Massey-Harris" was formed in 1903, with the former agent as manager.

The company's business in France accounted for 68 per cent of its European business in 1920, and so when European sales dropped from $9.2 million in 1920 to $3.4 million in 1922 it was obvious that the French position had to be investigated.[10] It was found that besides the usual cost disadvantages in freight and packing, the company as an exporter to France had suffered because of currency devaluation, import duties, and the emergence of local large-scale manufacturing. In addition, the company was prodded by the presence of the International Harvester Company as a manufacturer in France. Active consideration for establishing manufacturing facilities followed, including visits to France of the General Manager and a committee of engineering and manufacturing personnel; and it resulted in the formation of a French company, "Société Anonyme des Etablissements Industriels de Marquette" on October 27, 1925. This new company purchased a factory site of about 25 acres at Marquette-les-Lille in the north of France in November 1925, then built a factory to manufacture haying equipment, and by the end of 1926 its investment in plant amounted to $369,000. Expansion was soon undertaken to manufacture binders, since a new duty on these had prohibited imports, so that the investment by 1931 was $1,232,000, and over $4 million if current items such as inventories were included. The great

9 *Annual Address of the General Manager*, February 4, 1925, p. 18.

10 *Report of the President to the Shareholders*, March 15, 1921, p. 11, and February 16, 1924, p. 12.

bulk of that amount was apparently supplied by the parent company in Toronto. Marquette was chosen as the factory location because it was close to raw materials in France and Belgium, wages were relatively low there, and, being on the canal system, transportation facilities were good. International Harvester Company was already located there. The name of the company was changed to Cie Massey-Harris S.A. on January 26, 1927, and this it remained until 1954; in 1927 the company assumed responsibility for selling as well as manufacturing Massey-Harris products in France, and in 1928 for selling Johnston Harvester implements. An engineering department was also established, which for many years served all the company's European operations.

Production of the French company rose from $750,000 in 1928 to $1,600,000 in 1931, profits were made, and even at the depth of the Depression it appeared to the Comptroller of the parent company that "the French business appears to be on a sound footing if conditions improve."

Production naturally centred on those items that had previously enjoyed a strong market in France, principally binders and haying equipment such as mowers, rakes, and tedders. With the Depression, sales declined sharply from about $5 million in 1929 to $1 million in 1935, and losses occurred in each year of that period. But by 1935 improvement was visible, a small profit was made, and the worst was over. The factory had remained in operation throughout the Depression and was operating when war broke out in 1939.

Manufacturing operations in Germany began in a somewhat curious way, although, as with France, first associations had involved the sale of Canadian-made machines. The first export to Europe of the Massey company went to Germany in the 1860s but it was not until after 1886 that permanent associations were established with the German market through local representation. In 1897 a branch of the parent company, operating under German laws as "Massey-Harris Company Limited Filiale," was established in Berlin. This was replaced by a locally incorporated limited liability company, Massey-Harris G.m.b.H. in July 1902, a name retained until 1954. A large apartment block, in which the offices of the company were subsequently located, was purchased in February 1905 and operated as Industriestätte Charlottenburg G.m.b.H. As a selling organization the German company was well established by the beginning of the First World War.

Its first venture into manufacturing came by accident during that war. The interruption by the war of the flow of spare parts for Massey-Harris machines was regarded as harmful to German agriculture. The German government therefore built a small spare parts factory at Moabit, near Berlin, and placed the local Massey-Harris manager in charge. After the war the plant simply continued to operate, demand for its production expanded, and in 1925 additional premises were rented and some new machinery was installed. Production of new machines, principally mowers, trussers, and reapers, was undertaken in addition to the manufacture of spare parts. In 1927 the company's lease was about to expire, renewal was possible only for a portion of its property and this, at an increased rental; and since the buildings were not entirely suitable in any case the decision was taken to purchase another building. An unused war factory – previously employed by Messrs. Mannesmann to build lorries – with a floor space of about 5½ acres and land area of 36 acres was then acquired at Westhoven, near Cologne. Total cost was $155,000 and, with machines and equipment transferred from Berlin, production commenced in early 1928. The head office of the German company remained in Berlin until after the Second World War.

Things did not go well with the new venture. The low quality of the implements produced caused the company's reputation to suffer, and severe economic adversity soon depressed sales to very low levels. By 1932 losses were running as high as sales, in dollar value! Liquidation seemed imminent at that stage, and expenses for liquidation even appeared in the revised 1932 budget of the company. It was estimated that $1.0 million of the total parent company investment of $1.6 million would be lost if liquidation was pursued. But exchange controls of the new regime were disconcerting and rebuilding of lost prestige would be expensive. Yet in the end, it was decided to remain in Germany. By closing the plant at Westhoven for a period, and by sharp cuts in expenditures, losses were reduced so that even without any substantial recovery in sales a small profit was made by 1935. The German prohibition on the export of capital, however, induced the parent company to limit the business in Germany to the volume that the capital already invested there could accommodate.[11] Modest recovery continued but the parent company's contact with the company was lost in August of 1939 and was not regained until 1945. All in all, it had been neither a pleasant nor profitable experience.

In Australia, too, Massey-Harris began to fear the effects of proposed protectionist policies in the late 1920s, after many years of profitable business. A small number of harvesters had been shipped to that country in the early 1880s, although inadequate arrangements resulted in their not being sold. But as a result of the publicity received by Massey machines at the 1886 Indian and Colonial Exhibition in London, an Australian importer asked for a sample order of harvesting machines in that year. It was soon decided that Mr. Charles McLeod, of the Massey Manufacturing Company, should go to Australia with twenty-five binders and attempt to establish an appropriate selling organization. The Masseys themselves became interested in Australia and New Zealand. Walter and Fred Victor Massey visited there in 1887, established a branch at Adelaide with McLeod as manager, and appointed some agents. By the end of 1894 branch offices were operating in Melbourne and also in Dunedin, New Zealand, and by 1910 these had been extended to include Sydney, Perth,

11 *Annual Address of the President*, February 27, 1940.

Freemantle, and Brisbane in Australia, and Christchurch, Auckland, and Wellington in New Zealand.

This association with Australia had an important influence on the interest of Massey-Harris in harvesting machines. Its interest in reaper threshers or combines began when three Australians connected with the Massey-Harris Company sought to combine the essentials of the threshing machine with the American reaper. The first resulting reaper thresher or combine was made in Canada for the Australian market only and it was sold there in 1910. Many other reaper-thresher models were later developed by Massey-Harris. Interestingly enough it was also an Australian who designed the famous Massey-Harris self-propelled combine in the 1930s.

Ease of importation of Canadian-built machines had made it unnecessary for the company to consider establishing a manufacturing plant in Australia. This situation changed suddenly in April 1930, when the Australian government imposed severe restrictions on imports of farm machinery. Once more the company was faced with the inevitable alternative of either entering into local manufacturing arrangements or losing a valuable local market. Representatives of Massey-Harris went to Melbourne in August 1930 to investigate the possibility of local manufacturing. They came with the view that a suitable amalgamation of their sales organization with a local manufacturing company was preferable to becoming directly responsible for manufacturing in Australia. There were a number of reasons for this. Massey-Harris management did not want to add another plant to the eight that they already had, they were worried about managing a plant that was thirty travelling days and almost 10,000 miles away, they did not wish to add to their considerable investment in Australia, and they were concerned about their ability to find necessary capital in view of their uncertain financial position in Canada. In addition, they felt that in a period of low earnings their Australian investment would be unproductive for at least two years. The unfavourable labour and economic conditions in Australia were also disturbing, and they feared the financial consequences of any future lowering of Australian duties.

Soon after their arrival in Melbourne the representatives received a call from Mr. S. McKay, of H. V. McKay Proprietary Limited, Australia's biggest manufacturer of farm machinery at the time, and a meeting with him, with Cecil N. McKay, and with the McKay general manager was soon arranged. This was not the first time that Massey-Harris and McKay management had met, but it was the first time that discussion turned seriously to the possibility of amalgamating their Australian operations. Twelve days of negotiations ended in an amalgamation agreement. In one crucial respect it was an agreement signifying defeat and despair on the part of Massey-Harris management: according to the agreement, the company permanently sold its trade name in Australia to a company which it did not control. Twenty-five years later this deal caused moments of considerable anxiety for Massey-Ferguson management.

The basis of the amalgamation was that Massey-Harris would sell all its business assets and goodwill in Australia to H. V. McKay, and would buy stock in the H. V. McKay company amounting to 26 per cent of the company's total equity or 600,000 shares. This enabled Massey-Harris eventually to reduce its capital invested in Australia by about one-half while at the same time giving them an interest in a well-established Australian implement company.

Other important provisions of the arrangement were that the H. V. McKay company would be appointed sole agent in Australia and Papua/New Guinea for the sale of Massey-Harris agricultural implements for a period of twenty-five years; that H. V. McKay would have the right in perpetuity (it later seemed) to manufacture Massey-Harris machinery in Australia; and that the name Massey-Harris would be perpetuated in Australia by changing the name of the McKay company to H. V. McKay–Massey Harris Proprietary Limited. But the provision prohibited Massey-Harris permanently from operating in Australia through any other company bearing the Massey-Harris name. By an agreement of August 21, 1934, H. V. McKay–Massey Harris Proprietary Limited appointed Massey-Harris as their sole agent in New Zealand. This agreement could be terminated upon notice of one year.

In 1937 consideration was given by Massey-Harris to selling all or a portion of its investment in the McKay company as a way of earning capital urgently required to rejuvenate the Canadian and United States operations of the company. Massey-Harris was also concerned over the deteriorating exchange value of the Australian pound, and in 1938 they did sell 100,000 shares to the McKay family reducing their interest to 20 per cent or 500,000 shares. It remained at this figure until 1955, when Massey-Ferguson acquired the McKay company.

We have already seen that following the purchase of the Case company at Racine in 1928, and with it the Wallis tractor, the Massey-Harris company seemed to be in an improved position to make profitable progress in the United States market – something which had eluded it up to that time. The company opened additional branches in the United States, it added a two-plow tractor to the original three-plow Wallis and it continued to promote its reaper thresher. Sales of the United States company improved and amounted to a respectable $14.8 million in 1929.

Then came the Great Depression. While sales of the United States company declined only slightly in 1930, the deterioration in profits was staggering. In 1929 it had made profits of $435,000, but in 1930 its losses amounted to about $1.5 million. Sales fell to $5.8 million in 1931 and to $1.2 million in 1932, and by 1935 had recovered to only $3.1 million. We emphasize this development because it was one that was soon to cause a fundamental split on policy in the company, and almost led it to make what would certainly have been a fatal decision – to withdraw from the United States market and from Europe.

When the Depression began to hurt in 1930 the company had to retrench. The President of the company, Thomas Bradshaw, wanted to cease paying dividends on common and preferred stock in that year, but J. H. Gundy, a director and one of the larger shareholders, as well as the company's underwriter, objected to this. As a consequence Bradshaw resigned in the fall of that year. As it turned out, the company had no choice but to cease paying dividends in 1931 and Bradshaw's stand was vindicated.

The new President, T. A. Russell, then brought in B. W. Burtsell of the Packard Motor Company as General Manager. Burtsell's principal task at home and abroad was the unenviable one of ruthlessly cutting staff and wages, reducing inventories, collecting debts, and paying off loans. This he did most effectively. Even so the over-all annual losses on operations of Massey-Harris from 1930 to 1935 were $2.2 million, $4.0 million, $3.8 million, $3.3 million, $2.2 million, and $1.4 million, with the deficit on capital account in 1935 amounting to $22 million.

All the company's operations were experiencing great difficulties, but the difficulties of the United States company attracted particular attention. It had become a severe financial burden on the parent company. From 1920 to 1936 it suffered losses in fourteen out of sixteen years, and its net loss amounted to $8 million. The Johnston name in exports was no longer very useful, and sales of the company were running at a fraction of their 1928 and 1929 levels. Nor was the parent company any longer in a position to support its U.S. operations from profits made elsewhere. To the difficulties of the French and German operations in the early 1930s was added a debilitating experience in the market the company knew best – the Canadian market. Depression and drought reduced sales there from $13.5 million in 1929 to $3.1 million in 1932, and heavy losses resulted.

These were the circumstances that led the Board, apparently in early 1935, to decide to liquidate the company's business in the United States. It made plans for the Batavia plant to be demolished in order to save taxes, and orders were given for the machine tools to be shipped to the Massey-Harris plants in Canada.

Just about that time, in January 1935, James S. Duncan, who had earlier been sent from the company's European operation to reorganize the Argentine business, was asked to come to Toronto to become General Sales Manager. He had joined the company in Paris when he was seventeen, had worked for it in France, Canada, and Germany as well as Argentina and spoke several languages. Therefore he was thoroughly familiar with the international environment, in sharp contrast to the company's Board members. When he arrived in early March he heard about the decision to liquidate the U.S. operation under rather special circumstances. As he recalls it, two days after his arrival he was invited to lunch by Burtsell. In a suite at the Royal York, Burtsell told him of his plans to liquidate the United States operation, to close down the European business (except for England), including especially the factories in France and Germany, and to confine the company's operations outside Canada to a few export territories such as the Argentine and South Africa. Duncan directly and without undue tact expressed his opposition to such a move. Burtsell, in a rage, and after offering some intemperate comments, stalked from the room.

The President, of course, heard of the encounter, called Duncan in, and asked why he had quarrelled with the Vice President and General Manager so soon after his arrival. Duncan explained, as he had done to Burtsell, that he thought the worst was over in the United States and elsewhere. While still in the Argentine he had found a small market for surplus tractors in the United States which, he felt, was a sign of better times. Russell asked him to examine the company's United States business.

Duncan toured the United States operation, came back with his mind unchanged, and reported to Russell. He pointed out that while the net worth of the company in the United States in 1935 was about $6.4 million its salvage value was only about half a million dollars. Even this could disappear through fixed charges, such as taxes, if the salvage was not realized quickly. There seemed to be some indication that sales, along with farm income, were creeping off their Depression lows, and that given stringent control of costs the United States company would soon cease being a drain on the parent company.

The President then asked Duncan to prepare a budget for 1936. It showed that without closing any of its operations the company should be able to confine its losses to $250,000 in that year. Russell then asked Duncan to present his views and budget to the Board, which he did. Burtsell at the same meeting expressed his strong disagreement. The Board, following the suggestions of J. H. Gundy, agreed to accept Duncan's recommendations provided it was understood that if losses were in excess of $250,000 he would resign. Duncan accepted that understanding and the meeting ended. Telegrams were sent to Batavia halting the removal of machinery. In December 1935 Duncan was appointed General Manager. Burtsell resigned from the company. Losses in 1936 were much less than $250,000.

Those were momentous days. To have closed down the company's operations in the United States would not merely have removed Massey-Harris from what in normal times was the richest market for agricultural implements in the world, but it would also later have denied it very profitable war contracts; and perhaps most important of all it would have left the company without a tractor at a time when the tractor was becoming, indeed had become, of crucial importance to the success of large farm machinery companies. The company had not solved any of its immediate problems by not pulling out of the United States market, but the decision to stay in the United States and try again to become established must in retrospect be regarded as one of great importance to its development. It was a decision that owes much to the injection into the demoralized company at that time of the infinite self-confidence and optimism – some would also add arrogance and egotism – of the new and internationally minded General Manager, James S. Duncan.

The worst of the Depression as far as the company was concerned was over by the end of 1935, for its accumulated deficit did not increase after that and sales improved. But to make later developments more intelligible it is rather important to recognize that not all the company's difficulties resulted from the Great Depression. After all, its share of the Canadian market was only about 19 per cent in 1935, compared with about 50 per cent before the turn of the century. One must be wary of the tendency of management, in a period of economic adversity, to explain all reverses by pointing helplessly to external forces.

In this instance the truth was that Massey-Harris had failed to emerge successfully from its efforts to enter the age of power-farming. Its experience with manufacturing and marketing tractors is of particular importance because of the way this influenced its operations after the Second World War and because of the implications it had for the merger with Harry Ferguson Limited in 1953.

As we have seen, the company had been late in offering a successful tractor to farmers. Its attempt to build tractors at the Weston plant from 1919 to 1923 was a failure, and not till 1927 did it begin distribution of the three-plow 20-30 Wallis Certified tractor – fortunately a good tractor, with its distinctive U-shaped frame, having been developed by a company that had experience in that field since about 1912. (See illustration on p. 396.) This was followed in 1929 by the 12-20 Wallis two-plow tractor.

This was a period of swift development in agricultural tractor design. The United States manufacturers led the field, and the United States farmers led the change from animal to tractor power. Those farmers in the 1930s began to shift towards row-crop tractors (a tractor with high clearance and front wheels close together), a type which the International Harvester Company had successfully introduced in 1924, and to general-purpose tractors which could be used in some row-crop fields. In 1930 Massey-Harris, instead of moving right into the row-crop tractor market, introduced the fearsome-looking Massey-Harris General Purpose tractor with four-wheel drive and wheel settings to conform to a range of row spacings. (See illustration on p. 397.)

In spite of an improved version appearing in 1936 in the form of the MH-4W.D., sales proved disappointing and the tractor was soon dropped. Developmental expenses in relation to sales had been high indeed, at least twice as high as for the Wallis tractor. Another tractor, the MH-25-40 or model 25 introduced in 1931, a 3-4 plow tractor, did not seem to catch the fancy of the market either. Undeniably, in the early 1930s, that period of great uncertainty for the company, Massey-Harris tractor development lagged once again. The impetus it received from the purchase of the Case company had been lost.

Without row-crop tractors the company could not and did not sell an adequate range of row-crop implements. This denied it the rich market of the corn belt area in the United States where row-crop farming is important. Further limitations on the company's competitive position in tractors and implements resulted from failure to introduce high horsepower tractors, including crawler tractors. Nor had Massey-Harris developed a light narrow-cut combine as its competitors had done.

Besides all this, the company's United States subsidiary was in a demoralized state. The long period of uncertainty prior to 1936, an uncertainty arising from persistent threats to its existence, left it in a most unenviable position. Product development, as we have seen, had suffered severely. There does not seem to have been any sales leadership. Indeed there was not even a sales manager, and so the general manager had to serve as sales manager in addition to his other responsibilities. The extent of the demoralization of the sales force as the result of an inadequate product line and absence of sales leadership can be imagined, and this state of mind extended to factory management and personnel. Poor factory work was lamented. From 1933 to 1935 the Batavia plant was virtually dead. Besides all this, the United States Company suffered from frequent changes in management, and it was not always fortunate in the choice of its general manager.

Finally, there was the matter of company organization and personnel policy. Vincent Massey, commenting on organization during the period when he was President of the company, 1921-25, has written:

> . . . Looking back on this period it is extremely hard to realize that there was little, if any, departmentalization in the Company. There was no well-defined sales department, or department of production. Officers of the Company moved easily from one division to the other. No firm tradition of practice developed in any of them. This was perhaps less a defect in management than it was a tribute to the versatility of the men who had run the Company in the past. . . .[12]

And in speaking of the quality of the Board of Directors:

> . . . I cannot say that I was very happy with the board of directors of the day. They were, most of them, men whose experience had been narrow and, as a group, they were not equipped to direct the fortunes of a company that was about to become a great international corporation. I tried to interest my fellow directors in a plan for bringing into the business young men who had had a first-class education. There was no warm response to that. The principle I had in mind was that if a man brings a trained mind to his work in industry and commerce or, indeed, in almost any occupation, he will make faster progress than those who have not had that advantage. But the thought of taking a uni-

12 Vincent Massey, *What's Past is Prologue*, p. 79.

versity graduate and employing him in the business was something the board would not comprehend.[13]

As far as one can tell this essentially anti-intellectual attitude was not confined to the Board, but permeated much of the company for many years. So while there is no denying the cataclysmic effects on the company of the forces of the Great Depression, it is also apparent that management in preceding years had failed to meet a number of challenges.

In spite of all adverse developments and shortcomings, improvement in the company's position began to be visible after 1935. Persistent reduction in expenses had cut losses and with sales beginning to rise in all of the major international markets, profits returned to its manufacturing operations in Canada, United States, France, and Germany.

A new group of executives which was to run the company for the next two decades was coming to the fore. It was for the most part composed of men who, not unnaturally considering the state of management science at the time, lacked formal training, emphasized the superiority of the self-made man of practical experience, and as far as one can tell were enthusiastic, ambitious, and optimistic. By 1944 these men were to hold the following positions:

J. S. Duncan, President
W. K. Hyslop, First Vice President
C. N. Appleton, Vice President and Secretary
G. T. M. Bevan, Vice President in charge of Engineering and Research
H. H. Bloom, Vice President in charge of Sales
E. G. Burgess, Vice President in charge of Manufacturing
W. Lattman, Vice President in charge of Supply and Control
R. H. Metcalfe, Vice President and Comptroller
M. F. Verity, Vice President

Almost all had spent all their lives in the company, a few had engineering degrees, and all but Hyslop, who had retired by then, were still senior and influential executives in early 1956.

Improvement in farm income in the later 1930s lay behind sales increases, but sales were also assisted by the modest steps the company was taking, after a period of stagnation, in bringing new models of tractors, combines, and implements onto the market. In 1936 the Challenger, a 2-3 plow tractor with rubber or steel wheels and, significantly, in row-crop or standard model, was introduced, as was the standard model Pacemaker. The latter was subsequently marketed as the Twin Power Pacemaker ("twin power" involved increasing the horse power of the tractor by pushing a lever to increase the rpm's of the engine), and both were transition models for they retained the U-frame of the Wallis. They were displaced in 1938 and 1939 by the #101 and its successor, the #101 Junior and Senior. These models incorporated automobile-type engines

13 *Ibid.*, p. 76.

Peter Varley

Fall ploughing in Southern Ontario.

and so differed basically from preceding models. With them the U-frame disappeared, replaced by the independent frame. In 1941 the #81, a wartime tractor used at airfields, was added to the line, and temporarily a few 200-series tractors as well. No further basic model changes occurred until a new line of tractors was introduced in 1946.

One aspect of the company's tractor development may be commented upon in some detail, for it represented a new trend that was later to be rather troublesome. This was the trend towards subcontracting and so towards reduced control over sources of supply. Traditionally the company had manufactured a high proportion of the value of the products it sold, but the move to technically more complicated machines, which came swiftly when the tractor was introduced and motors began to be mounted on harvesting machines, posed new manufacturing problems. These the company initially circumvented when it purchased the Racine plant, for it had engine-making and other machining facilities as well as assembly facilities.

However, in the 1930s the trend away from the heavy tractor engines to the lighter automobile-type engines began, and in Massey-Harris this change was evident when it decided to put the L-head Chrysler-built six-cylinder motor into its new #101 tractor in 1938. Formerly its own engine had been used, but apparently the cost of developing an engine and of retooling for it at Racine at a time of low sales and low profits, not to mention the pressure of subcontractors anxious for orders, lay behind the decision to buy an outside engine. This trend towards subcontracting, we shall see, was accentuated in the war period when the company made the decision to assemble, rather than machine and assemble, tanks at Racine. By the end of the war the company had become quite accustomed to the subcontracting approach, and decisions, one-by-one, which increased or perpetuated the outside content of its products, came easily.

Product development in the mid 1930s was not confined to tractors. The company also developed the famous Massey-Harris Clipper combine, a straight-through narrow-cut machine which had a substantial influence on the trend in combine design. It was the clipper combine that put the Batavia plant back on its feet for that plant manufactured clippers from 1938 to 1952, an unusually long period for a basic model. A few other new implements also were introduced into the line at that time, including the semi-mounted mower and the power-take-off corn binder. But the most important machine developed by the company in the 1930s, indeed the most important machine that Massey-Harris ever developed, was the self-propelled combine – a machine that was to replace not only the binder and stationary threshing machine but, to a large extent, the pull-type combine as well.

The long period of development of the self-propelled combine, one in which various independent inventors participated, cannot here be traced. But the company's role in it may briefly be noted. The development of reapers and binders to cut grain, and of threshers to separate seed from straw, made great strides in the nineteenth century. Massey and Harris binders were part of that development. Attempts at combining the two processes of cutting and threshing were also made in that period. One such machine operated in Michigan in 1836 and other individual efforts followed. In Australia, a machine that stripped off the heads of grain and threshed them with a beater – the Ridley stripper harvester – had been used since 1845, and in 1884 H. V. McKay added a separating mechanism, patented the machine, and, for all its deficiencies, had in that way created a "combine."

Massey-Harris by the turn of the century had a good market in Australia, saw the need for a stripper harvester and, borrowing heavily from various Australian machines, designed one by 1901 for the Australian market. In 1910 it introduced its #1 reaper thresher which depended not on a stripping mechanism but on the North American type of cutter bar to cut the straw and a reel to guide the straw into the knives of the cutter bar. It was a combination of the North American reaper and threshing machine, and represented a giant step forward. (See illustration on p. 410.) A long line of successful Massey-Harris reaper threshers followed but all were pulled by animal or tractor. Initially sold extensively in Australia and Argentina, they were introduced into North America in 1917. In 1924 Headlie S. Taylor developed the 12-foot Sunshine auto header in Australia for the H. V. McKay company, which probably was the first self-propelled combine to be more or less regularly manufactured. It was sold in Argentina where it may have encouraged local experimentation with the self-propelled idea.

Tom Carroll, also an Australian, joined the Massey-Harris distributor in Argentina in 1911, saw local experimentation with combines and began a long career of designing such machines for Massey-Harris. His observations in Argentina of "blacksmith shop" attempts to build self-propelled machines led him to believe that a self-propelled combine designed to suit North American conditions might be successful. This induced him and E. Abaroa, General Manager of the Massey-Harris company in the Argentine after 1935, to attempt to convince management in Toronto to permit developmental work on such a machine to go forward. Management in general opposed the project, North American sales employees were completely skeptical, but Duncan, as General Manager, supported it, and approval was finally granted in 1936. Duncan, of course, had known Carroll well when he was manager of the Argentine Branch before 1935 and had personally seen that Argentine farmers were interested in the self-propelled idea. Carroll undertook the work in Toronto and had a prototype in the field in Argentina in eight months. The historically significant #20 self-propelled combine was born. (See illustration on p. 412.)

Eight machines were built, sent to Argentina, placed in the hands of outstanding farmers, and in spite of their large size (16-foot cut) and high cost, their performance convinced local observers that success was certain. A total of 925 of this #20 combine were manufactured but a lighter, smaller, lower-cost model was meanwhile developed. So emerged the famous Massey-Harris #21 self-propelled combine. In spite of the war, the company decided to put the new model into mass production, and the first units came off the line in 1941. It was a move that was soon to give the Massey-Harris company a greater lift than could ever have been imagined at the time, and that lift came first in the rich United States market which had eluded it for so long. For his work on combine development Tom Carroll was awarded the Cyrus Hall McCormick gold medal by the American Society of Agricultural Engineers in 1958.

As the Second World War approached the company's position was one of both strengths and weaknesses. It had accumulated a staggering deficit but also a self-propelled combine that was to prove remarkably successful; it was still an insignificant company in the United States market, where at least six other companies had a larger sales volume than it did, and there was still no real assurance that it could permanently operate profitably in that market; it had managed to persevere (but only just) with its manufacturing operations in the United States, France, and Germany, but political instability in Europe cast grave doubts on the company's prospects there; it had not won its place in the tractor market and consequently had lost a significant portion of the Canadian market, but it was staying in that race; its concept of management and organization had not changed, but it was beginning to be run by men who were optimistic – a not insubstantial factor considering that the shadow of Depression had hardly passed.

As significant as anything, it had accumulated wide experience in international markets. In 1939 Massey-Harris derived 35 per cent of its total sales from business in Canada, and over 40 per cent from its business outside of North America. There is no doubt that by the time of the Second World War the name "Massey-Harris" had meant something to many farmers in many parts of the globe for many years, and that income from export sales was of crucial importance to the company. Through its dealers, distributors and branches it had developed an international system of distribution and a channel of commercial intelligence that was to be most useful to it in the post-war expansion of the dimension of its international operations. The early decision of the Masseys to "go international" as far as selling was concerned was therefore one of the most important ones the company ever made, as was the decision to "stay international" in 1935.

CHAPTER 3

Impact of the Second World War

The Second World War affected Massey-Harris in a number of areas, particularly in its financial position. Payment of a dividend does not normally qualify as a momentous occasion for most old and well-established companies and yet by the end of the war circumstances had made it so for Massey-Harris. On April 15, 1946, the company paid its first dividend on common shares in sixteen years. Obviously things had changed. What things had changed and how had they changed?

The economic world had changed, as a result both of the disappearance of the 1930s Depression and of the economic consequences of the war. It is ironical that for many nations, many people, and we must add many companies the period of the Second World War provided more material rewards for effort and a more specific and deeper sense of purpose than they had experienced in many years; while at the same time the experiences of many other nations, organizations, and individuals sadly went to another extreme.

In a sense Massey-Harris, by being located both in North America and in Western Europe, experienced both extremes. But because its manufacturing facilities were at that time located mainly in North America, far from the threat of bombing, its loss of facilities and markets in Europe was minor in comparison to the benefits derived from its North American operations during the war.

Sales in 1939 were $21.0 million, but in 1946 they were $66.7 million (excluding war production of $5.7 million) and the number of employees had risen from 5,141 to 11,321 over that period. The company's assets had increased from $35 million to $52 million; its profits before taxes totalled $4.4 million or five times larger than in 1939. Sales of the company in 1946 were limited not by the volume of orders received but rather by the company's capacity to produce. There was no doubt that the resumption of dividend payments did reflect a material improvement, and that this improvement suggested a return to the prosperity which, apart from the war years, the company had not experienced since the 1920s.

Production at over-full capacity during the war, while of a very special and perhaps temporary character, nevertheless did inject into the company a feeling of optimism and enthusiasm that can come only from a growing concern. Prolonged and painful discussion by management of sales declines, close examination by them of collection reports, and endless rationalization of losses sustained had given way to discussion of material shortage, plant expansion, and profits taxes.

However, curiously enough, until 1941 it was the question of the company's capital reorganization, not its war effort, that attracted the greatest press and public attention.

Arguments in favour of a reorganization of the capital structure of the company arose simply because heavy losses during the Great Depression, including the loss of its German assets had, by November 30, 1939, created an accumulated deficit of $20.7 million. This amount had risen to $21.3 million by November 30, 1940, because of the need to write off properties under enemy control in France, Belgium, and Denmark. The prospects of soon removing that deficit through earnings were not very favourable, and as long as the deficit existed the company was legally forbidden to pay dividends on its preference and common shares. Preference shareholders became restive because they, like the common shareholders, had not received a dividend since 1930, and it was a group of preference shareholders that demanded action from management on capital reorganization.

Such reorganization was actively considered by the Board as early as June 1939, although it was urged by preference shareholders even before that. Action was deferred, however, and management justified further delay by pointing to the uncertainty which war reverses had cast over the value of assets and business on the European continent. This uncertainty was removed when France fell, for the company was then obliged to write off all its assets on the continent.

The company's next annual meeting was to be held on March 25, 1941, and a group of preference shareholders headed by James A. Gairdner, of the investment firm of Gairdner & Company, Toronto, prepared itself for a show-down with management. A letter sent to these shareholders and signed by A. G. Walwyn of Gairdner & Company, said that "it is time that some action was taken. . . . Renewal of dividends would not only give preferred shareholders immediate income but would assure much higher market values." Proxies were asked for in the names of E. J. Bennett of George A. Touche & Company, Toronto, Brian N. Barrett, an investment executive representing certain interests, and James A. Gairdner. Originally in 1929, a group headed by Wood, Gundy and Company had underwritten the Massey-Harris preference stock issue, and because of the financial crisis had had to take most of it up themselves. In the later 1930s Gairdner & Company apparently took over a large block of the issue from one of the original underwriters.

The annual meeting of March 25, 1941, was heavily attended by shareholders. The new President, James Duncan, explained that the company needed its earnings to finance its operations and that resumption of dividends would have hampered its development and restricted its operations. Action towards reorganization would be taken as quickly as possible.

Mr. Gairdner, representing a substantial group of preferred shareholders, argued for immediate reorganization. He had not accumulated sufficient proxies to force the issue, but he indicated that if management did not take action within ninety days he would seek subscriptions from sufficient shareholders to call a special meeting. Others pointed out that the deficit could easily be wiped out and dividend payments resumed by writing down the value of the common stock. With the proxies they had, the threat of the group of preference shareholders to call a special meeting was not an idle one.

Management took action. A plan was completed and circulated, and a special meeting of common shareholders was called for November 27, 1941, to consider it. Well before the meeting it became evident that the group of preference shareholders supported it – Wood, Gundy had assisted Massey-Harris in working it out – and also that Massey-Harris management with the help of their underwriters were actively seeking proxies in support of the plan.

It also became evident before the meeting that a group of common shareholders was going to provide some opposition to the plan. They called themselves the "Massey-Harris Co. Common Shareholders Protective Committee." This group, under the names of H. G. Stapells, K.C., and Avern Pardoe, asked common shareholders for proxies to vote against the plan proposed by management and supported by the preference shareholders.

The plan did have some interesting features with important implications for common shareholders. It envisaged a reduction of the share capital by $24.4 million through the exchange of 739,622 common shares standing on the books at $26.8 million for 732,508 new common shares valued at $4.8 million; and the exchange of 120,899 5 per cent cumulative convertible preference shares on the books at $12.1 million for 483,596 6¼ per cent cumulative convertible preference shares valued at $9.7 million. This reduction was sufficient to eliminate the accumulated deficit of $21.3 million, to cover other contingencies of $1.0 million, and to leave a surplus of $2.2 million. The payment of dividends on the new preferred shares would be cumulative with respect to dividends up to $1.25 a share for five years and three months, to the extent of net earnings in that period. The minimum accumulation in any year was to be 75 cents per share and payment of dividends would be mandatory to the extent of 75 per cent of net earnings. After November 30, 1946, the preferred shares were to become fully cumulative. The company could, during the five and one quarter years, pay dividends in cash or 5 per cent debentures, or both.

That part of the plan was not controversial. But the way in which arrears of dividends on the old preferred shares were to be settled and the nature of the conversion feature of the new preferred stock caused the organized group of common shareholders to become quite emotional. Since 1930 accumulated dividends on preferred shares had risen to $6.0 million. The plan envisaged settling this by giving preference stock shareholders 49.5 per cent of the new issue of common stock, leaving 50.5 per cent for the old common stock shareholders. Also each preferred share was convertible into two common shares as long as 322,399 or more preference shares were outstanding, and into one and a half common shares as long as less than that number but more than 161,200 preference shares were outstanding, with a one for one conversion after that. What this meant was that existing common shareholders would lose majority control of the company as soon as preference shares began to be converted.

Whether or not the common shareholders were being badly treated depended essentially on (*a*) the prospect for earnings, (*b*) the appropriateness of the valuation of the company's assets accepted by management, and (*c*) the impact on the company of continuing to carry the deficit and perpetuating the delay in paying dividends on preferred stock. While many things were said at the special meeting of November 27, 1941, the group representing some common shareholders emphasized that economic conditions had improved and would improve further, that assets in Europe might not irretrievably be lost, and that the company would benefit from being able to engage in war production – all of which led them to argue that an almost 50 per cent loss to them of the equity in the company and ultimate loss of majority control was too harsh treatment.

With this both management and the influential group of preference shareholders disagreed, emphasizing that without the reorganization many years would be needed to work off the deficit and that in the meantime the delay in resumption of dividends would hurt the company. In what way it was thought the Company would be hurt was explained at the meeting by James A. Gairdner:

> ... It is common knowledge that the action of the securities of a company and the condition of those securities affects the credit rating and the public goodwill of that company.
>
> The opposition have claimed that their re-organization, if put through or not put through, would have no bearing on the operation of the company. I would like to differ from them in this regard. You cannot have a company which deals with the public, whose bonds have been kicking around in the 80's and 90's,

with Preferred shares eleven years in arrears, and the Common kicking around the street at $1.00 and $2.00 a share, and have the good-will of the buying public.

The inevitable hour for the vote came and the results were made known at a meeting called for that purpose on December 2, 1941 – 433,962 in favour, 101,267 against, and a small number of votes in doubt. The required 75 per cent of shares represented and voted was exceeded, the legal order approving the reorganization was allowed in January 1942 in spite of opposing arguments, and in March 1942 resumption of preferred dividends began.

There was one interesting and historically important sidelight to the reorganization, for with it began the direct participation of E. P. Taylor and W. E. Phillips in the affairs of the Company. Duncan wanted to strengthen his Board and thought of Taylor. The two had been neighbours and friends in Toronto. Since both were in the service of the government of Canada as "dollar-a-year" men in the early war years, they had further opportunity to meet. Duncan returned to Massey-Harris in early 1941, and on one of his trips to Washington, where Taylor was located as chief executive officer of the British Supply Council, Duncan asked him to come onto the Board.

By the provisions of the reorganization the number of directors was increased from ten to twelve, three of whom were to represent the preferred shareholders, so room was created to bring Taylor in. At the annual meeting of shareholders of March 31, 1942, James A. Gairdner, who had spoken so strongly for the rights of the preferred shareholders, nominated E. P. Taylor, W. L. Bayer, and Gordon C. Leitch as directors of the company representing the preferred shareholders. Both Taylor and Duncan wanted W. E. Phillips on the Board as well, and he became a director representing common shareholders after the death of G. W. McLaughlin in December 1942.

Participation of Taylor and Phillips in the affairs of the company as directors and substantial shareholders – the latter after 1946 through common shares held also by Argus Corporation – was not to prove transitory, and Duncan was later to regret having brought them in. They soon controlled the company, something which no small group of shareholders had done since the Masseys sold their interest in the 1920s. Control was achieved through the appointment of directors friendly to them and through their accumulating stock of the company. The decisive change came at the annual meeting of February 12, 1947, when J. S. D. Tory, J. A. Simard, and H. J. Carmichael were elected to the Board, and J. F. Lash and F. K. Morrow were not re-elected. When the Executive Committee of the Board was formed in 1948 with the Argus group making up a majority of its members, it was clear that that group had gained firm control of the company. In later years the Argus influence resulted in a comprehensive reorganization of the company.

It may be wondered why Taylor and Phillips became interested in the company in the first place. During the period of the war their attention had begun to turn toward opportunities in the industrial field, and Taylor had ample opportunity to talk to Duncan about Massey-Harris. Phillips has noted that they were interested in what they regarded, in a rather abstract way, as "sick companies." He said that while their view of what constituted a "sick company" was never specifically defined, they felt at the time that one characteristic of such a company might be that its board of directors would be composed of individuals with only a nominal share interest in the company. Massey-Harris, one way or another, seemed to fit this definition.

In retrospect the reorganization appears to have been generous to the preference stock shareholders. At the time, management gave the example that thirty years of annual earnings after taxes of $1.5 million would be required to pay off the deficit, which may in a very general way be an indication of the order of prospects they felt were in store. Until the end of the war this proved to be a remarkably close estimate of results, for net annual profits from 1942 to 1945 averaged only $1.5 million because of war taxes. Then they began to rise steeply with the result that the figure of $45 million accumulated net profits after 1941 was exceeded during 1950 – that is, nine years after the reorganization.

The company did not hesitate to become engaged in war production after war was declared in September 1939. Its interest in war production had begun in the late spring of 1939, when it established a committee to consider problems of defence production and possible supply shortages; and in summer when its Chief Engineer, G. T. M. Bevan, spent some time in Britain as a member of the British mission sent by the Canadian Manufacturers' Association to bring to the attention of the British government the potentialities of Canadian manufacturing plants and to express the desire of Canadian manufacturers to lend their assistance at any time. All Canadian plants, of course, suffered from inadequate production volume.

On September 18, the General Manager of the company reported to the Board on a number of steps taken immediately following the declaration of war. They were not giant steps but they were a beginning: machine tools valued at $75,350 were purchased; the full staff of toolmakers at Toronto and Brantford was rehired; toolroom wage rates were increased from 60c to 70c per hour; the mechanical staff was strengthened; a skeleton staff to handle war orders was established; advanced material orders for $533,000 were placed; and cables were exchanged with foreign branches about price increases. In addition, fear that sterling might fall in value led management to sell forward about 80 per cent of the sterling they expected to receive up to November 30. Four days after Canada declared war Duncan and two senior officials went to Ottawa and informed the Canadian government of the facilities they had available for war production.

In December 1939 the company received the first munitions order placed in Canada by the United Kingdom – one for 40 mm. anti-aircraft shells. Production of these began in the Toronto plant in January 1940, to which were added the manufacture of bodies for trucks and personnel carriers in February, and 25-pound shells before the year was out. The company also received an order for manufacturing wings and spars for the Avro-Anson trainer from the Canadian Department of Munitions and Supply in December 1939. In 1940 it remodelled the Weston plant, which had been used only as a warehouse since a tractor project was abandoned in 1923, and began manufacturing the Avro-Anson wings. Production of all these items was facilitated by the stocks of materials which had been accumulated in 1939. In the midst of this new activity in 1940, a most significant decision was made – to plan for producing the new #21 self-propelled combine in volume in 1941.

But the year 1941 was also one of expanding activity in additional new lines of war production. The Department of Munitions and Supply built a factory of 475,209 square feet on the company's Weston property to permit Massey-Harris to undertake the production of wings for the Mosquito fighter bomber – which became the company's biggest single wartime project in Canada. Sixty pounder shells began to be produced at the Woodstock plant in August 1941, and in 1942 a new building for manufacturing mounts for rapid-firing naval guns for corvettes, frigates, and merchantmen was constructed at Brantford. Before the war was over the company's Canadian plants had also manufactured 4.2-inch bombs and cargo bodies at Woodstock, and radio trailers and airfield tractors at the Toronto plant.

More important in volume and in long-term significance was the company's war production activity at its Racine plant in Wisconsin. In November 1940 Massey-Harris received an order from the United States government for links for tank treads, and within a few months it was producing these at Racine. This was not a big project. However, in December 1941 two representatives of the Tank Division of the United States Army Ordnance Department came to Wisconsin to survey the possibilities of producing tanks in that state. Massey-Harris made its interests known, as did other companies, many of whom were however already occupied with other production. At a meeting on January 6, 1942, the Army Ordnance representative asked W. K. Hyslop, Vice President and General Manager of The Massey-Harris Co., of Racine, and C. P. Milne, his assistant, whether they were willing to take on a prime contract on tank production. They said that they were, and left the meeting virtually assured of the contract and with a feeling of elation; mental calculations suggested that 1,200 tanks per year at $70,000 each for a fixed 5 per cent managerial fee, added up to $4.2 million before taxes – more money than the United States company had hitherto been able to contemplate. "The whole thing may seem too good to you to be true, just as it was to me," wrote Hyslop to the President the next day. A contract was made with the government by way of a letter of intent dated January 11, 1942.

Immediate expansion of facilities was necessary, and on February 9, 1942, Massey-Harris purchased the Nash-Kelvinator plant in Racine for $675,000 – a plant which had been closed since 1936. The United States government was prepared to supply not only working capital but capital for machine-tool expenditures as well. The contract even called for $1.5 million of machinery. But Massey-Harris used only a small portion of that amount because management decided to subcontract most of its component requirements.

The decision before the war to assemble as much as possible, and manufacture as little as possible, resulted, it was said, from the unfortunate capital structure of the company which permitted no other policy. The decision in 1941 arose out of the view of management that the supply of toolmakers and other craftsmen was already short in Racine, as was the supply of housing, and that a faster way to produce tanks would be to rely on subcontracting. This proved to be the case: the first tank was scheduled for production by September 1942, but it was as early as June 9, 1942, that the first trial run took place. The environment in which the decision to subcontract was made was, necessarily, not one in which the relative costs of subcontracting and of internal manufacturing could carefully be considered.

A Purchasing and Subcontracting Department was organized at Racine by March 4, 1942, and literally thousands of firms were contacted as potential suppliers of tank components. It was in the ultimate refinement of subcontracting that the company took great pride; and Massey-Harris was pleased when a representative of the Ordnance Division referred to the tank project at Racine as "an example of subcontracting on a prodigious scale that was one of the most outstanding in the United States." Furthermore, it was prodigious in terms of the company's total wartime output and important in its impact on the company. Tank production was valued by the company at $133 million, almost 60 per cent of its total production of war items. It had resulted in a substantial increase in factory capacity in the United States, and had made management familiar with subcontracting on a major scale. Needless to say, it also lifted the U.S. company off the precarious balance that it had just been able to establish after the depressing experience of the early 1930s. Suddenly the company's name had become more familiar in United States industry than it had ever been before.

The growth of Canadian and United States operations during the war period is closely reflected in sales figures, since price controls prevented significant price increases. Sales from 1939 to 1945 inclusive in millions of dollars were $21.0, $24.3, $34.6, $58.2, $92.3, $91.0 $115.8; and the percentages of these sales accounted for by war production were 0 for 1939, then 3, 13, 49, 73, 55, 58. In 1946 sales from war production declined to $5.7 million from $67.7 million in the preceding year, or to only 8 per cent of total sales, and that decline brought total sales down to $70.1 million.

Operating at over full capacity throughout the war period, Massey-Harris was assured of great improvement in its financial position. Over the six-year period, 1940 to 1945, Massey-Harris net profits before taxes amounted to $25.6 million, half of which reflected profits from war production, and, with price controls on farm implements, profit margins on war production were about the same as those on implement production.

In spite of this substantial increase in profits the liquidity of the company, after the war, was not as high as one might have expected. Taxes took about 70 per cent of those profits, and the increase in sales volume from $21.0 million in 1939 to $70.1 million in 1946 demanded more working capital. Working capital rose from $23.5 million in November 1939 to $31.2 million by November 1945, yet the ratio of current assets to current liabilities declined from 6.3 in 1939 to 4.4 in 1945. It is true that the reduction of receivables and of inventories of finished goods to the vanishing point, and the increase in cash from $2.4 million in 1939 to $10.3 million in 1945, meant that the liquid assets became even more liquid as a result of the war. But it was also obvious at the end of the war that those assets would be needed to finance inventories and that they would not be available for capital expansion.

The impact of the war on corporate liquidity in general has probably been exaggerated. In the case of Massey-Harris the reasons why liquidity was not greater than it was can best be seen from a few figures that show on the one hand the company's sources of funds and on the other hand, how those funds were used. From November 30, 1939, to November 30, 1945, the company's net profits before taxes were $25.6 million. Depreciation allowances provided an additional $5.1 million, a long-term note issue of the U.S. company supplied $5.0 and other items a further $4.4 million – a total of $40.1 million. But of this amount taxes took $17.8 million, retirement of bonds and cash in escrow for bonds $4.0 million, preferred dividends $2.4 million, loss of foreign assets $1.0 million, leaving only $14.9 million for additional working capital and capital expenditures. Capital expenditures amounted to $7.2 million, leaving an increase in working capital of $7.7 million which, as we have seen, was not excessive for working capital purposes. It was therefore one of the facts of life of the company at the end of 1945 that further and substantial expansion would have to be based largely on outside borrowing and future retained earnings, and could not be based on past accumulation of earnings.

The figure of $1 million loss on foreign assets has behind it a story that could not be known to the parent company at the time. The loss referred to the German occupation of France, Denmark, and Belgium in which Massey-Harris had subsidiary companies. Contact with the Massey-Harris company in Germany had been lost in 1939 and the parent company wrote it off in its balance sheet that year. Yet operations of neither this company nor the other local European Massey-Harris companies ceased after they could no longer report to Toronto.

The last report received from the German company before the war was one for the month of August 1939. It is now known that at first the German company simply continued to manufacture implements and spare parts; steel was rationed and so the reduced production was not "sold," it was allocated. Almost immediately the company became subject to the control of the custodian for enemy property, and a gentleman by the name of Dr. Heinrich Troeger was the government's representative with whom the company had to deal. It turned out to be a most fortunate appointment from the company's point of view. R. A. Diez, the pre-war manager of the company in Germany, continued to be its manager. When France was occupied, Dr. Troeger was also responsible for overseeing the Massey-Harris company at Marquette, and Diez, through having gained Dr. Troeger's confidence, was able to maintain contact with the French company and render some assistance to it. The production of the German company was sold even beyond the borders of Germany. Surprisingly, the Massey-Harris company in Denmark received some of this production, and during the war that company even sent its sales representative periodically on business trips throughout the Scandinavian countries.

But on the night of May 30/31, 1942, a raid of the Royal Air Force damaged the Westhoven plant, destroying a warehouse and causing fire damage in the factory. The German government was anxious to restore the plant's production of spare parts and implements, and so compensated the company for the damage, enabling it to undertake necessary repairs. On the night of July 3/4, the R.A.F. again bombed the factory, this time causing very extensive damage, three times as much as on the first raid; once more the government financed repairs. These had just been made and the plant readied for two shifts when another raid ended the plant's operations for the duration.

The German authorities decided to salvage usable patterns and machine tools for spare parts production, and transferred them to Wels, Austria; other equipment and damaged implements were moved to Wipperfurth, not far from Westhoven. At the time the management of the company thought that Wels might be the location for a new start after the war. Further allied bombings and the general course of the war prevented the re-establishment of production, although right after the war a hand-to-mouth kind of spare parts manufacturing operation was undertaken at Wipperfurth.

Part of the Head Office at Berlin was moved to Welsleben when bombing became intensive. It was thought that the Russians would not go west of the Elbe, but when it became clear that Welsleben would become part of the Russian zone of Germany, the manager Dick Diez moved on to Wels which was in the United States zone of Germany. From there he began a trek in August 1945, by car, coal train, and on foot, to examine the company's properties at Wipperfurth and Westhoven. Nothing but rubble was found at Westhoven, and there was no sign of any contact with Toronto. After four weeks away he returned to Wels. In September James S. Duncan, President of the parent company in Toronto, and George Thomas, General Manager in the United Kingdom and previously General Manager in Europe, inspected the bombed plant and the small parts operation at Wipperfurth. It was a sad sight and the German company was to wait six years before its permanent survival as a manufacturing centre was assured.

In France the company's experience was quite different. Its branches at Juvisy, Nantes, and Lyons were destroyed, being located at important transportation centres, but the factory and office buildings at Marquette escaped severe damage although they were bombed by the Royal Air Force in the closing phase of the war. Production was continued at about half of the pre-war level. The remainder of the factory was run by a small group of German army officers who supervised prisoners of war in the manufacture mainly of electric lamp standards. When Duncan visited the factory in October 1944 he found that the Germans had removed some of the machine tools, and some bombs had fallen on the property. But the total damage was slight and, although the plant had badly depreciated through use, it was in operating condition. Increased production awaited only the appearance of raw materials. But Mr. S. S. Voss, who at the beginning of the war was the General Manager of Massey-Harris European branch, with head office at Marquette, and who had chosen to stay in France, was not there when the war ended; he had been arrested in 1942 and did not survive the concentration camps to which he was sent.

While the parent company in Toronto necessarily lost all contact with its European market, this did not happen in other of its foreign markets. Indeed, of its total sales of implements and spare parts over the period 1940 to 1945 inclusive, which amounted to $197 million, at least one-third was allocated to markets outside of North America. Sales in the United Kingdom were only $1.8 million in 1939, but the pressing need for local food production there had induced the government to import farm machinery, and so by 1945 annual sales of Massey-Harris machinery in the United Kingdom had risen to $7.3 million. Sales in South Africa, New Zealand, and Australia also expanded, and substantial shipments were made to the company's subsidiaries in Argentina, Uruguay, and Brazil. Canada and the United States received only a slightly larger proportion of the company's regular production (excluding war production) in 1945 than they did in 1939. However, if war production is included, the shift to North American markets was pronounced, amounting to 85 per cent of the company's total production in 1945 as compared with 59 per cent in 1939.

The substantial growth of exports to the United Kingdom was, in retrospect, the most significant development outside North America that the company experienced, since it led the company to undertake assembly and manufacturing operations in that country. Twenty years later, the company was to obtain about half of its production from its operations in the United Kingdom, and over a quarter of its total factory floor space would be located there.

All this began during the Second World War when conditions were created which induced the government of the United Kingdom to plan for import restrictions on farm machinery. Once again Massey-Harris saw a valuable market threatened by trade restrictions and in 1945 it started to assemble a few implements on its Manchester premises.

Massey-Harris associations with the United Kingdom were not new. It had established a selling branch in London in 1887. Warehouse space and office facilities were acquired on Bunhill Row. A private company, Massey-Harris Limited, with £50,000 capital was incorporated in the United Kingdom in 1908 (at 54 and 55 Bunhill Row) and, although its charter permitted it to carry on the business of "general manufacturers," in practice it confined itself until 1945 to selling agricultural implements, most of which it imported from the Massey-Harris factories in Canada. The company prospered until the 1920s. Land on Ashburton Road in Manchester was first rented from the Trafford Park Estates Ltd. in 1919, and a warehouse and bungalow were built on the new property in1920-21. But development of the company was severely hampered by a series of heavy losses during the period of economic stagnation after the First World War. Head office was moved from London to Manchester in 1931, and the London premises were used only as a sales office. To increase its sales volume the company entered into an agreement in 1933 with Blackstone & Company Limited, Stamford, to distribute Blackstone agricultural machines in addition to Massey-Harris implements. From 1932 on, the company was making profits again and the original building at Trafford Park was extended in 1933. Sales increased almost four-fold – to $1.8 million by 1939.

In 1940, as an emergency measure, the British Ministry of Agriculture took over the importation of all implements, with the result that Massey-Harris Limited became an agent of the ministry for the sale of implements of the Toronto parent company. This arrangement was not completely terminated until 1947. The old premises and stocks on Bunhill Row in London were severely damaged in air raids on September 11, 1940, and completely destroyed on December 29. The lease on the property was therefore not renewed and so ended a long association with that address.

Sales began to increase right after the outbreak of war on September 1, 1939, and the extreme urgency for increasing home-grown food supplies induced the Ministry of Agriculture to import large quantities of machines throughout the war. Sales of Massey-Harris machines and spare parts rose from $1.8 million in 1939 to $4.2 million in 1940 and they attained a wartime peak of $7.9 million in 1943. Even in 1945 over 7,000 units of drills, binders, plows, combines, tractors, manure spreaders, and potato diggers were imported.

But evidence of an imminent end to this import business appeared in 1944 when severe import restrictions began to be imposed, and when it became clear that the United Kingdom government intended that agricultural machinery be manufactured locally. Duncan had gone to England as early as mid-1944 and had been told by the British Minister of Agriculture of the possibility that import restrictions on farm machinery would remain after the war. Massey-Harris knew definitely in 1944 that imports in 1946 would be limited to a restricted number of combines, plows, drills, and binders. On January 10, 1945, the President had to inform his Board in Toronto that if the company's position was to be protected in the United Kingdom it would probably have to assemble, semi-manufacture, or even fully manufacture implements there.

Indeed by that time the parent company had already authorized certain minor stop-gap measures for the United Kingdom company, including an addition of 8,400 square feet to the Trafford Park, Manchester, premises and the purchase of machine tools amounting to $50,000 – in total an outlay of about $100,000. A total area of 48,000 square feet at Trafford Park was allocated for the assembly and semi-manufacture of several implements. In 1945, 163 of the #716 fertilizer distributors, 252 of the #721A hammer mills, and a few prototype #714 spring tine harrows, #730A tandem disc harrows, and #734 mowers were assembled at Trafford Park. Such was the beginning of Massey-Harris production in the United Kingdom.

But this token production was no solution to the problem and since it was clear by the end of 1944 that the agreement with Blackstone would not be renewed either (it was terminated as of October 31, 1945), the implements available to Massey-Harris distributors were further threatened. The President at Toronto instructed the British company to continue discussions with the Department of Agriculture regarding imports, to investigate possibilities of amalgamating with some British manufacturers, to study factory site proposals, and to compile information regarding costs of manufacturing in the United Kingdom relative to those in Canada.

Meanwhile the threat of further import restrictions did not diminish. The President had discussions with the Board of Trade and other government officials in London in 1945, the upshot of which was that if the company wished to continue business in the United Kingdom a manufacturing operation would definitely have to be inaugurated, even though production costs, it was estimated, would be 25 to 40 per cent higher there than in Canada. Duncan urged the Board to commit itself to manufacturing machinery in the United Kingdom, which it finally did in mid-1945. By then the post-war period had begun.

Heartening as the results of war production were, the long-run position of the parent company depended on the way it could adjust to peace-time conditions. This simple fact was forcefully illustrated by the decline in total sales from $115.8 million in 1945 to $72.4 million in 1946 through the disappearance of war production.

In this transition Massey-Harris had one important advantage over other North American manufacturers of farm machinery as a result of decisions it made during the war relating to its #21 self-propelled combine. As already noted, in 1940 it decided to put this combine into volume production, which gave it a tried and tested production model for a unique machine when basic designs were frozen for the duration of the war. It produced 606 of the machines in 1941, 491 in 1942, 765 in 1943, and 1,814 in 1944.

The increase in combine production in 1944 has behind it a story of success through imaginative initiative that does not come frequently to a company. Steel was still very short that year and normal allocations would not have permitted such a production increase; yet by that time year-to-year improvements had given the company a machine that was entirely ready for mass introduction into the huge and, as far as Massey-Harris was concerned, hitherto elusive United States market. Joe Tucker, Sales Manager of the U.S. company in 1944, began to develop his brilliant scheme of a "Massey-Harris harvest brigade."

It was thought that United States and Canadian governmental authorities might be persuaded to release more steel for Massey-Harris combines if they could be convinced that more bushels could be harvested with the same amount of raw materials going into self-propelled combines than into other harvesting equipment. It was also vital that the authorities should be convinced that the additional machines would fully be utilized. The scheme presented and finally accepted involved building an additional 500 Massey-Harris self-propelled combines and selling these to custom combine operators who would guarantee to harvest at least 2,000 acres with each machine. Starting from the Deep South of the United States in early April, and fanning out across much of the United States, the combine brigade reached the Canadian border in September with harvesting objectives achieved. The attendant publicity was invaluable, for Massey-Harris was soon viewed by the United States farmer as having no serious rival in self-propelled combines. The ground-work had been laid for an opportunity to become solidly established in the United States.

Just as Massey-Harris had at least one advantage over its major competitors as the war ended, the farm machinery industry enjoyed an advantage over some other industries in the period of transition that followed. By continuing to produce some of their regular products during the war, and in increasing volume after 1943 when government feared that machinery shortages were threatening food production, farm machinery companies could take advantage of the huge backlog of demand that existed; and even more important, it could see that farmers had the financial resources to express that demand. While many voices elsewhere expressed fear of early post-war economic reverses, Massey-Harris forecasts around the end of the war foresaw fairly healthy conditions until about 1950, although these proved subsequently to be conservative.

This is of some importance in appraising the performance of a company in the subsequent years when a seller's market prevailed. It is often felt that management cannot be given accolades for high sales in a seller's market. It seems that an appraisal of the performance of a company in such a period must concentrate on two things: first, on the way in which a company equipped itself to take advantage of that seller's market and second, on how it prepared itself for a return to more normal conditions. Massey-Harris was to be most successful in the first, but not so successful in the second.

The action the company took in preparing itself for post-war conditions related primarily to expansion of facilities, and much less so to product development. An internal Post-War Planning Committee was established and held its first meeting in early 1943. Its "planning" was restricted essentially to appraising the company's manufacturing facilities. In its appraisal the company could see that it had "inherited" substantial facilities as a result of its war production activity: additional land at Weston and the government-owned plant built on it, the Verity plant expansion, the purchase of the Nash-Kelvinator plant at Racine, and the F. F. Barber Machinery Company Limited of Toronto, were all related to the war effort and did not come out of post-war planning. The planning committee did, however, immediately think in terms both of modernizing and of expanding the existing facilities – the former out of the belief, engendered by the squeeze on profit margins imposed by wartime price controls, that costs would have to be reduced, and the latter as a result of the consensus in the company that business would expand in the immediate post-war years.

A survey of the state of its machine tools was undertaken in 1943. The Board was told in November of that year that of 1,900 machine tools in Toronto and Brantford, 417 could not be rebuilt, 653 could be rebuilt, and 830 could continue to operate as they were for about three to five years. Plans for improving and expanding facilities appeared in 1944. They had to be based on the assumption of a North American "common market," for in that year Canada abolished all remaining tariffs on farm machinery.

On September 13, 1944, the President outlined to the Board the major capital expenditures necessary if the company wished to remain competitive in the trade: a new combine plant should be built in Toronto at an estimated cost of $1.8 million and a new mechanized foundry at Brantford for $1.4 million. Purchases of machine tools, it was estimated, would cost $700,000 in excess of funds arising from depreciation in the succeeding year, material handling equipment – which had been primitive – would cost $200,000, moving to the Nash-Kelvinator plant from the old downtown plant at Racine would cost $575,000. In addition, a new machine shop was needed in Toronto, the Verity forge at Brantford had to be moved to the gun mount building, Verity plant required a new boiler and Toronto plant a new dry kiln – all of which was estimated at $300,000. A sum of $1.0 million, it was thought, would be needed to rehabilitate the European operations. In total, a programme of $5.9 million was envisaged, and the Board approved it subject to revision when final estimates would be available. It was a capital-spending programme that was to enable the company to improve its relative position in the industry, but one that still left it with a number of old, multi-storey factories. To obtain machine tools, engineers were sent to government war assets centres and war plants in the United States and Canada. A large number of new and slightly used machine tools were located and were installed in the company's Canadian and United States factories at relatively low cost.

The decision to proceed with building the combine plant was probably the most important capital-spending decision made at the time. It was important because of the role which the combine played in the fortunes of the company after the war and also because, by deciding to locate it on the company's original King Street site in Toronto, which over the years had become part of the central area of the City of Toronto, management for many years tied operations to that somewhat confined location.

Having completed its estimates, management recommended to the Board on December 26, 1944, that a new combine plant, estimated to cost $2.0 million, be built on the north side of King Street. In convincing the Board that this was a desirable move they emphasized that, assuming production of 4,000 combines annually, there would be an annual saving of $295,000 compared with previous costs of production. The Board approved the move.

Prior to that meeting eleven different plans had been considered. The possibility that the company might begin a gradual relocation of its Toronto facilities was not ignored. Sites at the existing facilities at Toronto, Brantford, Weston, Racine, and also completely new sites were appraised. The latter possibility was quickly discarded because it was estimated that completely new facilities on a new site would cost about $7 million, which was viewed as quite beyond the resources of the company – as it certainly was if other required expansion was also to go forward. It should be remembered that since common stock dividends were not yet being paid, and the capital reorganization of 1941 had not yet been completely forgotten by investors the ability of the company to obtain outside capital was somewhat limited. Higher wage costs would, it was estimated, more than offset the freight advantages of locating at Racine – a rather significant finding in that it suggested that this Canadian secondary manufacturing industry did not require tariff protection against United States industry. This was confirmed over fifteen years later when another new combine plant was located in Canada for the same reasons. The absence of adequate sites on company land or adequate buildings at Brantford, and a curious view that further concentration of company labour at Brantford was undesirable, resulted in that location being rejected. In the end the hardest choice to be made was the one between Toronto King Street and Weston.

There was the possibility that Massey-Harris would, after the war, be able to purchase the government-owned buildings at Weston, and cost studies showed that Weston was probably as desirable a location as King Street. Weston offered the great advantage of having substantial land available for future expansion. But it was rejected because uncertainty over termination of the war made it difficult to estimate when the airplane wing plant would be available for assembling combines, and also because of the disadvantages of having assembly operations at the Weston location and primary or machining operations at the King Street location.

The location studies were of a crude nature, but this was probably not of great significance since in the final analysis decisions were influenced more by the known limitations of capital resources – which prohibited the utilization of new sites – and by the uncertainty surrounding the course of the war, than by microscopic marginal cost analyses. So the Toronto King Street site was chosen and the new plant began producing #21 combines in mid-1946.

In total, capital expenditures had amounted to only $569,000 in 1943 and $891,000 in 1944, but in 1945 because of the projects already mentioned they increased to $3.3 million and in 1946 to $4.4 million. Reference has already been made to the extent to which these expenditures and increased activity had kept the liquidity of the company from increasing substantially in spite of the prosperity of the war. Outside capital was necessary. In 1945 its United States subsidiary arranged with a United States life insurance company for a term loan of $5 million in the form of long-term serial notes, $3 million at 3 per cent interest and $2 million at 2 per cent; in the same year it had to retire only $600,000 in funded debt. Additional funds were required in 1946 and the parent company issued $10 million of an authorized $20 million 3 per cent first mortgage serial and sinking fund bond issue, sufficient to redeem the $7 million 1939 issue of 4¼ per cent mortgage bonds and to provide new cash.

Plans relating to product development were more haphazard than those relating to capital expansion. Beginning in 1941, the idea seems to have developed within the company that its product line should be diversified in order to soften the impact of a future recession in the implement business. This led management to consider invading fields such as motor trucks, refrigeration equipment, and washing machines. In the end none of these items were introduced into its manufacturing programme, but it did for a period purchase milking machines, washing machines, home freezers, water pressure systems, water pumps, weed sprayers, and chemicals from subcontractors, and it sold these to its dealers. A new department, that of Engineering and Research, was created in the company reorganization of 1944 to assist in product development.

Since its #21 self-propelled combine made it well prepared in that part of its product line, the nature of the company's activity in planning for an improvement in the other major area of a full-line of products is of special interest. A new line of tractors was planned by Massey-Harris, but it was not available until 1947. What then came forward was a type of tractor that was not basically different from the #102 Junior and Senior tractors it replaced. Styling was much the same, the principle of an engine set into a frame was retained, and new features such as the "velvet ride seat" were not of fundamental importance. The Continental engine used in some of the new models was an improvement from a servicing point of view for, among other things, it had removable sleeves. The line of tractors, since it included the #55 4-5 plow tractor, the #44 3-4 plow tractor, the #30 2-3 plow tractor, the #20 2-plow tractor, and the Pony tractor for market gardening, had been extended both upward and downward. Unfortunately the #55 tractor revealed numerous faults after being introduced, which compromised the position of the company in that area, while in the lower range the revolutionary Ferguson tractor and its competitor from Ford were to play havoc with the company's tractor sales.

In addition to preparing a new line of tractors, Massey-Harris developed a self-propelled corn picker, a forage harvester, and improved various other implements. While the company did develop a corn picker, it did not plan specifically to introduce a range of machines suitable for the rich United States corn belt area. Its ability to improve its position there was not enhanced by the failure to have ready right after the war a successful tractor in the higher power range.

Matters of organization were not ignored during this period. In 1944 the company reorganized its distribution system and also its management structure. Massey-Harris dealers in Canada had previously been on an agency basis, but towards the end of 1944 one of its major competitors introduced into its dealer system the outright sales contract including a higher dealer discount. This induced Massey-Harris to adopt a similar contract, and since then it has proved to be the generally accepted approach to retailing in the agricultural implement industry in North America.

The company's organizational structure was also modified. On June 1, 1944, the President introduced a plan of reorganization which he felt was necessary because of the growth in the volume of business and the diversity of products. Under it there was a President, a First Vice President, and Vice Presidents in charge of Sales, Special Interests, Supply and Control, Manufacturing, Engineering and Research. The Secretary and Comptroller also were given the title of Vice President. The Vice Presidents were to report to the President through the First Vice President, and one purpose of the reorganization was to increase the authority of those in charge of the various divisions. At Racine the operating head was to be a Vice President who would also report to the President through the First Vice President. The position of General Manager of the parent company, previously held by the President, was abolished. Essentially the plan involved only a sharper functional division of the company's operations and, having in mind the relative absence of it in past decades, this was a sensible start. Walter Lattman, Vice President Supply and Control, devised and introduced a system for coding the many types and model variations of machines produced, a system still used, and one vital for proper production and inventory control. Unfortunately, the interest in organization and control that these first moves suggested was not sustained, and it was not for many years that important additional steps necessary for developing an efficient organizational structure were taken.

How well poised was Massey-Harris for launching itself into what was to be a period of unusual peace-time prosperity? There is no doubt that it had behind it a record of decisions which were greatly to improve its position in the farm implement industry right after the war. The accumulated deficit had been removed by the sure surgery of capital reorganization; the concerted move into war production had given the company a large, trained labour force and additional capital facilities; and the 1944 capital expenditure programme further improved its ability to take advantage of the seller's market that lay ahead. It had a long lead on the industry in the development and manufacture of the self-propelled combine, and it had the further advantage that this combine had already given the company invaluable publicity in the rich United States market. A new line of tractors, including the new Pony tractor, was being planned, but was not ready immediately, and it was not as adequate and complete as it might have been. The Engineering and Research Division at least held out the hope that necessary products would in future be developed, but it needed to prove itself. Reorganization of the management structure of the company appeared to indicate that management recognized the need for change in this area. The company in North America, while not entirely prepared, seemed reasonably capable of moving successfully into the post-war period of excess demand.

Still, compared to its major United States competitors, it was a small company. Its world-wide sales were less than half the sales of Deere and Company, and very much less than the sales of the International Harvester Company. About five companies had a larger share of the United States market than it had. Also, it manufactured a substantially smaller proportion of its tractors and implements than did its major competitors, substituting not so much whole implements manufactured by other companies as outside components for the implements and tractors it assembled. It still did not have its own engine for any of its tractors or combines.

Internationally its position was somewhat more uncertain than it was in the North American market. Before the war its manufacturing ventures outside of North America remained relatively small and their ultimate success had not yet been established. This uncertainty was magnified by the war, which made the Board of Directors of the parent company more reluctant to commit funds to such ventures than they had been in the 1920s. Nevertheless, the company's position in exports had not really been weakened by the war, and management remained fully committed to exploiting markets outside of North America. The President visited export markets and the company's plants outside North America just as soon as the war permitted it, beginning with France in September 1944. In the United Kingdom at the end of the war the company was on the verge of implementing a policy, first evident in the 1920s in its French and German operations, of protecting valuable export markets by committing itself to local manufacturing operations.

What followed was six years of dazzling prosperity, two years of accumulating difficulties, and then, in 1953, the merger with Ferguson, a development that changed the company's position entirely.

CHAPTER 4

Post-war prosperity

On February 25, 1946, the President reported to the shareholders of Massey-Harris that "There is an unprecedented demand both at home and abroad for the products of your company which would far exceed the capacity of its manufacturing facilities. . . ." But by 1953, conditions had reversed themselves and optimism about the company's affairs had diminished. Writing to the Vice President and General Manager of the Canadian division on October 31, 1953, the President had to use strong words to describe the company's position: ". . . We are spending a lot of our time cutting down expenses, reducing salaries, letting out personnel and doing all these distasteful things, some of which would not be necessary if those who are in charge of our senior departments . . . would take immediate remedial action when difficulties show up and not always assume that these are not very important and will probably work themselves out."

These contrasting statements reflect the over-all business experience of Massey-Harris from 1946 to 1953. In spite of the company's firm hold in the export market, its activity up to 1953 was heavily centred in North America. From a manufacturing point of view this was no change from the past. In 1938 about 88 per cent of the goods it sold originated in North America and while the disruption of war in Europe had increased this figure to 96 per cent by 1946, it was still 83 per cent by 1953. But from a sales point of view its activity in North America was different because of its successful penetration of the United States combine market. It had sold 18 per cent of its products in that market in 1938, and because of the war this had risen by 1946 to what must have appeared to be a temporarily high figure. But surprisingly enough, it still stood at

37 per cent in 1953. This was a higher ratio than for sales in Canada, and for the first time in its history the United States market was not merely an enticing opportunity: it had become vitally important to the company's over-all financial performance.

There were also historically important developments in its operations abroad, particularly in the United Kingdom, France, Germany and South Africa, as will be seen later; but the thrust of its prosperity, and then the diminution of it, was centred in its North American manufacturing and sales activity. A giant step forward had been taken in U.S. sales, and this was a historically significant event in the company's international development. There is no doubt that the company's extraordinary success in North America, and then its waning fortunes in that rich market, as well as the reasons for both, constitute the most important aspect of its history from 1946 to 1953.

In 1946 management of the parent company at Toronto realized that their previous forecasts of sales had all greatly underestimated post-war demands for tractors and farm implements, and that the ability to produce, not sell, would be the company's first major post-war challenge.

Sales, excluding defence goods, rose 36 per cent in that year and the scarcity of materials was uppermost in the minds of management. Plant capacity was never the problem that material shortages were, largely because major plant expansions were completed in 1946. The new combine and power plant in Toronto and the M foundry at Brantford were completed in that year or early 1947, as was the relocation of the tractor plant at Racine. The forge at Brantford and the addition to and re-arrangement of the Batavia plant were achieved in that year as well. At Woodstock tooling up for the one-plow Pony tractor, a new addition to the line, was under way and the new Engineering, Research and Sales building was under construction in Toronto. Even so, greater facilities were soon required and were added to the already expanded capacity.

Prime capacity of the new combine plant to produce the #21 was about 5,000 combines annually. This capacity reflected a forecast made in early 1945 that about 5,000 to 5,300 combines built at Toronto would be sold annually up to 1949. In 1946 the Toronto plant built 5,551 combines, not far from the forecast, but it was obvious to everyone that this number did not begin to satisfy the demand and merely represented the limit imposed by material supplies. A new forecast of demand was made in late 1946 which suggested that a demand of about 12,700 combines of the type produced at Toronto – mostly self-propelled – would exist in 1948. Capacity at the Toronto plant, after additional capital investment in equipment, had been raised to what was regarded as a top limit of 9,000 machines, much higher than the original 5,000 but still far short of expected demand. Therefore in 1947 a building on the south side of King Street, previously used as a foundry, was rebuilt and enlarged to provide space for a second combine assembly line. Combine production at Toronto in 1948 finally amounted to 12,931 machines, about three times as large as 1944 forecasts for that year.

While shortages of raw material and component parts were the major obstacles to production, for a time an acute labour shortage threatened to interfere with the output of the company's two factories and foundry at Brantford. It was overcome in a novel way. A housing shortage was at the root of the difficulty, and this is spite of the two staff houses accommodating 157 men, that Massey-Harris operated. After discussions with the government the Number 5 Royal Canadian Air Force Air Training Station at Burford was taken over, reconverted for family use, and soon 500 people were housed there.

Production was also enhanced by the company avoiding stoppages through strikes – the only major farm implement company in North America without a strike during the period of reconversion to peace-time operations. This record, however, seems to have had some hidden costs in the form of inadequate production efficiency, for the company was deterred from introducing a formal production control system. In the later war years outside consultants had been retained to introduce such a system into the Toronto works. Opposition from some groups in the factory, supported it seems, by at least one senior executive of the company, ensured its failure. Not only was an important opportunity for improvement in manufacturing missed as a result, but those favouring improvement lost influence in the company, and the control of management in the factories was probably weakened. But not until the early 1950s, when profit margins began to shrink, was this problem in the factories seen to be a serious one.

Material shortages appeared everywhere right after the war and it was fortunate that a separate Supply and Control Department had been established in 1944. In early 1946 the Vice President Manufacturing, E. G. Burgess, had to report a "very grim" supply situation at Batavia: some implements could not be shipped for lack of wheels, lack of sheet steel was limiting the production of hay loaders, corn binder production was two months behind schedule, the very important Clipper combine assembly line was shut down, lack of discs held up the production of the #20 harrow, and so it went. Attempts to place orders for 1947 steel requirements with the steel companies were meeting with some disappointment. Steel strikes in the United States and Canada aggravated the situation. Both the difficulties encountered and the frustration that resulted from them are illustrated in a report on tractor production to the President of the U.S. company by E. G. Burgess on November 1, 1946:

> ... We reviewed the production of tractors and I observed that we are daily going in the hole against our scheduled rate of 90. Something must be done about this. It is largely caused by a lack of materials and Mr. Kennedy told me of changing models last Monday five times, in order to keep the line going. On some models we are held for rims and wheels; on other models we are held for

magnetos, and the engine situation is indeed grim. Last week, only 73 engines for the No. 101 were received; there are 970 to come and we should have had them all by September 1st. On Tuesday and Wednesday of the current week, no engines were received at all. We are assuming that Mr. Lattman, through Mr. Krause, is initiating a more aggressive followup of materials and the use of expeditors. We simply cannot build tractors without materials.

But in spite of all the difficulties the volume of tractor production was

TABLE 1

Massey-Harris-Ferguson Tractor Production (1946–1954)

	1946	1947	1948	1949	1950	1951	1952	1953	1954
MASSEY-HARRIS									
RACINE, United States									
101 Sr.		537							
203		40							
20		4,367	3,564						
22			886	5,387	3,917	4,797	2,194	14	
23							257	3,088	207
21							100	1,569	
30		3,888	5,868	6,114	7,072	6,376	3,094	4	
33							1	5,614	2,605
44		4,147	10,442	16,364	16,955	19,942	12,534	4,404	6,894
55		1,293	2,627	3,235	2,968	2,999	3,454	2,390	1,787
Total Racine	N/A	14,272	23,387	31,100	30,912	34,114	21,634	17,083	11,493
WOODSTOCK, Canada									
11 Pony		1,314	9,073	5,106	2,834	4,543	4,087	962	705
14						33	17	24	
16 Pacer								164	1,444
MANCHESTER, England									
744			16						
KILMARNOCK, Scotland									
744				1,054	2,783	4,660	5,547	2,546	
745									2,952
MARQUETTE, France									
811 Pony						1,813	10		
812 Pony						587	5,800	8,133	8,758
TOTAL MASSEY-HARRIS	N/A	15,586	32,476	37,260	36,529	45,750	37,095	28,912	25,352
FERGUSON									
Ford	59,773	47,682							
Detroit			1,800	12,859	24,503	33,517	35,965	17,314	14,047
United Kingdom	316	20,578	56,878	38,689	51,381	73,623	69,567	48,060	62,483
France								2,007	4,496
TOTAL FERGUSON	60,089	68,260	58,678	51,548	75,884	107,140	105,532	67,381	81,026
Grand Total	N/A	83,846	91,154	88,808	112,413	152,890	142,627	96,293	106,378

quickly built up including, later on, production of the #744D at Kilmarnock and the Pony at Marquette. The volume and mix of tractor production from 1946 to 1954, including for the sake of interest and future reference, the production of Ferguson tractors, is shown in Table 1.

Conditions in the Canadian plants were no better than in the United States factories as the year 1946 ended. Because of general demand and because of strikes among steel, rubber and electrical workers, shortages were chronic. Should the plants shut down entirely or should they keep their productive labour output high by manufacturing components for inventory until shortages were overcome? The latter policy was pursued, one termed by the Vice President Supply and Control, as ". . . going full blast making left shoes." Inventory accumulation became rather frightening but the decision soon proved to be right.

In Canada, production of the self-propelled combines, first the #21, then the #26 and #27, and finally the #80 and #90, was of the greatest importance. How successful the company was in stretching its facilities to meet the demand is best shown by the rapid increase in machines produced. Until 1949 Massey-Harris depended entirely on its North American factories for combines, but then the company's new plant at Kilmarnock, Scotland, began to produce the #726 for the United Kingdom and export trade, and the French company first produced the #890 combine at Marquette in 1953. All this is shown in Table 2. No Ferguson combines are shown there because Ferguson did not sell combines.

Severe material shortages continued into 1947 but then began to ease. How well had Massey-Harris performed? Since Massey-Harris sales expanded more rapidly than did sales of its major competitors, it is apparent that the manufacturing and supply and control departments performed most successfully in supplying agricultural equipment to the strong buyers' market that existed in the years immediately after the war. Its first major post-war challenge had been met. What happened to costs of manufacturing in so doing later became worrisome. But a policy of maximum output through relative insensitivity to costs in a period of excess demand is an appropriate one, as long as it is followed by one of increasing control of costs in the succeeding transition period of decreasing excess demand.

Prosperity did not end when material shortages began to ease in 1947. Until about 1951 the company's sales and profit experience was most unusually encouraging, although scattered evidence of harder times ahead had begun to accumulate before then. Charts 1 and 2 show the trend of both, from 1946 to 1953, that is up to the time of the merger with Harry Ferguson Limited. But at least as significant as the total sales increase was the increased extent to which the company came to depend on the United States market. Sales rose in all major markets from 1946 to 1951 but not at the same rate, with the result that the United States market suddenly became much more important, and the Canadian market

TABLE 2

Massey-Harris Combine Production (1946–1954)

	Production Year (Aug. 31)						*Fiscal Year (Oct. 31)*		
	1946	1947	1948	1949	1950	1951	1952	1953	1954
ELF-PROPELLED									
ORONTO, Canada									
1 & 21A	4,601	4,678	9,139	10,279					
22		610							
6			2,792	5,311	4,616	5,209	5,797		
7				25	8,166	10,470	10,945		
0								200	795
0								1,499	
0							287	4,590	2,766
0								8,145	6,924
ATAVIA, United States									
.P. Clipper	750	750	1,000	1,000	875	635	670	196	
IANCHESTER, England									
22			395						
ILMARNOCK, Scotland									
26				1,321	4,100	5,000	4,636	1,150	
80								2,362	3,855
IARQUETTE, France									
90 Petrol								99	910
TOTAL SELF-PROPELLED	5,351	6,038	13,326	17,936	17,757	21,314	22,335	18,241	15,250
ULL TYPE									
ORONTO, Canada									
5 & 17	950	1,012	1,000	1,975	926	780	953	700	
0, 32, 33									
0								100	985
0									
ATAVIA, United States									
.T. Clipper	4,000	3,340	7,585	10,909	13,041	12,426	3,558		
0						100	850	3,670	3,912
ILMARNOCK, Scotland									
50								385	546
TOTAL PULL TYPE	4,950	4,352	8,585	12,884	13,967	13,306	5,361	4,855	5,443
Grand Total	10,301	10,390	21,911	30,820	31,724	34,620	27,696	23,096	20,693

somewhat less important than previously. The figures in Table 3, which exclude defence goods, show this in detail. The sales increase in the United States of just over 400 per cent to where U.S. sales accounted for almost half of the company's total turnover was certainly the most significant development of the period. It generated large profits, but it also made the company very vulnerable to changing market conditions in that highly competitive economy. Unquestionably, it was a major event in the history of the company's international development.

The $27 million net profit after taxes of the U.S. Massey-Harris company over that period represented about 40 per cent of the company's total net profits. Whereas earnings per common share of the parent company had been $1.58 in 1946 (excluding retained earnings of subsidiaries outside North America), in 1949 they were $9.71. Over that period the company's working capital rose from $31.5 million to $67.7 million, and the price of the common stock which had reached a high of $15.75 in 1945 rose to a new high of $24.00 in 1949, then to $44.50 in 1950 and,

CHART 1
Massey-Harris sales 1946-1953

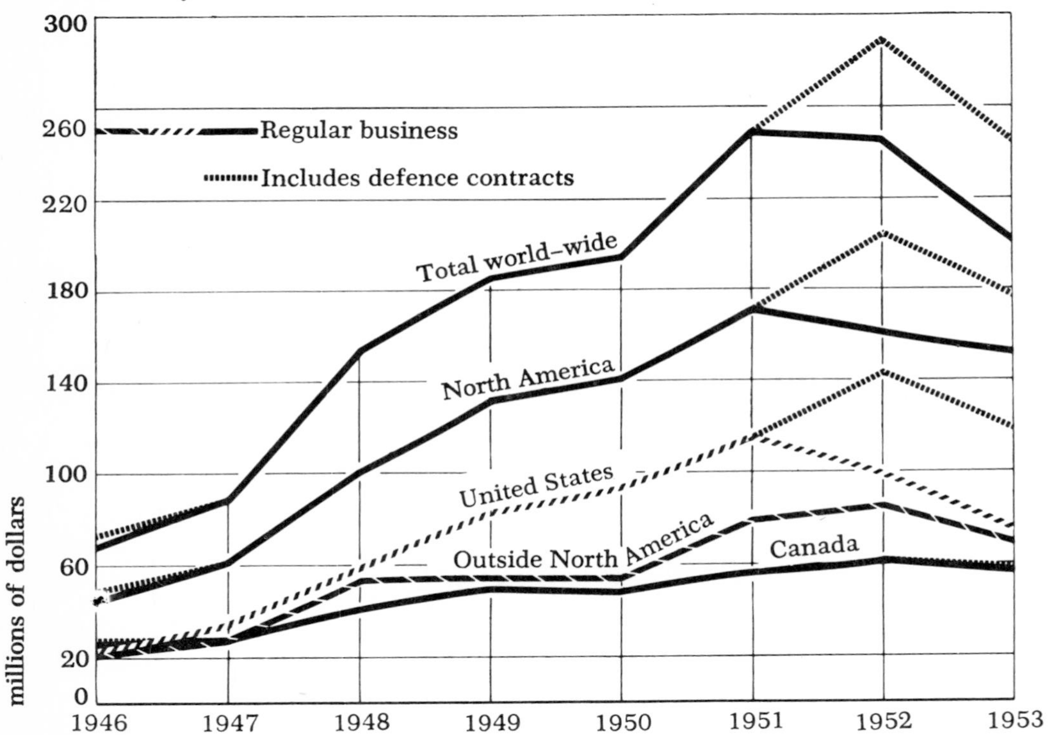

before a five-for-one stock split in 1951, it reached a high of $65.50. Its world-wide sales volume had expanded sufficiently to place the company well ahead of the other medium-sized North American companies, although still much smaller than those of Deere and International Harvester.

The key to the company's success in the United States was its self-propelled combine and, as already noted, this began with the Harvest Brigade during the war and was continued by the company's remarkable success in manufacturing large numbers of them. But the very success of that machine in the United States market created potential dangers, for it encouraged Massey-Harris to gear its operations to an unsustainable

CHART 2
Massey-Harris profits 1946-1953
(Net after taxes)

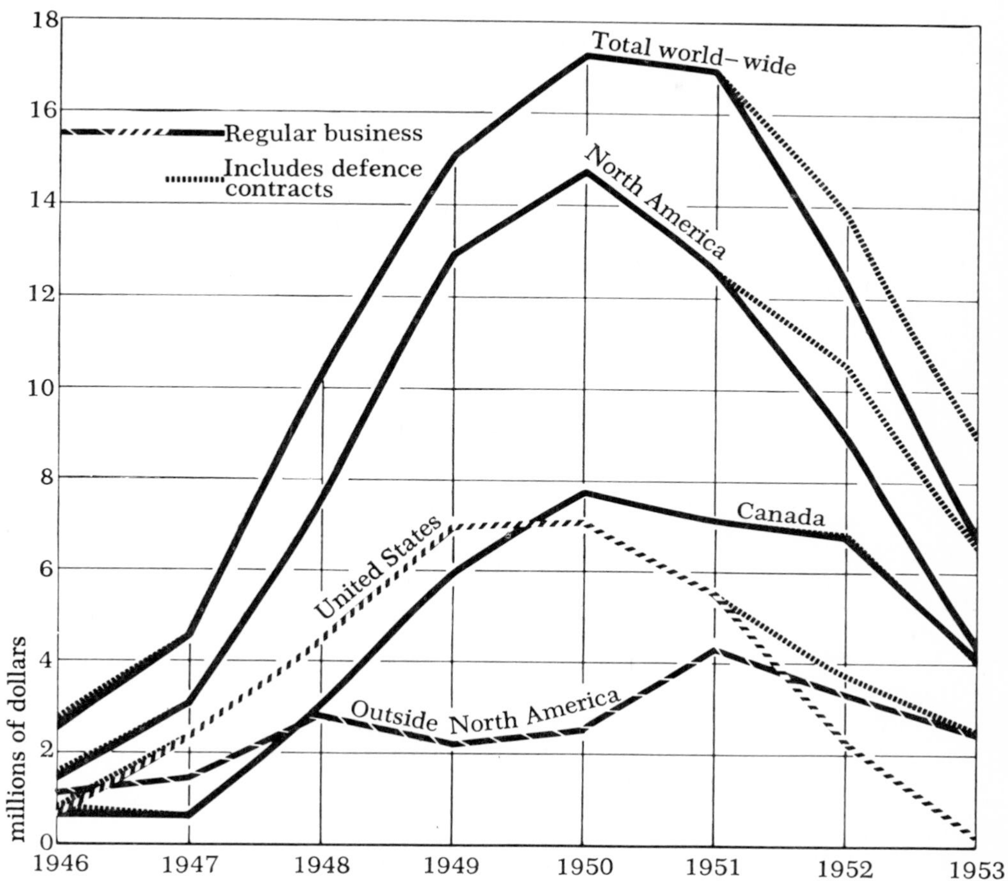

product sales pattern. The extent to which the United States sales pattern was distorted is sharply revealed by the share of the market the company enjoyed in the major machinery categories. Up to 1951 Massey-Harris enjoyed fully half of the United States self-propelled combine market,

TABLE 3
Massey-Harris Sales Increase (1946–51)

	Percentage Sales Increase 1946–1951	Percentage of Total Sales 1939	1946	1951
Canada	159	35	32	22
United States	406	24	33	46
Total North America	286	59	65	68
Total Outside North America	229	41	35	32
TOTAL	266	100	100	100

but it made little progress in penetrating the tractor and implement market. Taking the Canadian and United States markets together, the company's sales of grain-harvesting machinery, which included mainly self-propelled combines, accounted for 39 per cent of its total sales and for the United States market the figure was even higher.

TABLE 4

Massey-Harris Share of United States Market (1946–1951)
(per cent)

	Self-propelled combines (Units)	Wheel tractors (Units)	Implements other than harvesting machinery	Total
1948	52.9	3.1	2.1	4.1
1949	45.9	3.7	2.8	5.6
1950	51.1	4.8	2.5	6.2
1951	52.2	4.8	3.0	6.1
1952	45.7	3.7	3.4	6.0
1953	40.4	3.2		5.0

Source: Compiled from statistics in United States, Department of Commerce, *Facts for Industry*, and from company material.

Not that this represented a great change from the traditional Massey-Harris product mix, for its strength had always been in harvesting machinery. Its tractor sales from 1936 to 1938 averaged only 20 per cent of its total sales. But there was none the less one decisive difference. While its tractor sales in the United States in 1938 had accounted for about 45 per cent of its total sales in that market, this was only about 3 per cent of the U.S. tractor market; and its combine sales in this market were equally small relative to the total market and, of course, were all pull-type machines. Massey-Harris very obviously was no threat to anyone in the United States industry at that time. However, by 1948 its sales of combines were about 46 per cent of its total sales in the United States, its sales of self-propelled combines accounted for two-thirds of that and, what is more, these self-propelled combines accounted for 50 per cent of the total industry sales of that type of machine. There was no doubt that competing companies would strive hard to gain a larger portion of that market. When they did, the impact on Massey-Harris could be painful – providing alternative strength was not acquired.

How does a corporation work itself out of a precarious product mix? In the case of Massey-Harris, acquiring a more balanced strength involved mainly developing new and improved models of tractors and implements, manufacturing them at competitive cost levels, and creating an effective distribution system. In this the company was not sufficiently successful in the early post-war years.

Because of its historical strength in Canada, the company had the great advantage of a good system of dealers. This, unfortunately, was not so

David Fort

An MF tractor is observed at work from the steps of a temple near Bangkok, Thailand.

in the United States. A company with a small sales volume attracts few exceedingly good dealers and almost no dealers that will carry its line alone. Definite improvement in the company's United States dealer system was essential. Some steps in that direction were taken. The number of its branches in the United States was expanded to better service its dealer system and other branches were renovated. In 1946 it had only six branches there, the same number as in 1939. Then one or more were added almost every year until in 1951 there were fifteen, and in 1953 eighteen. Gradually it could dispense with the independent jobbers that it had used extensively in previous years and to some extent as late as 1949.

In addition to expanding the number of branches, there appears to have been emphasis on expanding the number of dealers. This was not difficult, owing to the strong demand for the Massey-Harris combine and the general shortage of machinery. Unfortunately many dealers appear to have become combine dealers who did not take great interest in penetrating the tractor market with Massey-Harris machines. The tractors they did sell were frequently competing and much better-known machines, for most dealers were not exclusive dealers. A determined policy on the part of Massey-Harris to ration its self-propelled machines to dealers taking a certain number of tractors and other implements might have helped, but the problem seems not to have been given priority or even explicitly understood. When sales softened in 1951 the difficulty was conspicuously and ominously centred in the tractor area.

To survive both the effects of slowing demand and the imminent attack on the company's position in self-propelled combines also demanded attention to product development. In combines the company did well in the 1946-53 period. At first the #21/21A self-propelled combine dominated the market because its design was well ahead of other companies, and because it had received thorough testing in owners' hands. Improved models were introduced and the lead the company enjoyed was, to some extent, maintained. The #26 and #27 models (illustrated on p. 412) appeared as early as 1948, and by 1950 had completely replaced the #21/21A models. Not that the new models offered many important new features but they confirmed that the company was concerned with evolutionary development. Developmental work was also begun on the #80 and #90 models, which were introduced in 1952. (See illustration on p. 413.)

To expand its product line the company purchased the plant of the Goble Disc Company of Fowler, California, for $375,000 in June 1948. A brief digression into the background of the Goble company may be interesting for the purchase of it gave the company several important items: an improved disc harrow, an additional factory in the United States which it still owns, and opportunities for experimenting with ways to meet competition from small, flexible, "shortline" companies catering to specialized markets.

Mr. W. E. Goble, the founder of the company, was born in Illinois in 1882, was raised on the farm, began farming on his own, but soon became interested in mechanical inventions. He and his family moved to California in 1907 where after sundry attempts at developing new machines he turned his mind to the improvement of the disc harrow. Older farmers will still remember that all disc harrows used to use block bearings of wood or cast iron. They will also remember that because of this dirt would inevitably find its way into the bearings until the discs would hardly turn. Power needed to pull the disc harrow was thereby increased and before long the bearings would wear out.

Goble began experimenting with an enclosed oil-bath bearing and eventually developed a successful one. This consisted of a two-and-three-quarter-inch piece of tubing as long as needed for a gang of discs. Into the centre of this tube he welded a cast bearing, and on either side of it, inserted in the tube, came an axle. A washer between the bearing and the axle took the wear that resulted from the thrust of the concave discs. Cast bearings were also placed at the outside ends of the axles. Grease was pumped into the hollow axle through a fitting at the end of the axle and worked its way to the inside bearings. It was simple in construction and had the great advantage that it kept dirt out of the bearings, reduced wear through friction and reduced the power required to pull the implement. Soon Pacific coast farmers knew all about the Goble disc, and Goble expanded his operation until he had about one hundred employees. But in 1948, with his patent on the inside bearing due to expire within a few years, he decided to sell the company, and Massey-Harris took over in May of that year. Massey-Harris thereby acquired the Goble disc and a factory in California. But while it exploited the disc fully, it did not pursue the idea of adapting the Fowler operation to meet the varied and changing demand for implements in the California area. Relatively little use was made of the Goble factory even though as an investment it proved to be profitable. Interest in it did not increase until after 1960.

It was in tractor development that the company lagged most sadly in the period before the merger with Ferguson in 1953. This was no minor matter. Long-run stability in its operations without strength in tractors was difficult to imagine. For most full-line farm machinery companies, the tractor is the most important product – accounting for about 30 per cent of industry sales, and the figure is higher for the major companies in the industry.

Massey-Harris did market new models of tractors. By 1948 the Massey-Harris #102 Junior and Senior tractors and the #81 had disappeared from the line and the company had committed itself to the #20 (succeeded before 1954 by the #21, #22, and #23), the #30 (after 1952 the #33), the #44 and #55, the #11 Pony, and also the #16 Pacer from 1953.

There is little doubt that the #44 tractor was the most successful Massey-Harris tractor of the group, but the big #55 designed to close to some extent the gap that had existed for so long in the company's line of tractors was at first plagued with defects. The one-plow Pony tractor, so successful in France, was suited in North America mainly for plot-size farms and market gardens. It was not big enough for the small mixed-farming areas, areas which bought thousands of the two-plow Ferguson and Ford tractors with their revolutionary features.

The new "Ferguson System" was the most significant development in the tractor market in the period 1946 to 1953, although it had first appeared even earlier than that. It involved integration of tractor and implement through a three-point linkage system and built-in hydraulics and had been developed by Harry Ferguson of Belfast, Ireland, between the two great wars.[1] Among other things, it permitted easy handling of mounted implements from the tractor driver's seat and, by transferring the weight of the implement down onto the rear wheels of the tractor, it increased its traction, giving it a performance comparable to much heavier and more costly tractors. The Ferguson tractor and the Ford tractor that was modelled on it were competitive in the two-furrow plow range.

Massey-Harris had the #20 tractor but it was of the old "lugging" type. In 1949 the "Hydraulic Depth-O-Matic Control" was available on the #22 as special equipment, and in 1950 the same feature was available as special equipment on the Pony, #30, and #44. But of course it was not the Ferguson System, since it did not include three-point linkage and weight transference, and the hydraulic attachment was external, not internal.

In 1951 a three-point hitch was provided for the #22 row-crop model tractor, but it did not operate satisfactorily, it did not include draft control, and it did not have many implements to go with it. The Ferguson and Ford tractors by that time had the fully developed internal hydraulic system and three-point hitch as standard equipment and they provided unrelenting competition in the smaller tractor market. After Massey-Harris made a survey of its branches in late 1951, a sort of crash programme to develop tractors of that type was instituted.

One result was the introduction in 1953 of the Colt and Mustang tractors, both two-plow tractors with hydraulics and three-point hitch. But again the hydraulic system was on the outside and the three-point hitch did not include a system of draft control. Cuts in the engineering budget in early 1952 had prevented the development of a three-point hitch of the Ferguson type and not until 1953 was the improved three-point hitch reinstated in the engineering programme. The company's research division was put to work developing a true "Ferguson" tractor, but while a prototype machine was built (internally called the Pitt tractor), it was a design failure and was abandoned. In brief the company's competitive position in the smaller tractor range was impaired because it once again

1 See below, pp. 96-9, for an explanation of the Ferguson System.

lagged in technological development (it misjudged also the appeal of the Pony in North America), and its competitive position in the market for larger tractors was compromised because of mechanical difficulties with the #55 tractor. Its technological lag in tractor development, to be sure, made it no different from most other North American tractor manufacturers but because of its distorted product line – including a very small share of the tractor market – it was inherently in a weaker position to ride out temporary difficulties and catch up with engineering development than were its major competitors.

The central significance of all these developments was that as the company entered the period of a buyers' market in North America it did so with a vulnerable sales pattern and a weak position in tractors.

The Korean War delayed slightly the change to a buyers' market. Sales in North America continued to climb steadily until 1951, and outside North America until 1952. But profits reached a peak in 1950 and, as far as the United States company was concerned, profits in 1950 were little better than in 1949. In 1951 there began a persistent decline in profits that did not end until 1957, by which time the company had undergone a major change in management.

Tentative signs of changing times appeared as early as 1949. Termination of the exceedingly fast pace of expansion of the 1946-48 period was clearly expected, indeed management frequently referred to such a possibility. But the worrisome aspect of the early signs was that they appeared in the area where the company had for long been weakest – its tractor line. In February 1949, E. G. Burgess, Vice President Manufacturing, expressed the feeling that because inventories of Pony tractors were piling up at the branches, production would probably have to be cut back from 50 to 40 a day at the Woodstock plant. The Pony tractor was expected to be one of the company's major achievements. The President had told the shareholders in 1946 that: " . . . we look upon the introduction of this small size tractor and its equipment . . . as one of the important agricultural developments of the post-war period." Now it was beginning to appear not only that its market appeal was disappointing but also that its costs of production could not meet those of its competition.

By June 10, 1949, the Woodstock plant had to be shut down entirely with an inventory of 1,514 Pony tractors. The President wrote graphically of his view of conditions on June 7, 1949:

> I have just returned from Western Canada and U.S.A., and conditions have changed radically. The customer is in the driver's seat again and is buying what he wants and when he wants it. Goods are piling up at the branches and in the agencies. We are taking drastic measures to cut back some of our programmes, including the closing-down of our Woodstock Plant for 3 months minimum. We are also revising the 1950 programme, with a view to establishing this on a more conservative basis.
>
> Some of our Directors feel that our present plans of 20% to 25% below 1949

> for Canada and, say 10% to 15% below the U.S.A. are too ambitious. Two of them suggested that a 50% reduction would be more appropriate. . . .
>
> The Company is quite prepared to face deliberately a smaller and less profitable 1950, but they are not prepared to get caught out on a limb for having followed ambitious plans which proved subsequently to be fallacious.

Conditions did not improve and in November 1,172 Pony tractors were still in stock. "The Pony is the first major problem which we have struck from a sales point of view, and I cannot see that we have been showing much initiative in handling it," wrote the President in December 1949. This appearance of weakness in Pony sales led the company in the United States to undertake the biggest sales campaign ever attempted. The President felt their objective should be to attain 5 to 6 per cent of the United States market in the one-plow and two-plow field. It was an unrealistic view, but then came the Korean War, and whereas a substantial decline in North American sales had been expected, they actually rose by 7 per cent. Still the experience with the Pony sales was sufficiently worrying to attract increased attention to the matter of costs of production, and after 1951 when Massey-Harris tractor sales fell much faster than its other sales, earlier worries about the strength of the company's position in tractors were confirmed.

Weakness in one direction not infrequently reveals weakness elsewhere. It did so on this occasion. While the basic difficulty facing the company was that it had again fallen behind in tractor technology, there was also the matter of production costs. In this latter respect the company's record was somewhat disturbing, particularly as it applied to production in the United States factories. Cost of sales of the United States company usually amounted to about 75 per cent of net sales from 1946 to 1951 which appeared to be rather above the costs of some of the major United States companies at the time, and above the costs of the Canadian operation. Since tractor manufacturing at Racine was the most important activity in the United States factories, attention naturally turned to tractor costs and to the possibility of reducing them. Here serious difficulties were encountered because tractor production at Racine was to a substantial degree an assembly operation. Among other major components, the company did not have its own engine – and this represented about 24 per cent of the factory cost of the #44 tractor.

It seemed that the only way to reduce costs and to increase control over supply was to increase the degree of manufacturing within the company's factories. This problem was recognized at the time. The difficulty was that positive steps to deal with it were not taken. It is true that the possibility of establishing a foundry at Racine for making tractor castings was considered in 1944. The Board even authorized it, and preliminary plans for one were obtained in early 1946. But the plans were not implemented. After visiting a number of factories of competing companies in early 1947, E. G. Burgess reported that:

... the information we now have clearly indicates two main reasons why our tractor costs are high: 1) Grey Iron Castings; 2) Manufacture of Engines. ... We did not see one tractor plant that was not producing its own tractor castings. ... With one exception ... every tractor plant visited manufactures its own engines.

Preliminary estimates were made of savings to be realized through engine manufacturing. The Research Division was also talking in 1947 of the possibility of developing a two-plow tractor inspired by the Ferguson tractor and equipped with enclosed Bendix hydraulics. The President quite rightly told his executives in December 1947 that:

In giving consideration to our tractor design and costs, we are dealing with the principal single problem of our Company, and I trust that Mr. Verity and Mr. Pitt are continuing to give the matter their first consideration. ... So far, the only concrete result is the agreeing to the Foundry, but this, of course, was accepted in principle before the present investigation started. We must, therefore, push this investigation energetically. ...

The resulting reports came forward in January and February 1948. They estimated that, based on the manufacture of 13,000 #44 tractors and 3,120 #55 tractors, $378,422 could be saved. However, if the company would furnish its own grey iron for motors, through building appropriate foundry facilities, the yearly saving would be $1,167,412. As for the ability of the company to proceed? One of the reports said:

It has been asked "Who will do the design engineering if we make these motors?" and in answer we would like to review the development of the H and J engines. In the first place, our Engineering Department drew up the original basic specifications, then Continental supplied us with sample engines of the specified displacement and we began to work out the bugs. It can be said that the development of the combustion chamber, manifold, water circulating system and the calibration of the carburetor was the work of our own organization. Of course, these items are all critical in the performance of an engine and therefore, we estimate that the ratio between the engineering work performed by Massey-Harris and Continental on the two motors was 50/50. (This proportion pertains to the gasoline and the distillate motors and not to the diesel development.)

M. F. Verity in sending the reports forward, recommended not to proceed with the project. Nothing was done. On November 17, 1949, E. G. Burgess repeated his earlier views in a memorandum to the President:

It is unnecessary to labour the point here that, if we are to become competitive, we must buy less and produce more. This is particularly and prominently true in the matter of tractor production.

He went on to outline and recommend a $6.5 million capital expenditure programme that would enable the company to equip itself with foundry, forge, and sheet metal plants at Racine, and allow it to renovate the Batavia plant, complete modernization of the south side of King Street plant, and tool up at Toronto for combine transmissions. Burgess knew what would be viewed as the major obstacle for proceeding, when he ended his memorandum by saying, "I know that this question of capital is one of availability and of fine judgment, therefore, before you make a final decision . . . I would appreciate having a very thorough discussion with you on this important subject." But the major aspects of this programme, those that would increase the company's degree of manufacturing, were not approved.

Finally there was the matter of quality. Its importance to the company was clearly understood. The Vice President Manufacturing, E. G. Burgess, in a memorandum to the President dated June 9, 1949, outlined a comprehensive programme, as he did several times later, designed to remedy the worrying problems of quality. He went on to say:

> I do not wish to be repetitious, nor to parrot what you said at the conference, but I know you realize that this post-war problem of quality in every commodity that is sold is not news..
>
> This whole question of quality has been accentuated because customers heretofore have bought anything, regardless of quality; we have had labour indifference and, possibly what is more important, the question of high volume has tended to emphasize quantity rather than quality.
>
> After four or five years of prevailing attitude with respect to workmen, it is going to take a lot of time, energy and determination to eradicate these practices, particularly now that we have the Unions to deal with.
>
> I am fully aware that any industry that does not realize quickly the need for re-education and re-dedication is in for trouble, and all I can say at this juncture is that we are all prepared to take whatever measures are necessary and we will see to it that these measures are carried out vigorously.

But things did not turn out that way. Earlier there had been embarrassing difficulties with the #1 baler and the #55 tractor. Then in 1952 when the new #80 and #90 combines were introduced numerous quality problems appeared almost immediately. The cleaning capacity of the new shaker shoes was inadequate; about half the engines tended to heat; some brakes caught fire; the variable speed pulleys wore too much; a number of the table rams began to leak; some hydraulic pump sieves collapsed; the knife drive was too noisy; and cracks appeared in the machine. "We are losing our prestige in the combine business," wrote the President in July 1953 to E. G. Burgess. "The 80 and 90 machines have proven not only disappointing, but, in many instances have given such dissatisfaction to the owners that they intend trading them in." Several months later he wrote to H. H. Bloom, President of Massey-Harris at Racine, that "it is most anomalous and distressing that as leaders and prime movers in the self-propelled combine business, we seem to be incapable of building a machine without having to recondition it the following year." These difficulties were then attacked and soon overcome, but they undoubtedly weakened the position of the Massey-Harris combine in the market at a time when tractor sales were already falling sharply.

Mechanical difficulties with the new combines, and before that with some of the tractors as well as the #1 baler, suggested that all was not well with the Engineering Department. In the reorganization of 1944 a Vice President had been placed in charge of Engineering and Research. That is, the new research department formed at that time had been made subject to the control of Engineering. This was changed in 1950 when the two were separated, the heads of both being responsible directly to the President. The respective roles of each were not clearly defined and the inevitable conflict that emerged between them in 1951 brought to the fore some of the difficulties in the company's approach to engineering development.

The Research Department tended to be engaged in "pure research" but yet became involved in projects such as engineering the Pony tractor which helped to expand the size, and cost, of its operations. The Engineering Department, some thought, had suffered from this misdirection of funds and so had not developed adequately as a department supplying the company with products that the market demanded. Comparisons with engineering departments of some other implement companies were not reassuring, it was felt. Still, the two groups were kept separate and remained so until after the merger with Ferguson – and internal concern over the quality of new machine models also persisted. It is therefore not surprising that at the time of the merger with Harry Ferguson Limited in 1953, senior management of Massey-Harris felt that the company would probably benefit greatly from the engineering staff and engineering projects of the Ferguson company.

All these disparate and troublesome developments in North American operations, and the indecisiveness at head office in dealing with them, when taken together, made the company's fortunes in 1953 appear somewhat blurred. The changing North American sales environment, problems of high costs and inadequate quality, a technological lag in tractors, market weakness in the United States corn belt area through inadequate tractors and complementary implements, and the inevitable decline in its share of the self-propelled combine market finally began to have a serious effect on the company's financial position. Only valuable war contracts from 1952 to 1956 could hide what was happening to the company's regular business. Table 5 shows what happened. Most significantly, by 1953 the United States company was making almost no profits on its regular business. The decline in those profits had begun in 1951 and continued year by year with frightening regularity – until the $14.4 million loss of 1957. To put it bluntly, the company's ability to survive in the highly competitive United States market was again seriously being questioned.

On the brighter side, the figures also revealed a fundamental characteristic of the company that it had enjoyed for many years, as it does today, the ability through its international operations frequently to offset, in a major way, losses in some countries with profits in others. For this

TABLE 5

Massey-Harris Net Profits after Taxes (1951–1957*)
(Millions of Dollars)

	REGULAR BUSINESS IN NORTH AMERICA		TOTAL NORTH AMERICA		Total	
	Canada	United States	Regular Business	Defence	Outside N.A.	World-Wide
1951	7.1	5.5	12.6	—	4.3	16.9
1952	6.8	2.2	9.0	1.5	3.3	13.8
1953	4.1	.3	4.4	2.2	2.4	9.0
1954	1.7	—1.5	.2	2.3	6.4	8.9
1955	1.1	—.9	.2	1.2	10.8	12.2
1956	—1.1	—6.5	—7.6	.2	10.6	3.2
1957	—1.1	—14.4	—15.5	—	10.8	—4.7

*After 1953, Massey-Harris-Ferguson

reason alone, the company's development outside North America over the years 1946 to 1953 is important.

To understand properly how the company proceeded with developments outside North America it is useful first to outline its approach to matters of organization and control in the 1946-53 period. Knowledge of that approach is also pertinent for understanding the difficulties the company was later to encounter.

Until 1950 there had been no change in the company's organizational structure from the one introduced in 1944. The United States company was headed by a general manager responsible directly to the president of the parent company. Operations in Canada had been the direct responsibility of the parent company. In the United Kingdom, the managing director, who also watched over the remnants of the German operation, was responsible to the president of the parent company. The European branch, created many years ago, continued after the war, headed by a general manager and with head office at Marquette. The French company reported to it, as did the Massey-Harris subsidiary companies (for sales only) in Denmark and Belgium.

By 1950 several things had happened to justify examining the company's organizational structure. The sales volume of the organization had grown substantially, the United Kingdom operation had emerged and had become important, and the French operation with its new Pony tractor project and additional expansion plans was also increasing in importance. How should these operations outside North America be controlled? The President considered this problem in a letter dated April 21, 1950:

> In the over-all European picture, one of the most important decisions to make is whether or not we should consider establishing top management, either in London or in Paris, who would be in charge of the British, French and Continental operations. If this were done, it would mean, of course, that we would

have to have a General Manager, probably a General Purchasing Manager and a General Engineering and Development Manager. I am assuming that the General Manager would be of the sales type and, therefore, a General Sales Manager would not be required.

. . . We must not overlook the fact that the ease and rapidity of communication between Toronto and Europe has materially changed the situation. You might, therefore, make a recommendation that we establish a European Management in Toronto.

This suggestion implied a very high degree of centralization, and it is difficult to imagine that it could have worked well. It was finally decided instead to establish a European division. A new organizational structure was announced on October 4, 1950, and it introduced the concept of the operating division, of which four were established: a United States Division (in practice it had been a separate operation), a Canadian Division, a European Division, and an Export Division (Organization chart 1). A high degree of parent company control was implied in this new arrangement, for the head of each division was a vice president of the parent company; the parent company appears to have felt it necessary to deal with the problem in a way that would not diminish its control over *operations*. The European Division, with Walter Lattman as General Manager, included the United Kingdom and French operations, and it was also responsible for sales in the European export areas, Near East, North Africa and Africa other than South Africa, the Rhodesias, and Katanga Province. Although its head office was in London, the General Manager had an assistant stationed in Toronto. The division's export department was also in London.

The Export Division was located in Toronto, headed by W. W. Mawhinney, and was responsible for sales in South America, Mexico, Central America, South Africa, the Rhodesias, Katanga, the Far East, New Zealand, and Australia. H. H. Bloom became General Manager of the United States Division at Racine and E. G. Burgess became General Manager of the Canadian Division at Toronto.

The 1950 reorganization also provided for a group of executives termed "Head Office Staff" who were responsible directly to the President (through the Vice President General Administration, R. H. Metcalfe), as were the heads of the operating divisions (see Organization Chart 1). But these did not include executives in the major functional areas of manufacturing and marketing, and in practice the distinction between "staff" and "operations" was not maintained. Within the Canadian Division the functional divisions were manufacturing, sales, engineering, supply and control, and that of the Comptroller, but no attempt was made to standardize the functional divisions of the various operating divisions.

The 1950 change in organizational structure was not accompanied by any attempt to define responsibilities, nor did it introduce new control concepts. Indeed, matters of organization were seldom discussed in the company. It may be that the high degree of operational centralization that the structure implied made it seem unnecessary to define responsibility in detail or to establish new reporting and control procedures. Control was based essentially on periodic submission of balance sheet and income statement data in traditional form; on annual budgets also submitted in that form, but revised during the year; and on periodic personal visits by the President, for which masses of material usually were specifically prepared. Frequently also, cable requests by the parent company for data such as sales forecasts, were employed. The basis of the budget was a sales forecast made by salesmen. There was no market research of a continuing nature.

Control through balance sheet and profit and loss data presented in traditional form, and usually giving regularly only year-to-year comparisons, makes it difficult to see trouble quickly. It was essentially that kind of data, together with a written report prepared in advance by the President and delivered orally, that formed the basis of control reporting to the Executive Committee of the Board, apart from data relating to capital expenditures. Because of this it is difficult to believe that the Executive Committee was adequately informed; but on the other hand the Committee seems not to have pressed for more adequate information on operations until serious difficulties had already appeared.

Now to turn to the company's developments outside North America, ones that were later to provide such welcome offsets to the financial distress in its operations in North America.

The company's interest in export activity was fully maintained after the war. Sales outside North America, as Table 3 shows, rose more than sales in Canada during the years of acute shortages up to 1951. Export markets were not starved in comparison to the Canadian market.

In the organizational structure of the company introduced in 1944, the Export Department of the company was a division of the Sales Department in Toronto. The department remained in this position until 1950 when it became a division at Toronto independent of the Sales Department but lost responsibility for export sales in Europe, and in certain other areas, to the newly formed European Division. Sales, as before the war, were made to local independent distributors in the export territories or, as in Argentina, Brazil, Uruguay, Denmark, Belgium, New Zealand, Union of South Africa, United Kingdom, Germany, and France, through local branch or subsidiary Massey-Harris companies.

In 1946, $24 million out of total agricultural equipment sales of $68 million were sold outside North America which is a measure of the extent to which export trade was resumed immediately after the war. The range of this activity is also indicated by the large number of countries in which sales were made. These included the United Kingdom, New Zealand, Australia, Union of South Africa, France, Belgium, Portugal, Denmark, Sweden, Norway, Ireland, Netherlands, Switzerland, Tunisia, Algeria, Palestine, Turkey, Iraq, Syria, Sudan, Morocco, Egypt, Madagascar,

ORGANIZATION CHART 1

Massey-Harris Company Limited/Canadian Parent Company
October 1940

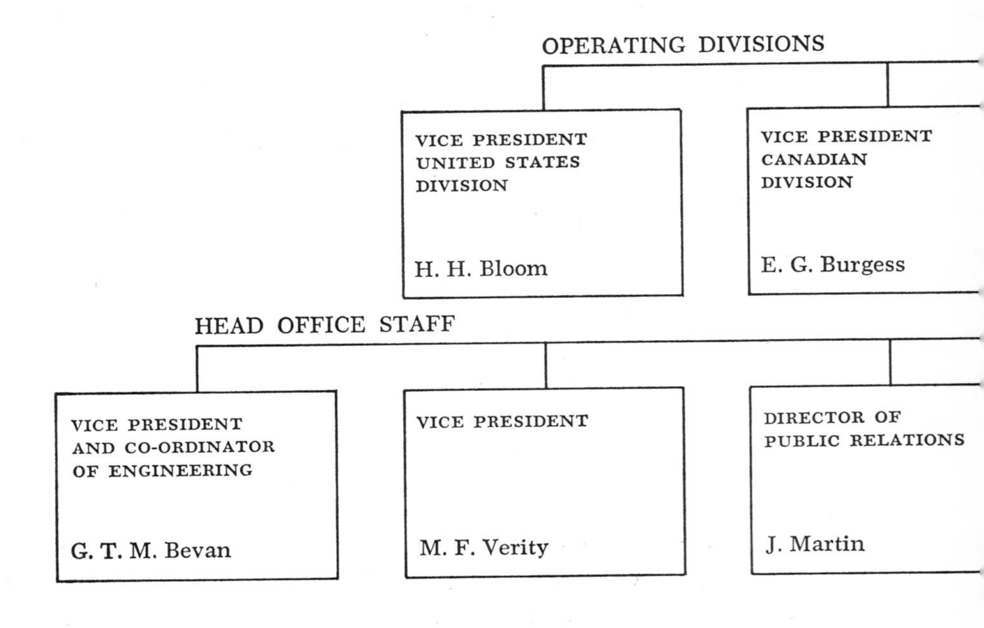

ADVISORY COMMITTEE: Appleton, Bloom, Burgess, Lattman, Mawhinney, R. H. Metcalfe.
EUROPEAN DIVISION: included Near East, North, East, Central, and West Africa apart from Kenya, Sudan, Rhodesia.

EXPORT DIVISION: included South and Central America, Mexico, South Africa, Rhodesia, Kenya, Sudan, Far East, Australia, New Zealand.

M. F. Verity was responsible for the Machine Tool Division.

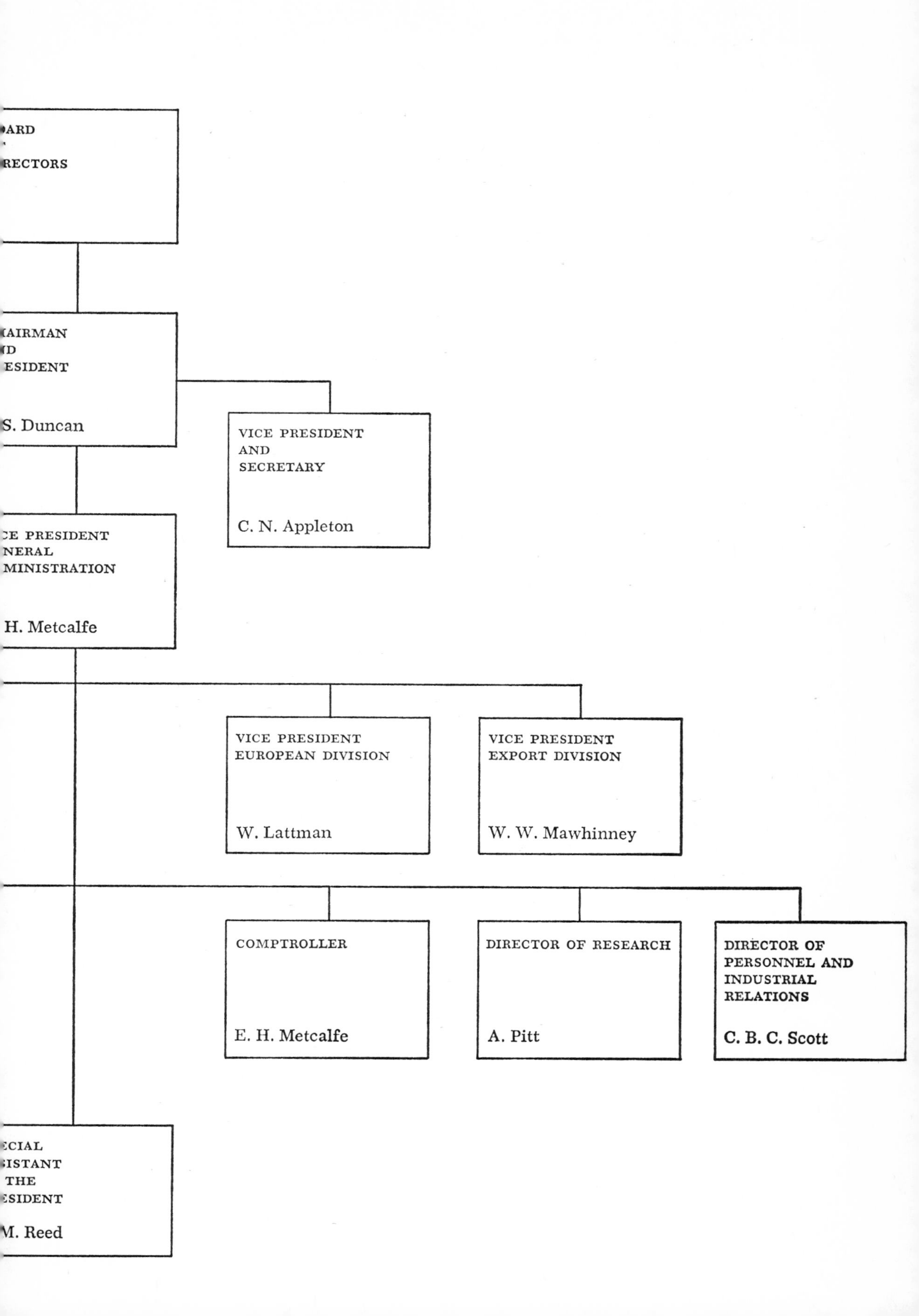
ARD
RECTORS
AIRMAN
D
ESIDENT
S. Duncan
VICE PRESIDENT
AND
SECRETARY
C. N. Appleton
CE PRESIDENT
NERAL
MINISTRATION
H. Metcalfe
VICE PRESIDENT
EUROPEAN DIVISION
W. Lattman
VICE PRESIDENT
EXPORT DIVISION
W. W. Mawhinney
COMPTROLLER
E. H. Metcalfe
DIRECTOR OF RESEARCH
A. Pitt
DIRECTOR OF
PERSONNEL AND
INDUSTRIAL
RELATIONS
C. B. C. Scott
CIAL
ISTANT
THE
SIDENT
M. Reed

Greece, Jamaica, Brazil, Ecuador, Chile, British Guiana, Colombia, Mexico, Peru, Cuba, Trinidad, Costa Rica, Bolivia, Puerto Rico, Venezuela, Argentina, Uruguay, Haiti, Ceylon, India, China. Sales were also made directly to the United Nations Rehabilitation and Relief Administration. In 1953, $68.7 million or 34 per cent of the company's total sales of regular goods were made in markets outside North America.

Although the company's manufacturing activity outside North America did not expand strikingly up to 1953, it was more significant in a historical sense than its relative size would suggest. This is because the three countries involved – United Kingdom, France, and Germany – remain to this day, centres of manufacturing activity of the company and, indeed, have become exceedingly important to it. Moreover, some of the difficulties that arose after 1950 or 1951, and which argued in favour of a merger with Ferguson, arose in the company's manufacturing operations abroad.

First there were major developments in the United Kingdom. By agreement with the United Kingdom Board of Trade and the Ministry of Aircraft Production, Massey-Harris leased and occupied a war factory of 196,368 square feet on the Barton Dock Road, Manchester, about one mile from their existing facilities on Ashburton Road. The plant was taken over on January 1, 1946, toolage was installed, and in 1946, the company was able to produce 1,383 #716 fertilizer distributors, 774 #714 spring tine harrows, 500 #730A T.D. harrows, 950 #734 mowers, 368 #721A hammer mills, and they also made a few of the #711 manure spreaders for the first time. In all, 3,981 units were produced, nine times as many as in 1945, and by 1948 this number had increased to over fifteen thousand, including now also a C.M.A. rake, the #726 plow, and more important still, the #722 combine and a few #744 P.D. tractors. Whereas in 1945, 430 units had been produced and 7,036 imported, in 1948, 15,242 were produced and only 3,506 imported.

The way in which the company financed its operations is rather interesting for it contrasted with the way it financed the German and French operations in the 1920s, and to some extent set the pattern for financing of such operations in later years. Expansion in 1946 was financed by a $500,000 bank loan guaranteed by the parent company. The parent company extended credit in the form of materials and implements, and this debt was satisfied in 1947 when the British company was transformed into a public company (January 28, 1947) and the whole of its increase in issued share capital of £280,000 was issued to the parent company. However, the bank loan, other capital expenditures and necessary addition to working capital were met by a local public issue of £400,000 3½ per cent debenture stock, guaranteed by the parent company and maturing in 1972. By leasing the factory and securing substantial local finance in the form of debentures, the parent company had minimized the use of its own funds and yet had retained complete ownership and control over the local operation.

Events leading to the local production of tractors are significant in that they illustrate the predilections of management at that time for subcontracting and they explain why the company became involved in its next manufacturing venture in the United Kingdom – the factory at Kilmarnock, Scotland.

The glaring gap in the production schedule of the Barton Dock Road factory in 1946 and 1947 was that it did not include tractors and combines. For most implement companies these items represent a very large proportion of total sales, but their production also involves heavy capital expenditures. The President of Massey-Harris was reluctant to take that plunge in the United Kingdom and sought alternatives. He wrote to the Nuffield Organization in February 1947 offering them an arrangement whereby Massey-Harris would cease distributing its own tractors in the United Kingdom and Ireland in exchange for exclusive selling and distribution rights in those countries for the Nuffield tractors; exclusive selling rights in other territories where restrictions prohibited the sale of Racine-built Massey-Harris tractors; and selling rights in certain other European territories where, however, the Nuffield tractor would be sold by Massey-Harris alongside its own tractors. This proposal was declined by the Nuffield Organization, which the President found " . . . disappointing, if not unexpected." Similar arrangements were offered to the David Brown Company with the same result.

It was this failure to become distributors for a competitive line that left the company no alternative but to begin manufacturing tractors in the United Kingdom. Had it succeeded in obtaining distribution rights it would have meant that the company would have taken a further step away from being a manufacturing company – a step that it appeared quite willing to take at the time.

A decisive incentive to begin manufacturing tractors appeared in the later part of 1947 when Massey-Harris began to see the possibility of supplying increasing quantities of tractors and implements for the Ground Nut Development Scheme in Tanganyika, and also the proposed future developments in Gambia and French West Africa – provided it could locate these in the sterling and not the dollar area of the world. The ground nut (peanut) project arose out of the need to increase the United Kingdom's supplies of edible fats. In March 1946, the United Africa Company, a subsidiary of Unilever, presented a proposal to the Minister of Agriculture of the United Kingdom government for a 2.6 million acre project for growing ground nuts on the vast undeveloped plains of Tanganyika in East Africa. The scheme involved intensive mechanization. The government was enthusiastic, sent out a mission, and chose Kongwa as the prime site for operations – unfortunately a low rainfall area.

Machinery was scarce, a motley assortment of war surplus crawler tractors was used to help clear the land, and finally even converted Sherman tanks were pressed into use. It was in this kind of environment that scarce dollars were allocated to buy proper farming machinery from Massey-Harris.

The company signed a contract with the United Africa Company Ltd. in January 1947 to supply them with approximately $5.0 million worth of tractors, implements, and spare parts for their project in Tanganyika, and by October 1947 some of these had been delivered from North American factories and were at work in the Kongwa part of the project. It then became clear that the Overseas Food Corporation, a British Crown corporation, would assume responsibility for the ground nut project. While this meant that the project would continue to go forward, it was by no means certain that Massey-Harris would be permitted to supply the large number of implements and tractors that the project would need. The company did have one advantage over others, in that it had assisted the venture in its early stages when farm machinery had been exceedingly scarce. But the problem was that it had no tractor manufacturing facilities in the United Kingdom, and because of the shortage of dollars this was officially regarded as a serious obstacle to the company's participation.

G. T. M. Bevan, Vice President in charge of Engineering and Research, was requested by the President on August 5, 1947, to prepare a general report on "Part Manufacture and Assembly of Tractors in the United Kingdom." The report, completed in November 1947, concluded that such a project was feasible. An engine was found to be available in England which could be installed in the Massey-Harris standard and row-crop #44 tractor chassis without major modifications, and certain other parts could be purchased or made by Massey-Harris in England; but the transmission, rear axle, and some other special parts would have to be imported from Racine. The engine Bevan chose was the Perkins P.6 Diesel, after having investigated numerous others. He was clearly very impressed with F. Perkins Ltd. of Peterborough, and of their engine production facilities; he wrote, "I have seen nothing to equal it or come anywhere near it yet. It could be put down in Canada or the U.S.A. and be a show factory." Many years later when Perkins was acquired by Massey-Ferguson, its product and its operations were familiar to the company.

When the company began to estimate the likely demand of the ground nut operation for tractors and implements, and when to this it added the strong demand for its farm machinery, particularly its combines, in the United Kingdom and export markets, there seemed little doubt that its facilities in the United Kingdom would be inadequate. Unfortunately expansion at Manchester on the Barton Dock Road was prohibited by the government since it was not classified as a development area, something which had not earlier been expected. However, during a visit by Bevan to the Permanent Secretary of the Board of Trade in October 1947, the company's attention was directed to the possibility of locating in Scotland with government assistance. This proposal was investigated and accepted.

The Scottish Industrial Estates Limited, a non-profit division of the government, operated in conjunction with the Board of Trade, facilitated the construction of two buildings, one of 150,000 square feet, the other of 30,000 square feet, which were rented to the company on a 21-year lease without option to purchase. The major attraction of the arrangement was not just that it relieved Massey-Harris from having to provide funds for the construction of a factory but also that the rent and taxes were exceedingly low, amounting in total to only about £9,750 annually.

The apparent feasibility of manufacturing the #744D in the United Kingdom and the availability of substantial assistance in financing the project at Kilmarnock, strengthened the company's position in maintaining its association with the ground nut project. After lengthy negotiations with the Overseas Food Corporation a letter of intent was signed on April 22, 1948. Production of the #744D at Kilmarnock began in 1949. As for the ground nut project, it was a fiasco and was abandoned. But Massey-Harris obtained valuable business and to this day retains distributors in Africa that were established indirectly as a result of the project.

Other machines were also introduced into the Kilmarnock plant: the #722 self-propelled combine in 1948, the #726 in 1949, the #780 in 1953, the #701 hay baler in 1951. There was the usual difficulty associated with introducing new machines into production, but difficulties with the #701 baler were somewhat more than normal. No sooner had a large number of them gone into the hands of farmers than complaints came in from all directions. The difficulties lay in the general construction and design of the machine, particularly with respect to welding, and with the packing and knotting mechanism. In the end almost all the machines sold were taken back into the factory or reconditioned in the field at a substantial cost to the company. What started as a disaster turned out much better than it might have done, for the new manufacturing establishment had demonstrated its determination to stand behind its product. But there was the nagging question of the quality of the company's engineering staff, which had permitted the difficulty to arise in the first place.

While the profits of the United Kingdom operation began to decline after 1952 and a number of other difficulties appeared on the horizon, the operation was without doubt very successful from a financial point of view. Head Office in Toronto did not directly increase its 1947 investment of £330,000 in the form of share capital, but through retained earnings and various reserves its investment had tripled by 1951.

Those years also revealed something that had not originally been expected – that costs of production in the United Kingdom would permit the company to source some of its exports from there rather than from North America. The 1949 devaluation of sterling was an important factor in making the United Kingdom operation internationally competitive.

But there were problems, the most important one being the competitiveness of the tractor produced at Kilmarnock. While it is true that the #744D tractor was originally intended for the export market, its total failure in the United Kingdom was probably unexpected: in 1952, 5,225 were sold in export markets and only 68 in the United Kingdom, a year in which industry tractor sales in the United Kingdom amounted to 33,188, of which 17,335 were Ferguson tractors. It is unlikely that a company can continue a tractor operation in a country where it has failed to penetrate the local market. Nor did the #744D appear to have much future in the export market. A forecast made in 1953 for 1956 sales of the tractor foresaw sales of only about 3,000, little more than half of 1952 production. The tractor plant itself had not changed from an assembly to a manufacturing operation: in 1952 about 95 per cent of its costs were accounted for by purchases of direct materials. Sales prospects for 1953 were disappointing as well, and in the end only 2,546 were produced in that year, compared with 5,547 the year before. The #744D project was in jeopardy.

The fundamental position of combine production at Kilmarnock was much stronger. Penetration of the United Kingdom market was most successful, and that penetration had a reasonable chance of being perpetuated for the company was planning to replace the #726 self-propelled with the new #780 combine in 1953. Still, in 1952 sales did begin to decline, production was cut to 4,636 combines from 5,000 the year before, and to 3,897 in 1953, so that excess capacity and labour "redundancy" was becoming a problem here as well as on the tractor assembly line.

The matter of labour redundancy at Kilmarnock was a serious one for it brought into the open certain difficulties with labour that the company had experienced for some time. These problems stemmed from a number of related circumstances: a new factory located in an area where redundancy had not been common, somewhat limited alternative opportunities for employment, and a company with a strong desire, until sales declined, to maintain production even at the expense of compromising with sustainable labour relations policy. One possible way out was to increase production by taking on subcontracting projects. Massey-Harris began actively to explore that possibility and in so doing encountered Harry Ferguson Limited, an encounter that was to change decisively the character of Massey-Harris. To that story we shall turn in the next chapter.

In France, the company's experience up to 1953 was rather different than in the United Kingdom, for there it already had a plant. The President was able to report to the Board of Directors in Toronto on December 26, 1944, that during the German occupation the Marquette plant had operated on a fairly normal basis, that the financial position of the French company was reasonably good, and that it would probably not require funds for rehabilitation from the parent company. This was good news considering that capital was short and that the Board was worried about investing funds abroad. The French company simply continued its production, but at an increasing rate. In 1945 it manufactured 1,333 mowers, 762 mower bars, 80 tedders, 311 binders, and 1,011 reapers, as well as a few rakes and disc plows – a total of 3,502 units, many of them of the horse-drawn type. By 1948 this had risen to 30,203 units, including now also some fertilizer distributors and a greatly increased production of rakes. But that number is less impressive than it appears. Sales of Marquette-made machines in Canadian dollars amounted to $3.1 million in 1948, which was only 23 per cent of the French company's total sales, and actually less than the plant had produced in 1931. Virtually all the rest of the machines sold came from the North American factories of Massey-Harris.

There were two major obstacles to further significant expansion of the French company; one was the scarcity of capital which inhibited extension of manufacturing operations, the other was exchange restrictions which reduced imports. The Board of the parent company had starved the French company for capital equipment, so that its condition in 1949 was rather like the condition of the North American factories in 1945. Exchange restrictions made it increasingly difficult for the company to be supplied with implements from North America. Something had to be done if it was to survive on a profitable basis.

Meaningful expansion of local production meant tooling up for a tractor. Incentive for doing this came from the large demand in Europe for tractors and from the policy of the French government to limit tractor imports from hard currency areas – at that time North America.

Massey-Harris considered establishing a tractor plant in France as early as 1946, and thought in terms of an annual production of 5,000 tractors by 1950-51 and of capital expenditure of $3.8 million, all to be raised in France with government assistance. But events moved slowly, and not until the latter part of 1948 were satisfactory proposals presented to the French authorities. The approach favoured by Massey-Harris was one which confined their contribution to design engineering and technical "know-how," with all the capital being raised locally.

Massey-Harris preferred producing a 20-horsepower tractor in France, but the "Monnet Plan" officials of the French government would not permit it since International Harvester already had a tractor of that size in production. So Massey-Harris executives argued in favour of the Pony and demonstrated it before French government officials. The demonstration showed that it could pull a one-furrow 12-inch plow, to a depth of 8 to 10 inches in all normal French conditions and it could draw a wagon of 3½ tons in the field. Fortunately the 98 Pony tractors that the company had been permitted to import from Woodstock in 1948 were giving good service. And it was also fortunate for the Pony proposal that the famous Ferguson TE-20 tractor was not in production in France. So finally, after prolonged negotiations, agreement between Cie Massey-Harris S.A. and the French authorities was reached, and by means of a loan emanating from counterpart funds of Marshall Plan imports, expansion of the Marquette plant was undertaken in early 1950. Involved was a twenty-year 7 per cent mortgage loan amounting to Fr. 700,000,000, and it was to be followed by a further loan of Fr. 100,000,000 in 1951 – all from the French government through the Crédit National.

In addition a medium-term loan from French banks for Fr. 140,000,-000, guaranteed by the parent company, local overdraft facilities for Fr. 200,000,000 and a $450,000 ten-year loan from the parent company, were arranged to finance the manufacturing programme. In this way Massey-Harris was able to retain full ownership of its French operation with a minimum dollar outlay, almost exactly what it had wanted all along. In spite of worrisome delay in receiving the two instalments of the Fr. 700,000,000 loan the project went ahead, and it included implements for the Pony tractor as well as the tractor itself.

An engineering building, steel storage shed, and the Pony tractor works were built – the first capital expansion since 1927 – and soon production of Pony tractors was under way. This was without question a milestone in the development of the French operation. It gave its manufacturing operation a new lift, it assured the company of a place in the tractor market in France, and it greatly improved the French company's profits, for the Pony proved to be a resounding success in France. In 1951, 2,400 tractors were produced, 5,810 in 1952, and 8,133 in 1953, along with harrows, plows, and cultivators. Before the Pony (in various models) was dropped in 1960 over 92,300 had been manufactured at Marquette.

Even before the Pony tractor operation was in full swing the French company turned its mind to the possibility of producing combines at Marquette. Walter Lattman, then Vice President European Division, submitted a plan for such a project to the Commissariat Générale du Plan de Modernisation et d'Equipement in March 1952. The plan envisaged a total capital investment of Fr. 789,000,000 and therefore a long-term loan of Fr. 800,000,000 was thought to be necessary, and a production target of 2,000 machines annually by 1956 was decided upon. Cie Massey-Harris S.A. would arrange for its own working capital.

That this plan should be accepted became a matter of some urgency because political instability in France made it progressively less likely that the parent company would commit funds to the project. Writing to Lattman on March 5, 1952, the President pointed out that:

> If our Directors were somewhat reluctant six months ago to put Canadian money into France, you can well imagine how reluctant they would be today at the very thought of making such a move. On the other hand, I am just as convinced as ever of the absolute essentiality of manufacturing combines in France, and the object of my letter is to ask you to try to bring this matter to a head. . . .

With this Lattman agreed, but he had to explain that frustrating delays were being encountered. In the end a loan for Fr. 800,000,000 was received from the government through the Crédit National in three instalments, and the project went ahead after the first instalment was received in 1953.

In addition a medium-term loan for Fr. 390,000,000, backed by the Crédit National, was received from French banks and as before an investment of $450,000 – in the nature of a gesture – was received from the parent company in Toronto. A combine plant was then built at Marquette and also additional workshop facilities. The combine chosen for production was the new and large #890 self-propelled, a bigger combine than the ones produced at Kilmarnock at that time, for this was what the French market seemed to require. By the end of 1953, 99 of these self-propelled machines had been turned out, followed by over 900 in 1954. Production of combines by the company at Marquette has continued to the present day.

In both the tractor and the combine projects Massey-Harris made great efforts to minimize capital spending – an approach necessitated by their policy of minimizing, even attempting to avoid entirely if possible, the investment of dollar capital and yet retaining ownership of the local company. In later years the investment policy of the company in France had to change somewhat.

As a consequence of the tractor and combine projects, Massey-Harris became leaders in France in both those fields. Sales rose steadily and continued to do so in 1952 and 1953 when sales for the rest of the company, including those in the United Kingdom, were declining. Building up production at Marquette, and sales in France generally up to 1953 was not achieved without difficulty. Besides exchange restrictions, capital scarcity, and shortages of materials, there was the continuing problem of labour unrest at the Marquette plant and major difficulties with quality. These difficulties however deepened the company's knowledge of the prerequisites for operating successfully outside of North America, and by 1953 solid accomplishment was evident. Important decisions for building up the French operation had been made, the name Massey-Harris was firmly established in France, and the French company was making quite satisfactory profits.

In Germany there was little hope for the local company, Massey-Harris G.m.b.H., until late 1951, for the parent company appeared to have no definite plans for its resurrection. The Westhoven plant, as we have seen, was completely destroyed and management in Toronto was forced to admit that it probably would never be rebuilt. But people must eat, and a few of the old employees of the German company went back to the factory site to see what might be done. A local contractor wanted some of the plant salvage. He agreed to pay a small sum for it and he also agreed to clear about 10 per cent of the factory space. After the area was cleared, some bricks were sorted out of the rubble and were used to build walls, girders were straightened and after a difficult search for necessary materials, a roof was built. When it was finished the building still had the appearance of one that had been through a fire or earthquake.

Machine tools, equipment, spare parts and materials that had been moved to Wipperfurth, Germany, and Wels, Austria, during the war, were returned to Westhoven and placed in the improvised structure. Later, as it became evident that business with farmers in the eastern part of Germany would not develop, the spare parts stored in Berlin were gradually moved to Westhoven. This ended an association with Berlin that had begun around the turn of the century, although actual sale of the Industriestätte property did not occur for some years.

The local staff in 1948 was seriously concerned with obtaining enough food to eat, and the income from sales of spare parts provided welcome assistance. Even so, in early 1948, the Managing Director of the United Kingdom company, G. H. Thomas, who had also been asked to watch over the German interests of Massey-Harris, recommended to Canadian management that regular shipments of food parcels be increased. Inhibiting any real economic progress was the everpresent probablity that the German mark, hopelessly inflated, would be devalued. Finally it was, in June 1948, and some order in commerce began to emerge from the chaos that had prevailed for so long.

Activity at the Westhoven plant was at first confined to making spare parts, but arrangements were made for subcontractors to build some mowers, rakes, and tedders. Thought was even given to manufacturing the old Junior binder at the Westhoven plant. But this essentially was a pre-war approach to product line, for it involved horse-drawn implements, and it could not hope to be successful. Disappointing sales of subcontracted machines could be explained in various ways, but the hard fact soon emerged that if there was to be a German operation it could not be built successfully on pre-war concepts.

Then in 1949, almost by accident, the German company began its first manufacturing operation. The United Kingdom company had encountered a serious roller chain bottleneck at its Kilmarnock combine plant. Its Managing Director, on an inspection tour in Westhoven, met a former East German roller chain contractor to whom he told his troubles, and who then suggested that they co-operate to produce it at the Westhoven plant. This was finally done after the compilation of necessary technical information from memory and in spite of capital costs that ran higher than had been expected. To this day the German company has manufactured roller chain. However, the venture was not at that time part of any definite plan to reconstruct the German company.

At first the quality of the chain was not up to the standards expected by the United Kingdom company and problems of delivering chain on time harmed its relations with the North American Massey-Harris companies. But these difficulties were slowly overcome, and for a period a large portion of the German company's manufacturing activity was centred on the production of roller chain for the United Kingdom company.

In the meantime, signs of hope in German agriculture were reappearing. Other farm machinery companies were moving more aggressively than Massey-Harris, although it must be said that Massey-Harris was the only major implement manufacturer in West Germany that had had its production facilities destroyed completely during the war. Rather interesting is one comment by G. H. Thomas in his report on the first post-war Agricultural Exhibition held at Hanover in early 1949:

Ferguson, Coventry, had an excellent display. I was informed that they were holding demonstrations throughout Germany and I was rather curious, in view of the fact that at the moment it is impossible to export tractors to Germany. I was advised by their Representatives that although it is true imports of tractors are forbidden, the company felt it was desirable to show the German people the most modern development as far as tractor manufacturing is concerned.

Massey-Harris did not show similar enthusiasm in Germany at that stage. Their factory had been destroyed, their important East German business was gone, and also, since companies such as International Harvester and Fahr were intact in West Germany, they faced stiff competition in that part of Germany. Their pessimism at that time, as well as their policy, is revealed in a letter from the President, J. S. Duncan, to G. H. Thomas, in March 1949:

Basically, my views with reference to the German situation are that our business has deteriorated there to such an extent that it would be very difficult for us to build it up again exclusively on borrowed capital. On the other hand, none of us believes that the situation in Germany justifies our investing Canadian dollars at the present time, and even if we did, the process of building up our own plant and selling organization would be a slow and laborious one, and would have to be probed very carefully before we decided to go ahead with it.

All of which brings us to the conclusion that probably the best way to re-enter the German market would be through the front door, namely through some tie-up with existing companies, such as Fahr, or perhaps others in whom you might become interested, and that if we were prepared to risk Canadian money in Germany, this medium would be more satisfactory than the slower process of building up our own organization.

Attempts were made to become associated with existing German companies. Discussions were held with the Fahr company in early 1950, first with the thought of forming a new company, and then on the basis of Massey-Harris investing in the Fahr company using blocked marks. The inevitable conflict that would arise in the export market as a result of having to market two lines – Fahr and Massey-Harris – was recognized but accepted. In May 1950, W. K. Hyslop, Vice President of the parent company, reported to the President as follows:

. . . I am not in favour of opening up a factory in Germany, of going into a 50/50 arrangement with the Fahr Company to manufacture Combines, or of investing money in the Fahr business unless under very favourable circumstances with blocked marks.

I recommend that we make arrangements to buy Tractor Binders from Fahr, to buy a Tractor (30 H.P.) in Germany, preferably from Fahr. . . .

But no agreement was reached. In early 1951 Massey-Harris approached the Hanomag company to manufacture their R.28 wheel diesel tractor as a Massey-Harris tractor to fill the gap between the Pony, being built at Marquette, and the bigger #744D tractor being built at Kilmarnock. But again negotiations proved to be unsuccessful, as they had also been, it will be remembered, with the English tractor manufacturers a few years earlier.

Nor did the company in 1950 feel that it could recommence manufacturing operations with government assistance. In 1950 Massey-Harris officials had a meeting with Dr. Heinrich Troeger, now Deputy Finance Minister of the Rhineland government at Düsseldorf; it will be remembered that he was custodian of Massey-Harris G.m.b.H. during the war, and had co-operated closely with R. A. Diez of the German Massey-Harris company. Dr. Troeger advised them that he could perhaps help them to obtain a long-term loan to establish manufacturing facilities in Essen, a distressed area. This was not pursued. Not until the later part of 1951 did management begin to take a more positive view on the future of their German company. Some additional work was done on the Westhoven factory to facilitate increased production of spare parts, and the chain manufacturing operation was reorganized, but these were minor moves.

Then in March 1952, Diez, who had been with the German company during the war, then with parent company in Toronto from March 1948 until the end of 1950, and finally with the European Export Division in London, was transferred to Germany as Managing Director. The guidance given him by the President as he remembers it was: "I don't know what you are going to make in Germany. I don't know how you are going to make it, but we will give you every support except money." This indeed constituted a clear statement of the policy of the Massey-Harris Board with respect to manufacturing operations outside North America at that time, in that it confirmed that a more positive parent company view on the German company's future did not include significant financial support.

But the first move towards the permanent establishment of a new manufacturing operation in Germany occurred through curious circumstances. Executives of the French company had discovered that to increase their sales of combines, they had to sell a straw press attachment that could be put on their combine and would permit farmers in one operation to retrieve threshed straw for bedding and fodder. Before the war the company had been distributors in France of a popular low-tension straw press built by the Raussendorf company in what, after the war, was the Russian zone of Germany. Hermann Raussendorf lost his factories after the war and attempted to re-establish himself by becoming associated with the John Grebestein company in Eschwege. The French company officials met Raussendorf and asked him to develop and manufacture a mounted press, which he agreed to do. But the Grebestein company failed to meet delivery schedules and Diez was asked by the French company to investigate. The problem appeared to be one of lack of finance, and on the recommendation of Diez, the French company advanced them DM100,000 in April 1951, and additional amounts after that. Having become involved in this way Diez naturally took a close interest in the press manufacturing company at Eschwege, and began to think of it as a possible nucleus for a new start for the German company. Then when the Grebestein company was unable to overcome its difficulties it appeared that, both to assure their source of supply of presses and to protect their investment, Massey-Harris should perhaps acquire all the assets of that company.

The difficulty was the old one – the scarcity of capital. The parent company would not supply it. Diez went to the government of the Land of Hesse, the province in which Eschwege is located, and there discovered that Dr. Heinrich Troeger was now Minister of Finance of that government. It also developed that the government was most anxious to increase employment opportunities in Eschwege. An agreement was made whereby the Land of Hesse would lend Massey-Harris G.m.b.H. DM600,000 at a favourable rate, for the purpose of buying the machine tools, patterns, installations, raw materials etc., and would rehabilitate the buildings (part of an old airport) and rent it to the company. The latter was not unlike the arrangement at Kilmarnock. Raussendorf was employed by the new owners. So Massey-Harris now had two small manufacturing operations in Germany – one at Eschwege and one at Westhoven. It was an important step forward for the German company.

Perhaps the first really significant and visible evidence that management at Toronto was beginning to consider seriously the development of its German operation came in 1953 when it was decided that a small combine would be produced in Germany. The President of Massey-Harris gave permission to the German company not merely to manufacture but also to design that combine, a decision that lifted the German company into a new role within the structure of the organization. Tom Carroll was brought in from Toronto and he, together with Hermann Raussendorf, designed the #630 self-propelled combine. But actual production did not begin until 1954, so the period up to 1953 ended for the German manufacturing operation with new hope for the future rather than with much ground for satisfaction with the past – a past extending back to the late 1920s.

Looking back on the experiences of Massey-Harris over the years 1946 to 1953, many things come to mind. Until 1951 the company had experienced exhilarating success and that success was explained by four factors: the remarkably correct decisions relating to the development, the manufacture, the marketing and the improvement of the self-propelled combine; the success of the company in stretching its production facilities to satisfy post-war demand; the resumption of export activity right after the war; and the backlog of demand for agricultural machinery created by depression and war, by the shortage of farm labour, and by high farm income and low farm debt.

The parochial fear that abolition of the Canadian tariff would harm the company proved groundless, and in a relative sense production actually shifted into Canada. But a tone and temperament of caution and conservatism pervaded the company in its international operation. The Board was fearful of making investments abroad and the progress made in establishing operations in the United Kingdom, as well as in expanding continental operations had to be sponsored by the President in a positive fashion. Yet the Board did approve the moves, and its cautious attitude induced the company to depend heavily on local funds for financing operations outside North America. Long-sought penetration of the U.S. market was achieved with a major product, a most significant milestone in the development of the company as an international corporation.

The company did not make any significant progress in reducing its dependence on outside suppliers, and indeed it was in the area of manufacturing that management showed great reticence. Production control had made little progress either. As for the company's organizational structure, it was evolving slowly, but neither the company's basic approach to organization nor its systems of control changed visibly with increased size of operation. Control remained highly centralized.

In terms of relative size and rate of growth, Massey-Harris had done well. For many years it had been a large company in Canada, but in comparison with major United States companies it had been small – in fact almost insignificant. By 1953 its world-wide sales of agricultural equipment and repairs had, without question, moved well ahead of the sales of the medium-sized North American companies among which it had found itself for a long time – Case, Minneapolis-Moline, Oliver, and Cockshutt. Also it had grown more quickly than the major United States companies – Deere and International Harvester. Its sales volume had been much below that of Deere before the war; it was still under half of the volume of that company in 1946, but by 1953 it amounted to over 60 per cent of the volume of sales of that company, which in dollar terms meant $249 million. A considerable change from the $72.4 million of 1946 and the $21 million of 1939, even after adjusting for higher prices.

But as the year 1953 unfolded Massey-Harris was faced with some harsh realities. The United Kingdom factories needed output and the possibility of finding subcontracting work came to mind. Tractor sales in North America and in most other markets were dishearteningly weak and the company's rush programme to close an important technological gap was not at all certain of success. Past company experience in tractors, after all, was not encouraging. Combine sales, so long based on a sort of technological monopoly, were beginning to suffer, partly because the company's technological lead in combines was diminishing. Quality problems in several product areas cast doubt on the strength of the Engineering Department. Finally, the disappearance of profits in the regular business of the United States operation raised an age-old question: could the company really become soundly established in that market without greatly improved strength in tractors? Also, could its traditional strength in export markets be maintained without improved strength in tractors?

In the light of these problems, a merger with Harry Ferguson began to make great sense. Above all else Ferguson had a strong name in tractor development and unusually high penetration of the European tractor market, as well as a much higher share of the United States market for two-plow tractors than did Massey-Harris.

But who was Harry Ferguson? By 1953 many people outside of the farm machinery industry knew who he was because of his "David and Goliath" lawsuit with the Ford Motor Company and, after a four-year legal battle, of the award in 1952 to him of $9,250,000 in payment of patent royalties. Within the industry he was well known for his revolutionary "Ferguson System," involving a new approach to the integration of tractors and implements; and because of the astonishing commercial successes he had scored in marketing his TO-20 and TE-20 Ferguson tractor. That remarkable little tractor is still referred to by thousands of farmers and machinery dealers around the world, with a tone of affection that is not often accorded a mechanical device, as the "Fergie."

CHAPTER 5

Harry Ferguson's system

Harry George Ferguson was born on November 4, 1884, on a small farm in County Down, Northern Ireland. He and his brother, both without benefit of much formal education, opened a garage for servicing cars in Belfast when Harry was sixteen. That Harry was a rather unusual young man was soon shown. At nineteen he built a motor cycle to his own design, then he successfully engaged in motor-car racing, and by the age of twenty-five he had built and flown his own monoplane.[1]

These exploits and accomplishments attracted a good deal of attention, and during the First World War the Department of Agriculture of the Irish government offered Ferguson a job which involved supervising the operation and maintenance of tractors in use on farms in Ireland. Harry Ferguson accepted it. It proved to be the most important decision of his life, for from then until almost the end of his life in October 1960 Harry Ferguson devoted himself more to the development of agricultural machinery than to anything else. In his lifetime he, more than any other person, was responsible for developing and popularizing an innovation in tractor and implement design – still widely referred to as the Ferguson System – which has probably been as important to the evolution of that area of mechanized agriculture as the self-propelled combine was to harvesting. During the celebrated Ford-Ferguson lawsuit of 1948-52 the matter of patent rights and origin of inventions involved in the Ferguson System became crucially important, and made it desirable for the defendants to argue that Ferguson's role in development had not been an important one. A former Ferguson engineer, William Sands, was even willing

1 For details of the life of Harry Ferguson, see Norman Wymer, *Harry Ferguson* (London: Phoenix House, 1961).

to permit the impression to be created that he, and not Ferguson, was primarily responsible for the Ferguson System. In the environment of bitter legal debate, history becomes distorted. Comments of essentially historical interest given from memory, but not tightly documented, are not merely questioned by the opposing side but dismissed with scorn if they in any way cast doubt on favoured lines of arguments. An observer must tread carefully.

Ferguson's mind ran to mechanical inventions and experimentation was in his blood. All his life he spoke of "invention" even in cases where "development" would have been a more appropriate word. Because of his lack of formal training and his lack of confidence in books and in advice emanating from professionally trained engineers, he favoured a trial-and-error approach. He could not draw blueprints, found it difficult to read them, and favoured model demonstrations. But he was fascinated with novel mechanical devices, loved experimentations, enjoyed unusual perception concerning the potential usefulness of novel devices, and excelled at demonstrating them to commercial advantage.

In his government position during the First World War he was required to take a direct interest in farm machinery, and of course he already knew something about farming. At that time he seems to have begun to understand and appreciate the full significance of a rather simple fact – that hitching a tractor to implements previously drawn by horses introduced new problems, and that these problems detracted from the advantages of farm mechanization.

There was the obvious element of inconvenience in adjusting the tractor-trailed implement and also in lifting and lowering it. With a horse-drawn implement the driver's seat, at least, was on the implement not the horse. Bringing the implement closer to the tractor, indeed attaching it to the tractor, would ease this problem. A tractor-trailed implement also involved a new element of danger. If a stone or root was hit by a plow – and Ferguson had known many stony fields – the implement could be damaged, and if the front end of the tractor lifted off the ground, the driver might be hurt. The hitch geometry was the cause of the difficulty. Older tractors were heavy and Ferguson believed, rightly or wrongly, that such weight on the land interfered with the growth of crops and in any case wasted power. In spite of these disadvantages Ferguson was convinced that appropriate mechanization of agriculture was highly desirable because it would reduce the cost of food by freeing land used to raise fodder for horses.

As Ferguson began to think about these things, Henry Ford began to attack the problem of tractor size and efficiency. He developed the Fordson which was lighter than other tractors but which tended to tip up in front if care was not taken. Production of this tractor began in 1916 at Dearborn, Michigan. To increase wartime food production the British government requested Ford to consider manufacturing Fordsons in England, and in June 1917 Charles Sorenson came to England for Ford to investigate its feasibility. Scarcity of facilities, the need for speed in agricultural mechanization, and the problem of costs led Ford to decide against it. Instead Ford agreed to supply 6,000 tractors from Dearborn at an early date and at a satisfactory price.

Harry Ferguson heard of Sorenson's 1917 visit to England, made a point of meeting him, and later he recalled that he talked to him about the "one-unit system." This involved developing a light plow and linking it directly to the tractor, so that the two units, in effect, would become one. Ferguson has also recalled that at about that time he began thinking about using tractor power instead of manual power to lift and lower the implement. These ideas were the ingredients of what later became known as the Ferguson System, but it took until at least 1936 to develop them to the point of their being acceptable for practical application.

The development of the Ferguson System is best appreciated if the system itself is first understood. Its essential features are seven in number: integration of implements with the tractor; weight transfer from implement to rear wheels of tractor; implement penetration without excessive implement weight; continuous automatic control of the implement in the soil; quick and simple attachment and detachment of implements to and from the tractor; protection of the implement from all normal hazards of operation; effortless control and operation of the tractor and implements. These features were realized through the development of a suitable linkage arrangement between tractor and implement, of a system of hydraulic controls, and of an appropriate tractor.

When horses were used to pull implements power was derived from the neck or shoulders several feet above the ground and transmitted by traces to the load, as shown in Figure 1. The illustration shows that locating the hitch point in a straight line between the load point and the pulling point was depended upon to maintain the plow in equilibrium at the desired breaking depth, supplemented by the exertion of the plowman on the handles. Elementary depth control was later obtained from a depth gauge wheel added to the front of the plow beam. When tractors were introduced the linkage arrangement remained unchanged in principle as can readily be seen in Figure 2.

However, new problems were introduced with the appearance of the tractor. In heavy pulling, the torque of the engine created a lifting effect at the front of the tractor, tending to rotate the tractor around its rear axle, sometimes to the point of its tipping over backwards. To offset this tendency tractors tended to depend on built-in or added weight. But this approach ignored the hitch geometry that was the real cause of the problem. The Ferguson System does not depend on weight to overcome the problem of torque reaction. It depends on novel hitch geometry.

Rather than trailing an implement behind the tractor, Ferguson linked it directly to the tractor so as to make one integrated unit; in effect the implement was mounted right on the tractor and was lifted and lowered

Ferguson System

Figure 1
Typical hitch with horse-drawn plow. Power is derived from neck and shoulders of the animal several feet above the ground.

Figure 2
There is no improvement in method of hitch when tractor merely replaced animal to pull a trailing implement.

Figure 3
The Ferguson System integrates the tractor and the implement. The two bottom links of the three-point hitch draw the implement forward, while the top link transfers force applied by the implement into a forward pressure, activating the draft-sensing device in the tractor hydraulics. The effect is to add weight, and therefore more traction, to the tractor's rear wheels.

hydraulically. The linkage system, referred to as "three-point linkage" involved two links at the bottom that pulled the implement and one link at the top, as can be seen in Figure 3. The top link served two principal purposes. First it served as a rigid brace between tractor and implement which in fact made it impossible for the front end of the tractor to lift high off the ground. Second, as the implement pulled back on the bottom links when in the ground it pushed the top link forward which in turn activated a hydraulic mechanism that tended to exert a lifting force on the implement. This lifting tendency added weight to the rear wheels and increased traction. It also acted as a counter weight to the front end, shifting front weight to the rear wheels where useful traction was obtained. Excessive or dangerous lightness and lifting of the front end was prevented by the top link serving as a rigid restraining brace. By applying these implement reactions to the tractor the Ferguson tractor could be made shorter and lighter while actually gaining increased stability. That was one of the important aspects of the Ferguson System.

Another was the method for lifting and lowering implements and for controlling the depth of implements in the ground. This was provided by a hydraulic mechanism built right into the tractor which was controlled both manually and automatically. Manual control was furnished by a small lever which caused the built-in hydraulic mechanism to lift, lower, and adjust working depth. As already noted automatic control utilized the resistance of the soil, as the implement was pulled forward, to activate a control spring through the pushing or compression forces set up in the top link. The control spring in turn governed the hydraulic mechanism, causing it to react automatically, and so governed the working depth of the implement as pre-set manually. In this way, for example, increased soil resistance would not stop the tractor. Rather it would push the top link forward which in turn would cause the hydraulic mechanism to exert a lifting tendency on the implement; and this in turn would increase tractive weight on the rear wheels, thereby adding to the pulling ability of the tractor, and furthermore, to the extent that the implement was momentarily lifted partially out of the ground, allowed the tractor to go forward even under hard-pulling conditions.

Since increased pull on the implement led to increased forward push on the top link, the Ferguson System automatically acted to increase the amount of weight transferred to the tractor at just the time when changing soil conditions required the tractor to have added traction. When no such traction was required all the advantages in economy of a light and highly manoeuvrable tractor and implement were automatically realized.

The hydraulic mechanism was also utilized to avoid, or reduce, the danger to implement and driver arising from the implement hitting a solid obstruction. An overload release action automatically reduced the traction or weight on the rear wheels and increased it on the front wheels when such an object was hit. Instead of extra weight on the rear wheels

MF self-propelled combines harvest wheat in Saskatchewan.

causing the front end of the tractor to lift, there was reduced traction causing the rear wheels to spin.

Clearly, the Ferguson System was a combination of mechanical linkage and hydraulic mechanism for controlling the operation of farm implements and it involved a number of distinct actions: traction without excessive tractor weight; hitch geometry to keep down the front end of the tractor; finger-tip and automatic hydraulic control; implement penetration without excessive implement weight; protection for the implement, tractor, and operator from the consequences of hitting hidden obstructions. Being light and manoeuvrable, the Ferguson System tractor was ideally suited for the small fields that predominated in most countries of the world. It could easily be backed against a fence or hedgerow so that all the ground in small fields could, for once, be plowed with tractor power. In recent years the Ferguson System has successfully been adapted to serve large tractors with mounted, semi-mounted, and pull-type implements.

It took Harry Ferguson and his team of engineers and mechanics a long time to perfect the mechanism involved in the Ferguson System. Up to 1920 he concentrated his experiments on the development of a new plow that could be linked directly to the tractor – that is, with developing a one-unit system. Even then, indeed for several years before that, he was assisted by a few mechanics and engineers, including William Sands. The plow he developed, the "Belfast plow," was first linked to the Model T Ford, but then was designed especially to fit the Fordson tractor. Ferguson and Sands went to the United States in 1920 to show the still far from perfect model to Ford, returned to England to improve it further and in 1921 went again to see Ford. Ferguson later said that Ford refused to provide manufacturing facilities and instead offered him a position in his organization to develop the plow further. Feeling that this would result in his losing all his patents and ideas without compensation and preferring to remain independent, Ferguson went away, deeply disappointed in Ford, to seek out other manufacturers.[2]

An arrangement was made with the Shunk Company of Bucyrus, Ohio, to manufacture the plow, and later the business was shifted to the Vulcan Plow Company of Evansville, Indiana. In 1925 Ferguson, together with Eber and George Sherman, formed Ferguson-Sherman Inc. to distribute the plow in the United States – Sherman supplying funds and Ferguson a licence under his patents.[3] This was the first of many future arrangements by which Ferguson was able to launch his machines onto the market with almost no investment of his own capital in manufacturing facil-

2 *Ibid.*, p. 36; Civil Action no. 44-482, United States District Court, Southern District of New York, Harry Ferguson Testimony, p. 2484.

3 *Ibid.*, Amended answer of Ford.

ities. The Shermans were distributors of Fordson tractors and the plow, as mentioned earlier, was designed to fit that tractor. In 1928 Ford decided to stop manufacturing tractors in the United States, which effectively put an end to the market for the plow sold by Ferguson-Sherman. The latter company therefore ceased to operate, although liquidation of its inventories took a long time.

Harry Ferguson himself spent little time in the United States during the 1920s. His time was spent in Belfast with his research team which included William Sands and Archibald Greer. Development work was necessary because the Ferguson-Sherman plows being built during the 1920s were still some distance from incorporating the present Ferguson linkage system, and the hydraulic system on the tractor was still very much in the experimental stage. But the three-point linkage system in embryonic form was already there. First a light plow was simply attached directly to the tractor, with one link, much as before, and an attachment was added to enable the tractor operator manually to lift and lower the plow from the tractor. But the front end of the plow tended to dig into the ground and the back end to tip up. To correct this, a second link, directly above the first one was added which tended to keep the plow firmly in the ground. But these two links were on the plow, not the tractor. Numerous difficulties remained and experimentation continued.

In the later 1920s a decisive move was made when the links were fixed to the tractor, and when two upper and one lower instead of one upper and one lower were used. Tackling the problem of how to attach these links to the plow marked a great step forward for the Ferguson linkage system, and, apart from details, it was virtually completed when it was decided in the early 1930s to improve torsional stiffness of the linkage by moving the two links to the bottom and the single link to the top. All this involved placing a cross-shaft at the base of an inverted V-strut on the plow, attaching the two bottom or pulling links to the cross-shaft and the top or control link to the peak of the inverted V-strut so that when the two bottom links pulled the plow, the top link tended to push forward.

Experimentation leading to a satisfactory system of linkage was combined with experimentation leading to a satisfactory system of hydraulics to lift and lower implements with ease from the driver's seat and to control automatically the draft on the tractor. While much experimental work on hydraulics was undertaken in the late 1920s, it was not until the early 1930s that Ferguson faced the fact that the proper development of his system required not only satisfactory hydraulics and linkage but a suitably designed tractor as well. The Fordson tractor he had used in experimenting with his hitch and hydraulics was not adequate. With the assistance of Archie Greer, William Sands, and John Chambers, Ferguson therefore designed and built a tractor by 1933, calling it the Black tractor for the perfectly good reason that it was painted black (see illustration on p. 404). That tractor had incorporated in it for the first time a relatively successful system of linkage involving two converging bottom links and one top link, and a successful system of draft-responsive hydraulic controls. Since then the Ferguson System has changed only in certain details and in refinement, even though patents protecting it in later years related to design refinements that were yet to come. The experimental work was financed by Harry Ferguson's old Belfast company.

With the success of the experimental Black tractor the next major task was to convince some manufacturer to produce it. A number were approached without success. In some cases negotiations broke down because manufacturers wished to incorporate the Ferguson System into their own tractors, whereas Ferguson insisted that his own model had to be built. He then met representatives of the David Brown Company of Huddersfield, England, and personally demonstrated for them the Black tractor with plow, tiller, ridger, and row cultivator – one of many demonstrations that he was to plan and supervise with infinite care and unusual skill.

Ferguson's considerable powers of persuasion, as well as his skill with demonstrations, undoubtedly lay behind the decision of David Brown to build the first Ferguson tractor. For that purpose David Brown formed David Brown Tractors Limited, but Ferguson retained control of distribution of the tractor through Harry Ferguson Limited, Huddersfield, formed in 1936. This set the pattern for his approach to selling tractors and implements, a pattern involving his remaining aloof from manufacturing, but retaining control of distribution and of patents and engineering development. Here was the secret to his being able to build an empire with surprisingly little capital, and he compromised with it only in 1947 when failure to go into an assembly operation at Detroit would almost certainly have lost him the United States market.

The Black tractor, for purposes of production, was modified with improved brakes, hydraulic controls, linkage arrangement, and steering. Its weight, length, and general arrangement remained virtually unchanged. When David Brown began producing it, it was variously referred to as the Huddersfield tractor – after the location of the Brown factory – or the Ferguson-Brown tractor, and also as the Irish Ferguson tractor.

It is perhaps not surprising that the tractor with its revolutionary design, relatively high cost, and small appearance did not sell well when first introduced into the market. In order to improve sales a new corporation, Ferguson-Brown Limited, was formed in 1937 as a successor company to Harry Ferguson Limited, Huddersfield. The name "Brown" was better known than "Ferguson" and it was undoubtedly believed that it would help to sell tractors. Brown was Chairman of the Board and Ferguson was Managing Director and a major shareholder. The new company did not manufacture tractors or implements, these being supplied by the David Brown Company as before, but rather was engaged in distributing the Ferguson-Brown tractor with accompanying implements and in designing farm machinery.

This somewhat closer relationship between Brown and Ferguson soon deteriorated into fundamental disagreement. Ferguson never worked easily in partnership with anyone, except for a few years with Henry Ford, and he even quarrelled with him in the 1920s. The disagreement with Brown, in the words of Ferguson, centred on Brown's plans to design his own tractor which would conflict with the Ferguson tractor promotion; on Brown's plans to open his own tractor depots instead of working through dealers (which Ferguson felt would reduce the profitability of the Ferguson-Brown selling company); and on price, with Ferguson favouring no price increase in spite of the small volume of sales. John Chambers, who had taken a leading part in developing the Black tractor and who had continued as an engineer with the David Brown company, resigned from that company in November 1938 giving as his reason, in a letter to David Brown, that he had lost confidence in the manufacturing abilities of the David Brown people. David Brown reminded Chambers of the major design defects in the ramplates and transmission housings of the tractor which, he felt, were more serious than manufacturing defects.

Suspicion and distrust between the "Ferguson people" and the "David Brown people" had by then overtaken their relations. Agreements were terminated by the end of 1938, although for a short while Ferguson-Brown still had a non-exclusive and non-assignable licence to sell the Ferguson-designed machines.

In an immediate commercial sense, the Irish Ferguson tractor was not much of a success, for only about 1,250 were sold; but from a longer point of view its success was undoubted. With the appearance of the production model, the Ferguson System as such had definitely passed out of the experimental stage. There is evidence that various companies in the United States were beginning to show some interest in the Ferguson System with the result that the commercial value of the ideas and patents that it embodied had probably become very great. It is true that some defects in design and manufacturing were encountered in the production model, but these proved more troublesome than they should have done simply because of inadequate after-sales service. Even so there were many satisfied customers, in Norway, Sweden, and Denmark, as well as the United Kingdom. What restricted the sales volume of the tractor was the inadequacy of the marketing organization, the natural reticence of farmers to adopt a revolutionary machine, particularly one that was rather more expensive than its competitors, and the unsettling effect on the whole organization of the rift between Ferguson and Brown. Desperately needed now was a much-improved marketing organization and an efficient manufacturer capable of volume production, as well as a greater degree of harmony between the two.

Harry Ferguson did not wait until the final break with David Brown to investigate new arrangements. He had over the years maintained friendly contact with the Shermans at Evansville, Illinois, and had even demonstrated his tractor to Eber Sherman at Huddersfield in July 1937.[4] Eber Sherman, in turn, was a good friend of Henry Ford, his company still being the distributor for English-made Fordson tractors in the United States. He kept Ferguson informed about Ford's activities and undoubtedly kept Ford informed of Ferguson's activities. It was Eber Sherman who, after seeing Ferguson and viewing the latest developments in the Ferguson System in England in 1937, arranged for Ferguson to demonstrate the Ferguson tractor and implements for Henry Ford, Senior.

Ferguson arrived in the United States in October 1938 and brought with him a tractor, two plows, a tiller, a ridger, and a row cultivator, all of which had been manufactured in England for the Ferguson-Brown company. A private demonstration on the grounds of Ford's residence was arranged without difficulty. Although Ford had taken an interest in tractor development since 1906, he had produced no tractors in the United States since 1928; but he did experiment with various models, which he rejected, and he may well have regarded Ferguson's visit as an opportunity to achieve, at least indirectly, the success in the farm tractor industry in the United States that had eluded him up to that point.

There were a number of demonstrations and finally after one of these when Ferguson and Ford were alone, both standing, debating, arguing, planning, and discussing tractor design and the future of world agriculture, these two unusual men agreed that the Ferguson System had a great contribution to make. They decided, without scratch of pen or pencil in evidence, to work together in mutual trust as gentlemen. The famous "gentlemen's agreement" was born.[5]

Following that demonstration Ferguson returned briefly to England and, in his absence, the Ferguson tractor was disassembled at the Ford Airport Building on Henry Ford's direction, after which Ford had his engineers build two light tractors resembling the Ferguson tractor in their basic features.[6] Ferguson returned to the United States in February 1939 bringing with him John Chambers and William Sands, although Sands did not stay long and returned to Belfast where he, and other engineers, worked on the development of implements during the war. The design drawings for the new tractor, called the 9N tractor, were soon made and prototypes were built. Ferguson always maintained that he had retained full control of tractor design, and so did his engineers, but during the lawsuit of later years Ford management denied this. Ferguson undoubtedly contributed the unique aspects of the design, in particular the hydraulics, linkage, and (in the Black tractor) the concept of a tractor designed especially to suit the Ferguson System, whereas Ford contributed much to production engineering. It was Ferguson's contribution that made it a

4 *Ibid.*, Harry Ferguson Testimony, pp. 2538, 8766-8.

5 *Ibid.*, Harry Ferguson Testimony, pp. 2310-12.

6 *Ibid.*, Plaintiff's Memorandum, August 6, 1951, p. 2, Appendix.

unique machine and engineers of Harry Ferguson Incorporated also worked with Ford engineers in developing the 9N model.

Field tests proved to be successful and on April 27, 1939, newspapers carried Henry Ford's announcement of the new tractor, and also a most unusual tribute by him to Ferguson, which read:

Now, we have tackled power farming from a new angle. Instead of a new tractor weighing about 3,000 lbs. we expect ours to come out at about 1,700 lbs. . . . we have been able to do this by the application of a principle developed by Mr. H. G. Ferguson of Ireland, with whom I have been in touch for a number of years. His System will not only revolutionize agriculture, but will put his name alongside those of Edison, Bell and the Wrights. . . . [7]

The details of the "gentlemen's agreement" cannot be known, but apparently involved was an understanding that Ford would manufacture the tractor and would finance its manufacture; and that a new corporation would be formed to distribute the tractor, to subcontract implements to other companies, and to engage in developmental work. Ferguson-Sherman Manufacturing Corporation was formed for this purpose on July 1, 1939, but the Sherman association was short-lived and the company was renamed Harry Ferguson Incorporated in 1942. The agreement with Ford could be cancelled for any reason at all, at any time, by either side with notice.

Manufacturing began before the end of 1939 and until July 1, 1947, Ford was the sole supplier to the Ferguson company of a tractor widely known as "Ford tractor – Ferguson System" but in various places, and at various times, also referred to as the Ford 9N tractor (or 2N during the war when a model with steel wheels instead of pneumatic tires, and magneto instead of generator and battery was sold), or the "Ford tractor with Ferguson System." As a matter of policy, from which it did not deviate, the Ferguson company chose to work through a number of independent distributors who in turn built up a system of dealers.

The experience with the Ferguson-Brown tractor had clearly demonstrated the need for an educational programme as part of the company's promotional work and for adequate after-sales servicing. The novelty of the tractor made the demonstration approach to sales promotion ideal, and Harry Ferguson was the man to develop such a programme. Herman Klemm, for many years Chief Engineer of Harry Ferguson Incorporated, has commented on this aspect of Ferguson:

I remember . . . the opening of the Southfield tractor assembly plant in Detroit when Mr. Ferguson drove the first tractor off the assembly line. After leaving the assembly track, he applied the right rear wheel brake and wheeled the tractor in a circle at high speed to the astonishment of the visitors.

In a field demonstration, he would never operate the tractor for any operation at the highest permissible speed, but would deliberately select a lower speed to impress his audience with his effortless ease of operation. He was also ex-

7 *Ibid.*. Plaintiff's Memorandum, August 6, 1951, p. 3.

ceedingly resourceful in the event field conditions did not favour his demonstration.

For instance, Mr. Ferguson once gave a demonstration to a large audience near Washington, D.C. and it rained the night before leaving the surface of the ground wet and slippery. As Mr. Ferguson plowed around the field, he noticed that the left rear tire, which was running on unplowed ground, was slipping excessively in spite of the weight transferred from the plow to the tractor. As he approached the assembled spectators, he stopped the tractor and called for Mr. Checkley, who was for years his right-hand field man, and asked him to sit on the left fender to "keep the wind away from him." The left side was, of course, the side where the tire was slipping and Mr. Checkley's 210 lbs. added weight to the left tire reduced the slippage to normal and Mr. Ferguson drove along with a smile. None of the spectators, of course, understood his real intent but assumed that Mr. Checkley was merely a wind-break.

Demonstrations were always carefully rehearsed and the equipment had to be in first-class condition and perfectly adjusted. If, during the rehearsal, the engine of the tractor would idle at too high an r.p.m., Mr. Ferguson would get off the tractor and adjust it to his liking. He always wanted the engine to idle at the lowest possible r.p.m. so as to be almost noiseless. He applied the same care to implements and would often get off the tractor and make the fine adjustments himself and, at the same time, give a lesson to his staff.

Tractor and implement adjustment during a demonstration was never allowed. If a demonstration called for a change to another implement, the detachment of one and the attachment of the other had to be made by one man completely effortless. It was sometimes necessary for the man to practice for days to bring his skill to perfection.

Special attention was paid to product education. Pamphlets, manuals, films, and other training aids were widely used in training distributors and dealers. Distributor personnel were encouraged to attend the company's training school at Ypsilanti, Michigan, and schools were also being held during the winter in the south. From the start intensive product education played an important part in the Ferguson organization, an approach made necessary by the novel features of the Ferguson System and one favoured by Harry Ferguson.

Ferguson personally participated in many public demonstrations of the tractor, and these he planned and executed with utmost care and skill. He made frequent speeches in which he extolled the benefits of the Ferguson System. In these speeches Ferguson developed the theme that his system would be of great benefit to the whole of humanity. He began to fulminate against "the system of ever-increasing wages" which he felt meant ever-increasing prices and advocated instead "the price reduction system."[8]

How the Ferguson System came into this he never tired of explaining:

If we are going to bring the benefits of price reduction, we must give the farmer

8 *Address* by Harry Ferguson, International Food Conference, Ford-Ferguson Field, Bethesda, Maryland, June 5, 1943.

something which will cut his costs, cut them in pieces . . . all over the world. . . . Agriculture, let me tell you, is more important than all the other industries in the world put together . . . yet it is the only industry left in the world that is still being conducted by antiquated methods. Why, gentlemen? Because until now nobody has come along with equipment which will do everything, on all farms of the world that animals now do – and do it at a fraction of the cost.[9]

Endlessly Ferguson preached about this "price-reducing system" taking full-page advertisements in newspapers, writing to government leaders. As time went on his obsession with this "system" seemed, if anything, to increase, and increasingly he saw it as the answer to most of the world's ills. When communism appeared as a threat, the Ferguson System was seen by him as the instrument for combating it by helping to rid the nation of "ever-increasing prices." "Under a Price-Reducing Economy we would head for security and prosperity and full employment" he wrote to the British Chancellor of the Exchequer, R. A. Butler, on February 7, 1953, after he had already written a letter to the Prime Minister. Nor were his letters short; indeed they normally ran to many pages sub-divided by carefully chosen and underlined sub-headings. A letter to the Chancellor was sub-divided into "Our Glorious Opportunity," "Foreign Competition," "What Causes a Slump," "Real Wealth," and the next one to the Chancellor into "Our Most Urgent Need," "The Wrong Approach," "The Right Approach," "A Serious Warning," "International Co-operation," "How to Give a Practical Lead," "National Savings Movement," "H.R.H. The Duke of Edinburgh."

Now, to an economist, the view that a technological innovation would increase productivity or reduce costs through permitting substitution of capital for labour is a rather straightforward one. To Ferguson it seemed to be revealed truth, and how wonderfully effectively he preached it – stretching the theme so far that it became embarrassing to the economically literate to listen to him. In associating the Ferguson System with the economic salvation of the world he had a theme that attracted unusual mass publicity. Accompanied by Ford, he was permitted, after the war had started, to demonstrate his system to President Franklin D. Roosevelt and to several other leading politicians, and thereby succeeded in obtaining a priority of steel for the production of his tractor. Even President Roosevelt's own farm at Hyde Park soon operated under the Ferguson System.

But the most astonishing thing about Ferguson's "price-reducing" views was that he applied them to the operation of his own company, particularly after the war, with the result that his profit margin on tractors was exceedingly small. He seemed not to grasp the fact that general inflationary forces far exceeded the opposite pull on prices of productivity increases, and that Ferguson alone could do nothing to overcome them.

To increase interest in its tractor in the United States even further the

9 *Ibid.*

Ferguson company financed the National Farm Youth Foundation for three years. This foundation trained boys and girls in farm management, farm engineering, and agricultural methods. In addition, of course, the Ferguson company built up a system of 33 distributors and 2,876 dealers in the United States after 1939 to give it a comprehensive nation-wide distribution system. This distribution system was quite different from that of the other farm equipment companies for most of them did not sell through independent distributors at all. This system was also one that subsequently did not stand the test of time. Ferguson was undoubtedly greatly assisted by his association with Ford – that magic name – in developing his distribution network.

And what a development it was. The expansion of the company is seen by the spectacular behaviour of its sales and profits. Table 6 gives some of the details. Obviously the growth of the company until 1942 was quite remarkable. But then on March 25, 1942, production halted. Wartime prohibition on the use of rubber tires on tractors, the rigid price ceiling that prevented a changeover from rubber to steel wheels, and the squeeze on profits created by rising wage levels forced this move. Production was not resumed until after the Office of Price Administration granted a price increase on September 17, 1942. Ford was permitted to charge the Ferguson company $60.00 more per tractor but curiously Ferguson passed only $32.00 of that increase on to the farmer. As a result, Ferguson made very little profit on his sales of tractors and depended for the most part on implements to generate profits. This strange attempt to keep down the price of the Ferguson tractor and to compensate for it by profits on implements, was a policy of Ferguson's even in the Ferguson-Brown days and remained his policy until the merger with Massey-Harris in 1953. After the disappointing experiences of 1942, sales again improved and

TABLE 6

Harry Ferguson Incorporated Sales and Profits (1939–1947)

	Tractors*	Implements*	Total sales $ millions	Total profits $
1939	10,233	9,730	4.9	170,030
1940	35,742	47,774	18.2	270,802
1941	42,910	85,219	26.9	426,601
1942	16,478	48,896	14.1	258,697
1943	21,062	56,393	20.2	1,155,152
1944	43,443	174,176	50.9	813,482
1945	28,749	176,022	42.0	632,930
1946	59,773	238,142	79.4	4,273,253
1947	47,682	169,367	70.5	2,164,811

***Based on Units produced**

by 1946 the company's sales volume amounted to $79 million, which included 60,000 tractors and 238,000 implements.

But the company's 1942 experience revealed the dangers involved in depending on subcontractors as a source of supply for tractors and implements. Various incidents, such as having to vacate hurriedly its quarters in the B building at the Fort Rouge plant of Ford in 1944, and Edsel Ford's feeling that his company should perhaps cease building tractors, began to make life increasingly uncertain for the Ferguson company. The Ferguson people were also concerned over the prices that they were having to pay for the Ford tractors, and they worried about costs of production at Ford.[10]

When the first moves were made which subsequently lead to a break between Ferguson and Ford, they arose more than anything else from new policies introduced by new management at Ford. Henry Ford II assumed leadership of the Ford company in September 1945 and he faced monumental difficulties within his company – indeed the Ford Motor Company was tottering on the brink of disaster.[11] In its over-all business operations during the last quarter of 1945 and the first two quarters of 1946, the company seems to have lost over $82 million.[12]

Among Ford's smaller difficulties was its relationship with Ferguson. The tractor manufacturing operation apparently was not generating profits, partly because of the Office of Price Administration price restrictions and, no doubt, also because of the generally unsatisfactory organization and administration of the Ford company. But in addition to the matter of losses, there was the unusual, and for Ford perhaps somewhat irritating, arrangement whereby the tractor that it was manufacturing in North America, its home base, was being sold by a company in which it had no participation. This difficulty could be resolved only through Ford marketing a tractor itself in North America.

As early as June 12, 1944, Henry Ford II suggested to Roger Kyes, President of Harry Ferguson Incorporated, that the verbal agreement or "gentlemen's agreement" between Ferguson and Ford was unsatisfactory and that there should perhaps be a written contract.[13] But nothing happened for some months. However, Roger Kyes, on December 12, 1945, wrote to Harry Ferguson and reported to him that Henry Ford II had reiterated his dissatisfaction over the absence of a written agreement and had expressed the view that Ford should not be making a product with the Ford name on it which they did not sell through their dealers.[14]

In early January 1946 a Ford representative, speaking for himself, not

10 Civil Action no. 44-482, Plaintiff's Memorandum, pp. 10 to 22, Appendix.

11 Mira Wilkins and Frank Ernest Hill, *American Business Abroad: Ford on Six Continents* (Detroit: Wayne State University Press, 1964), p. 338.

12 Civil Action no. 44-482, Plaintiff's Memorandum, p. 42, August 6, 1951, Appendix.

13 *Ibid.*, p. 28.

14 *Ibid.*, p. 30.

the company, expressed the view that the Ford family should perhaps have 30 per cent interest in the Ferguson company. Then on January 14, 1946, Ernest Kanzler of Ford, speaking personally, suggested that the Ford Motor Company might take the position that it might be advisable for it to acquire a controlling interest in any company established to distribute the tractor.[15]

Kyes indicated that there was little chance of Ferguson agreeing to Ford participating in a company established for distributing the tractor. A committee, referred to as the "Watt Committee," was appointed to consider all the problems of the relationship between Ferguson and Ford, including the matter of a written agreement; it held its first meeting on May 21, 1946, and adjourned in late June because of vacation schedules. Then on July 18, 1946, in a meeting between Henry Ford II, Ernest Breech (who had recently joined Ford), Roger Kyes, and Horace D'Angelo (of Harry Ferguson Incorporated), Breech indicated that Ford wanted 51 per cent of a new company.[16] It seemed to the Ferguson management that this was quite a change from the hopeful atmosphere of the Watt Committee meetings. They could explain it only by the possibility that Ford wanted a new company so that he could offer his new senior executives a chance to make capital gains through investing in it (which they could not do in the Ford Motor Company itself), and that he wanted majority control so that such capital gains could be assured through Ford, in effect, taking over the Ferguson business. Henry Ford II subsequently testified that he had from the time of the January meeting made up his mind that he wanted majority control, which, if true, casts some doubt on the sincerity of Ford representatives on the Watt Committee.

What is certain is that the position of the two parties had become finally fixed and permanently in conflict, Ford insisting on majority control and Ferguson on complete independence. It was to be expected that both would begin to consider what each would do if a final break came, even though Ferguson management somewhat naively continued to believe, for some time, that an agreement with Ford might still be reached.[17]

Ford management began to think of forming their own new corporation, which subsequently turned out to be Dearborn Motors. Ferguson representatives had discussions with Continental Motors Corporation over the possibility of Continental manufacturing tractors for Ferguson if a break came,[18] and these discussions went to the point of draft contracts. But they ended in September 1946 when Continental felt that they could not take on a programme of the size contemplated, and nothing else was done.[19]

By October 1946 Roger Kyes appears finally to have concluded that if the Ford Motor Company insisted upon control of the Ferguson company and refused to enter into a written agreement, a break with them would be necessary. Even so Kyes met with Kanzler on November 4, to try to

15 *Ibid.*, p. 33. 16 *Ibid.*, p. 49. 17 *Ibid.*, p. 60. 18 *Ibid.*, p. 49. 19 *Ibid.*, p. 50.

work out an agreement involving profit-sharing but excluding surrendering control of the Ferguson business. Horace D'Angelo, of the Ferguson company, later testified that on November 6, in a meeting between Henry Ford II, Breech, and Eaton of the Ford company, and Kyes and himself from Ferguson, Ford suggested that a new company should be formed, that it should not include the name Ferguson, that Ford would have 70 per cent control of the company, of which 40 per cent would go to the Ford Motor Company and 30 per cent to the Ford executives. In addition Breech was reported as saying that Ford would want to have a say in who was to receive the 30 per cent allotted to the Ferguson company, and no value was to be attached to the patents, licences, and goodwill contributed by the Ferguson company.[20] And what was the reaction of the Ferguson company? Its minutes for November 27, 1946, read: "This preposterous proposition, which was made an ultimatum backed by a veiled threat that they would usurp our patents, was so ridiculous that there was no choice but to break with Ford."

Ford established Dearborn Motors Corporation in November 1946 as the distribution organization for its tractor and implement division. A number of Ferguson employees were hired by Ford, and in the evidence of the subsequent lawsuit Ferguson argued that Ford used a variety of tactics to win over Ferguson distributors, dealers, and suppliers. By 1948 Dearborn Motors was producing approximately 100,000 tractors annually. The tractor that Ford chose to produce, named the 8N was fundamentally the same one that it had been producing; and Ford did not obtain patent rights held by Ferguson, presumably believing either that they were invalid or that Ferguson would not, or could not successfully, contest its action.

Harry Ferguson Incorporated needed a new source of supply for tractors, and quickly too, if it was not to lose its position in the United States market. But instead of moving purposively forward the company then entered a period of considerable internal confusion. The curious lack of leadership within the company that was to exist during 1947 could not have come at a worse time. Fortunately for the future of the Ferguson organization it was at about this time that Ferguson began to have tractors manufactured by the Standard Motor Company in the United Kingdom. Coventry, not Detroit or Belfast, soon became the focal point of the Ferguson company and Coventry opened the door for Ferguson to export trade around the world.

20 Evidence of Horace D'Angelo, and *ibid.*, p. 61.

CHAPTER 6

Ferguson System in mass production

At the beginning of the Second World War, Harry Ferguson had wanted Ford to convert its Dagenham plant in England from the production of Fordsons to the production of the Ferguson tractor. Charles Sorenson of Ford at Detroit agreed, for after seeing the Ferguson tractor he told Lord Perry, head of Ford in England, that it " . . . will so outperform the latter (the Fordson) that it is almost pitiful."[1]

Dagenham executives refused the idea, giving as their reason the restricting conditions imposed on them by war demands, but it was also true that the English Board of Ford had little use for Ferguson.[2] It is not unreasonable to believe that the English Board viewed Ferguson's ideas as coolly as they viewed his person. Ferguson always felt that his manufacturing agreement with Henry Ford was world-wide and he interpreted this action of Ford as a contravention of it. So after that, Ferguson terminated it in writing, as far as the eastern hemisphere was concerned, and this left him free to seek other manufacturers.[3] In the end Sorenson's and Ferguson's appraisal of the Ferguson tractor proved correct, for in 1948, after being in production in England for just two years there were more Ferguson tractors produced than Fordsons and by 1951 Ferguson production was double that of Fordson production. The little TE-20 had become "the tractor" of the post-war period.

But before that could happen Harry Ferguson had to find a manufacturer in the United Kingdom. He moved back to England in May 1945,

1 Mira Wilkins and Frank Ernest Hill, *American Business Abroad: Ford on Six Continents*, p. 312.

2 *Ibid.*, p. 313.

3 Civil Action no. 44-482, Harry Ferguson Testimony, pp. 9206-7.

and his chief engineer, John Chambers, went with him. Harry Ferguson Limited, a private company, was incorporated under the United Kingdom Companies Act on November 15, 1945, with the ridiculously low capital of £50,000. This company was to be the principal distribution and engineering organization for Ferguson in England, but not a manufacturing company, and so a supplier was again required.

Ferguson went to Sir John Black of the Standard Motor Company but that company was reluctant to take on the project at a time of severe steel shortage. Only the government could overcome this shortage for the company. Ferguson therefore went to see the Chancellor of the Exchequer, Sir Stafford Cripps. He made one major point to the Chancellor: if the Ferguson tractor could be manufactured it would prove to be an important dollar earner, and he pointed to sales of the tractor in the United States as evidence to support his view. Sir Stafford Cripps must have been convinced for he sanctioned the necessary steel and instructed the Standard Motor Company that they could take on production of the tractor.[4] This they did in a government-owned wartime shadow factory on Banner Lane, Coventry. Here again Ferguson's enthusiasm, which so often appeared to many (sometimes rightly) as the fulminations of a crank, proved to be correct. Exports from the United Kingdom of the Ferguson tractor soon reached a large volume and the Coventry plant is still a huge exporter of tractors.

An agreement with the Standard Motor Company for the manufacture of tractors in England was dated August 20, 1946. Standard was also given first right of refusal to manufacture Ferguson tractors anywhere in the eastern hemisphere. The first tractor built was quite similar to the 9N or Ford-Ferguson tractor built at Detroit until June 1947, but it differed in that it had a four-speed transmission whereas the 9N had a three-speed transmission. Its motor had been designed by Continental with about the same displacement as the one on the 9N, but it was a valve-overhead motor, was considerably more rugged and powerful, and could plow at 3½ to 4 m.p.h. instead of 2½ to 3 m.p.h. Incidentally, when Herman Klemm, Director of Engineering of Harry Ferguson Incorporated, Detroit, after September 1945, told Ferguson this, Ferguson said angrily that no one in his right mind would plow faster than 2½ m.p.h.! He also objected to its styling, but it was pointless to tell him that when it was designed he himself had been in charge at Detroit. Styling of the new tractor was different but the hydraulic and linkage systems were almost identical with those of the Detroit tractor and major components were interchangeable. But there was no doubt that this tractor, designed by Harry Ferguson Incorporated engineers was a distinct improvement over the Detroit tractor. It was the famous TE-20 (illustrated on page 405).

By an agreement of December 4, 1945, Continental Motors in the United States supplied 25,000 gasoline motors for the first United King-

4 Norman Wymer, *Harry Ferguson*, pp. 62-3.

dom Ferguson tractors produced by Standard Motors. Almost immediately Harry Ferguson Limited began to appoint distributors in various countries and in this the company was materially assisted by the Standard Motors connections just as in the United States it had been by the Ford name. Many of the distributors had sold Standard Motors' cars in the past years. Ferguson, however, always insisted that a distributor establish a new and separate organization to handle tractors and implements.

For the most part Ferguson distributed its products through independent distributors around the world rather than through locally incorporated Ferguson companies. When local companies were formed by Ferguson in France, South Africa, India, and Australia they were intended essentially to meet the threat posed to company markets by restrictions favouring local production, although the approach taken by these subsidiary companies was different than that taken by the Massey-Harris companies. While the Massey-Harris subsidiary companies became directly involved in the manufacture and assembly of products locally – as in France, Germany, and the United Kingdom – the Ferguson companies were established primarily to arrange for local subcontracting and generally to supervise the activities of independent local distributors. The progress of the operation in the United Kingdom is seen in the volume of production of tractors at the Coventry plant of Standard Motors and in the sales and profits of Harry Ferguson Limited shown in Table 7.

TABLE 7

Harry Ferguson Limited Production, Sales, and Profits (1946–1953)

	Production of tractors at Banner Lane (Units)	Total sales tractors and implements (£ millions)	Total profit (£)
1946	316	.1	—79,765
1947	20,578	7.3	150,892
1948	56,878	19.0	181,812
1949	38,689	14.3	—1,589
1950	51,381	18.8	278,848
1951	73,623	28.2	684,465
1952	69,567	30.0	363,535
1953*	48,060	20.9	221,385

*Ten months

By 1948 just over 56,000 tractors were being produced. While this generated an encouraging profit it was much smaller than one might have expected because of the small profit on the 25,000 tractors that were sent to the United States to keep the distributors of the company there in business until the new Detroit plant could come into operation. However, before the year 1948 was out the effects of the dollar shortage, and in some countries the sterling shortage, were being felt. Sales were definitely on the decline.

Eric W. Young told the Ferguson Board on February 23, 1949, that more than 4,200 tractors were in stock and that he had already informed Standard Motors that another reduction in the production schedule might be necessary. This schedule was eventually cut from an original 350 tractors a day to 150 tractors a day. By the end of March the production of a number of implements had been stopped outright and the problem of financing inventories had become acute.

Before long the company had reached its borrowing limit with the bank and subcontractors had to give additional credit. The bankers had informed them that they feared a trade recession which precluded additional facilities being provided. Inventories of tractors continued to rise. To add to the difficulties, Sir John Black of Standard Motors indicated to the company that with the reduction in tractor production at Banner Lane, Coventry, he might have to recommend that Banner Lane should become the centre of car production.[5] Then came the 30 per cent devaluation of sterling which almo[illegible]vernight fundamentally changed the position of the company. By O[illegible]49 sales exceeded forecasts and they continued to do. In ear[illegible] restricted only by the lack of availability. Profits s[illegible]ented levels and the financial worries of Harry Ferg[illegible]er, for a while at least.

Even thou[illegible]guson Limited began to decline in late 1952, and 1[illegible]ose of 1952, the company's profits held up remark[illegible]ar of the merger. Indeed, in spite of crisis after crisis betw[illegible]ack of Standard Motors and Harry Ferguson, the United King[illegible]guson company had become a great commercial success. For the org[illegible]ization as a whole this was not the case, and the year 1953 saw a marked turn for the worse. Whereas profits (excluding those for Harry Ferguson Holdings Limited and Harry Ferguson Research Limited) in 1952 were £1.7 million, in the first ten months of 1953 they fell to £0.2 million, almost entirely because the United States company moved from profits of £1.4 million in 1952 to a small loss in 1953. This was virtually what was happening at that time in the Massey-Harris organization as well.

Certain developments within the Ferguson company, however, tended to detract from its commercial success and long-term viability. From the very early beginnings of Harry Ferguson Limited, it was clear that Harry Ferguson was determined to introduce the "price-reducing system" into the operations of that company. The price of the tractor was actually reduced and the ratio of profits to sales was kept at an unusually low level in spite of repeated appeals by management at Board meetings for price increases. It became the policy of the company to make profits on its implements which induced it to emphasize the rapid development of a range

5 Harry Ferguson Limited, Board Minutes, September 7, 1949.

1 / The first Ferguson tractor produced at Ferguson Park, Detroit, on October 11, 1948. Left to right, H. G. Klemm, Director of Engineering; Robert G. Surridge, Director, Secretary and General Counsel; Philip C. Page, Director, Executive Vice President; Horace D'Angelo, Director, Executive Vice President; R. W. Hautzenrueder, Chief Tractor Engineer; Nils Lou, Tractor Plant Manager; Albert A. Thornbrough, Director of Procurement.

2 / Henry Ford and

3 / Harry Ferguson and James S. Duncan, President of Massey-Harris Company, sign the agreement for the merger of their two firms in 1953.

4 / The "million-dollar" tossed coin and cigar box.

5 / Group attending the Massey-Harris-Ferguson merger signing at Abbotswood in August 1953 are, left to right: Tony Sheldon, M. W. McCutcheon, John Sonnet, Colonel W. E. Phillips, H. D'Angelo, Trevor Garbett, J. S. D. Tory, I. J. Wallace, Harry Ferguson, Mrs. Duncan, L. G. Reid, John Peacock, James S. Duncan, R. H. Metcalfe, Mrs. Ferguson, M. F. Verity, H. H. Bloom, G. T. M. Bevan, J. A. McDougald, J. D. Turner, D. Morton.

6 / The President of Massey-Ferguson Limited, A. A. Thornbrough, presen a duplicate of the Massey-Ferguson Educational Award and Medal to the Rt. Hon. Vincent Massey, PC, CH, Presi- dent of the Massey-Harris Company 1925-1927.

7 / The 1,200,000th MF tractor at the opening of the new Detroit tractor plant on May 7, 1958. Left to right: Herman G. Klemm, Vice President Engineering; Senator M. W. McCutcheon; A. A. Thornbrough; Harol A. Wallace, MF Director and Vice President Manufacturing; W. Eric Phillips, Chairman of the Board; and John A. McDougald, Director and member of the Executive Committe of the Board.

of implements. The ratio of implement sales to tractor sales consequently came to represent a very important rule of thumb in the operations of the Ferguson selling organization. When adversity came, the organization did not have wide profit margins to cushion a decline in sales.

Ferguson had also made little progress in building up a line of tractors. As early as October 1948 the Board recognized the need for developing a diesel engine for their tractor. Discussions were held with Standard Motors. The latter agreed to develop such a motor and indicated that it would obtain a plant for its production. However, developmental progress appeared to be unusually slow and by early 1950 Ferguson management began to think of going elsewhere for a diesel engine and they even had exploratory discussions with F. Perkins Limited of Peterborough. But, as it turned out, the Ferguson company remained unusually patient and not until about the time of the merger did Standard have a prototype model ready.

There had been much talk and no progress over developing a larger Ferguson tractor, the TE-60, until competition from the Ford Major tractor finally began to inject a note of urgency. Drawings were ready by October of 1951 and the first tests of the large tractor were made in May 1952. In August 1952 the Standard Motor Company agreed to manufacture the tractor, planning to use an existing gasoline engine and the diesel engine that was in the process of being developed. Standard began to order necessary equipment. But as late as April 1953 the design had not finally been released because of delay in the development of the diesel engine. Astonishingly enough, a letter announcing the big tractor was sent to all distributors on May 13, 1953, and orders were solicited even though the tractor was still a long way from the production line. In the spring of 1953 a prototype of the TE-60 was sent to Detroit where its unsuitability for corn harvesting was immediately obvious because a corn picker could not be mounted on it; and it was also unsuitable for row-crop cultivation, since it could not accommodate mid-mounted cultivators. Harry Ferguson did not like tricycle tractors and the mid-mounted implements they carried, and he failed to understand why farmers needed them. But then came the merger and after a bitter controversy within the company the TE-60 project was cancelled, to the deep consternation of many Ferguson people.

Harry Ferguson Limited had only just begun to arrange for the manufacture of the TE-60 tractor and implements in countries other than the United States and the United Kingdom prior to the merger of 1953. Not long after beginning to export from Coventry, Ferguson encountered precisely the same threats to its markets that Massey-Harris had encountered and had therefore also to consider manufacturing operations in other countries. The matter first arose in France. On April 28, 1948, the Board agreed in principle that a French company should be formed, and that its operations should include the purchase of tractors from Harry Ferguson Limited and the manufacture of Ferguson implements under licence.

Progress in the formation of this company was very slow and by December 1948 the Chairman felt that the time was approaching when it would be necessary to manufacture tractors in France as well as implements. The question now arose whether it would not be better for Ferguson to amalgamate its proposed company with the interests of Compagnie Générale des Machinise Agricole (COGEMA), its French distributor, headed by J. A. Bouilliant-Linet. But negotiations did not move swiftly, and in the meantime COGEMA began to arrange for the local production of disc tillers, reversible and vineyard plows, of which samples were sent to Coventry for approval. Finally it was decided that it was undesirable to combine management of COGEMA in anticipation of joining the new French Ferguson company.

Harry Ferguson de France was formed at last in 1952 with J. A. Bouilliant-Linet as President. However, this company was not intended to engage in manufacturing, but to supervise subcontracting and distribution. In 1953 the Standard Motor Company acquired a 50 per cent interest in a new company, Société Standard Hotchkiss, formed for manufacturing Ferguson tractors in France. By that time, of course, the merger with Massey-Harris was imminent.

Towards the latter part of 1948 Eric W. Young of the Ferguson company had a discussion with the High Commissioner for India, Krishna Menon, who in spite of reservations over the Ferguson tractor permitted thirty such tractors to be sent to India for experimental and demonstration purposes. Two Ferguson distributors – Escorts (Private) Limited at Delhi, and Kamani Engineering Company Limited at Bombay – were appointed to import Ferguson tractors and implements from England and distribute them to their own dealers. In late 1949 Standard Motors announced its intention to build tractors in India, which led Harry Ferguson to form Harry Ferguson (India) Limited. This Ferguson company entered into a contract in 1950 with Standard Motor Products of India Limited, the latter importing completely knocked down (C.K.D.) tractors, assembling them and supplying them to the Indian Ferguson company. However, by the time the merger came about the Indian operation had not had sufficient time to move away from the simple C.K.D. stage into a manufacturing operation. The Indian Ferguson company even found itself in financial difficulties, which were only just being straightened out when the merger was effected. Standard Motors was still importing some C.K.D. tractors and continued to do so after the merger.

In South Africa, Ferguson products began to be distributed by Tractors and Farm Tools Limited in 1948. Toward the end of 1949 the subsidiary company, Harry Ferguson of South Africa Limited, controlled by Ferguson, was formed and its functions were the familiar ones of subcontracting locally for the manufacture of implements and supervising, although not being directly responsible for, local distribution of Ferguson tractors and implements.

It was not until 1952 that Harry Ferguson of Australia was formed as a wholly owned subsidiary of Harry Ferguson Limited. Its formation was prompted by the Australian balance-of-payments situation which indicated that there would be strong emphasis on local manufacturing. The need for local manufacturing of implements became apparent and this activity began in 1952. A further reason for becoming formally established in Australia was that it would give sharper identity to the Ferguson name there. All of these operations outside the United States and United Kingdom were relatively unimportant, however, at the time of the merger.

The remarkable growth of the United Kingdom company, assisted by exports to many countries, was just beginning in 1947 when things went from bad to worse in Harry Ferguson Incorporated of Detroit. Coventry rather than Detroit became the heart of the Ferguson company: in 1949 Detroit was to produce 12,859 tractors, Coventry 38,689. It was Coventry's cost advantage in many international markets that made the difference.

But in spite of the emergence of Coventry as a source of Ferguson tractors, developments in Detroit following the break between Ford and Ferguson in November 1946 were none the less a serious matter for the company. That it seemed to involve losing the rich United States market was bad enough, but even worse was the fact that this was happening at a time when the profitability of the United Kingdom operation had not yet fully been confirmed. Also, with Ford marketing a tractor that included features closely resembling the Ferguson system, there was the risk that even the English Ferguson tractor might lose its distinctive selling features in the minds of buyers around the world.

After November 1946, Harry Ferguson Incorporated began frantically to search for new manufacturers. When no suitable ones could immediately be found, management decided, somewhat hastily, in February 1947 to purchase a large plant of 800,000 square feet in Cleveland, Ohio, for $1.9 million from the War Assets Administration. It had been operated during the war by the Cleveland Pneumatic Tool Company. An engineering firm was retained to make the plant operational. Almost immediately a distributors' meeting was held at the Cleveland plant, for it was vitally important for Ferguson to show them that Harry Ferguson Incorporated was going to be in business.

While the Ferguson company had $6 million of capital funds at the end of 1946, substantial capital expenditures involved in the Cleveland manufacturing project required outside financing. After exploring private financing with various underwriting institutions, an underwriting agreement for public financing was signed in February with F. Eberstadt & Company Incorporated of New York, whereby Ferguson was to receive approximately $8 million through the sale of common and preferred stock. Eberstadt arranged for McKinsey, Kearny & Company to prepare a report on the Ferguson business, which report appeared in April and was favourable to the company. Then on April 22, 1947, Harry Ferguson Incorporated suddenly faced another crisis. Roger Kyes, President of Harry

Ferguson Incorporated, after discussions with his underwriter, made the decision to withdraw the financing programme. Foremost in the mind of Kyes when he made this decision were disturbing reports circulating publicly that the Ferguson company was losing a significant proportion of its distributors. In a letter to Harry Ferguson he attributed this to the efforts of the Ford organization. Market conditions, too, were not particularly suitable for a new issue, but the underwriters were very worried about the future of the Ferguson company if it lost its distributors.

As soon as this financing had been cancelled, Harry Ferguson began to retrench. Distributors were told that plans for production of tractors at Cleveland had to be abandoned and they were advised that they should feel themselves free to accept Dearborn franchises. The Cleveland plant was put up for sale and was finally sold to General Motors for $2.0 million, and negotiations were begun to reduce commitments to suppliers. Albert A. Thornbrough of Harry Ferguson Incorporated had the thankless responsibility of negotiating these reductions. In the end the reductions, which might have been very costly to Harry Ferguson Incorporated, turned out to involve a relatively small loss.

The company now was back where it had been months before. Many thought it was finished. James S. Duncan, President of Massey-Harris Company Limited, writing in May 1947 to one of his Vice Presidents expressed that view: "Have you heard that the Ford Ferguson of the U.S.A. are folding up? They have been unable to finance their manufacturing projects, and they are through." Much time had been wasted, and evidence continued to accumulate that the Ferguson distribution system was falling apart. Ferguson had had thirty-three distributors, but twenty-six of these became Dearborn distributors as of July 1, 1947, and all but two of the remaining seven subsequently sold their buildings or inventories to Dearborn distributors. With them Ferguson lost most of their 3,300 local dealers. Many dealers cancelled orders placed with distributors and it was suggested that they were being warned not to buy Ferguson products.[6]

The supply of Ford manufactured tractors ceased in mid-1947, and July sales of Harry Ferguson Incorporated dropped to $2.1 million from the June figure of $8.3 million. Even these reduced sales included corn pickers which would be lost to the Ferguson company in early 1948 as a result of Ford buying Wood Brothers who had manufactured them for Ferguson since 1943. Meanwhile the production of Ferguson-System tractors by Ford rose quickly to reach about 400 per day, while Ferguson was not even in a position to announce a tractor programme. More and more the question of taking legal action against Ford began to be discussed, and the Board, on August 27, 1947, approved the action of management in employing the New York law firm of Cahill, Gordon, Zachry and Reindel to

6 Minutes of the Board of Directors, Harry Ferguson Incorporated, July 24, 1947, p. 867.

make a preliminary survey of courses of action.

As if all this was not enough, serious management and personnel problems began to develop within the Ferguson company. The personnel problem was one of vanishing morale. Many individuals were torn between staying with the uncertain Ferguson organization or going to Dearborn. When the Cleveland project was being contemplated many employees sold their homes and property and were preparing to move to Cleveland. Then suddenly the project was cancelled, and plans had to be reversed. The difficulty was heightened when, toward the end of June, Ford gave notice to the Ferguson company that it would have to vacate immediately space that it occupied in the Ford Highland Park building. New space had hurriedly to be found. Thornbrough located some in an old building on Milwaukee Avenue and a move involving all departments was made within three weeks. The unhappy contrast between the accommodation there and the accommodation that employees had been accustomed to only increased the pessimism within the organization.

Discord among management did not make matters any better. In May 1947 Harry Ferguson seems to have lost confidence in Roger Kyes and a split was beginning to occur within the Detroit management. Harry Ferguson and Horace D'Angelo generally favoured a more venturesome policy than the somewhat conservative approach counselled by Roger Kyes. The conservative approach that Kyes favoured took the form of a continuing search for subcontractors instead of a revised programme for manufacturing and assembling. Just prior to the break between Ferguson and Kyes in November 1947, the two differed also over the implement programme that should be planned for 1948. Kyes, pointing to all the difficulties, favoured retrenchment into a liquid position whereas D'Angelo, backed by Ferguson, favoured expansion. It appeared to some at the time that the President of Harry Ferguson Incorporated wished not to risk his investment in that company by directing it into a new manufacture and assembly operation.

By the fall of 1947, therefore, the management of Harry Ferguson Incorporated was split. The President had lost the confidence of Harry Ferguson, sales were running less than $2 million a month and threatened to decline further in 1948, no subcontractors for tractors had been found, there was no implement programme for 1948, employee morale could not have been lower, and many key employees had left the company.

These were the conditions that greeted Harry Ferguson when, in November of 1947, he came personally to Detroit. On November 17, 1947, he announced to the Board as its Chairman that the resignations of Roger M. Kyes and Charles R. Vincent, Jr. had been requested. The Chairman felt that, "... the course on which Mr. Kyes appeared to have been guiding the company for the past 6 months seemed deliberately calculated to bring about the liquidation of the company whereas what was needed was courageous leadership to restore faith and confidence in the company and aggressive action to rebuild its business and find a tractor source."[7] Earlier Harry D. Myers had resigned as Vice President in charge of procurement.

Harry Ferguson assumed the position of President and Horace D'Angelo became Executive Vice President and Treasurer. An operating committee of D'Angelo, Klemm, Page, Surridge, and Thornbrough was formed. Almost immediately a remarkable spirit of enthusiasm replaced pessimism, and the decision was made to import 25,000 Ferguson tractors from the United Kingdom pending arrangements for a tractor source in the United States. In January 1948 an option was taken on 72 acres of land on Southfield Road, Detroit, in the name of "Mr. Harlan" – a curious procedure prompted by the mysterious and sudden failure of another land transaction a few weeks before, which management felt was related to the dispute with Ford. The option was exercised and a contract was let for the construction of a factory for completion by July 31, 1948. Ground was broken in February 1948. Negotiations were entered into with the Detroit Gear Division of Borg-Warner Corporation for the production of transmissions, rear axles, and hydraulic pump assemblies, while the engine that had previously been designed by Continental for Ferguson and which had powered the first 25,000 tractors built at Coventry was available to the new Ferguson Detroit project. Because it was generally a much smaller project than the abortive Cleveland one, Harry Ferguson Incorporated could finance the Detroit installation itself.

By April 8, 1948, Thornbrough, in charge of procurement, could report to the Board that only a few component parts remained to be procured, and that the project generally appeared to be ahead of schedule. It was not easy to win the confidence of some of the suppliers because by that time the rift between Ferguson and Ford was in the litigation stage. To go with Ferguson could prejudice a company's business with Ford. In spite of all this, Harry Ferguson drove the first tractor off the assembly line of the new plant on October 11, 1948. Before the end of the year Harry Ferguson Incorporated was assembling tractors at Detroit at the rate of 100 per day, some distributors and dealers had been located, and a market was being re-established. It is difficult, when one recalls the state of the company in November 1947, to regard this achievement as anything other than remarkable.

Sales and profits soon began to rise and the figures in Table 8 show the magnitude of the improvements. In terms of profits the improvement was modest. In 1949 the company had to embark on a decisive cost-reduction programme to trim the excessive cost incurred in hurriedly moving back into business. Competition, particularly with Ford, was continuously hard. Litigation expenses, as we shall see, were beginning to be heavy. The market was not immediately as buoyant as management had expected it would be. But all this was suddenly changed with the outbreak of the

7 Minutes of Adjourned Special Meeting of the stockholders of Harry Ferguson Incorporated.

Korean War. Production availability, not demand, became the limiting factor as far as sales were concerned. These sales were helped by the fact that the Ferguson tractor with Continental valve-overhead engine had quite a considerable superiority in torque performance over the old L-head engine in the Ford 8N tractor – a superiority which could be put to very good use in field demonstrations.

TABLE 8

Harry Ferguson Incorporated Production, Sales, and Profits (1946–1953)

	Assembly of tractors at Detroit (Units)	Total sales ($ U.S. millions)	Total profits after taxes ($ U.S.)
1946	—	79.4	4,273,253
1947	—	70.5	2,164,811
1948	1,800	33.1	540,968
1949	12,859	32.6	—368,251
1950	24,503	41.0	296,943
1951	33,517	64.6	406,956
1952	35,965	71.0	4,009,859†
1953	17,314	35.7*	—104,341*

*Ten months

†Includes net effect of litigation settlement. Net profits from operations were under $2 million.

After July 1950, official allocation of materials was introduced and a production cutback was threatened. Allocation was to be based on historical production data, which placed a new company like Harry Ferguson Incorporated at a great disadvantage. Through formal presentations to government and considerable manipulation of sources of materials, a cutback was avoided and tractor production exceeded 33,000 units in 1951, 35,000 in 1952. When Coventry failed to bring out the long-promised bigger tractor, Detroit introduced the slightly more powerful TO-30 tractor in the fall of 1950, which also contributed to the company's success. The TO-30 improvements included increased engine power, improved steering, improved hydraulic system control (including two-way control through the top link, positioning control, sensitivity control), an improved seat, clutch on left-hand side, and brakes on the right-hand side.

Detroit engineering activity, including the design work on the proposed TO-35, recognized the increasing demands of a sophisticated market, whereas at Coventry work on the larger tractor began with emphasis on utmost simplicity – roughly a TE-20 increased in weight and power. It was most difficult to convince Harry Ferguson that the United States market made special demands on tractor manufacturers and in later years it was the unsuitability of the TE-60 for that market that resulted in its being completely rejected.

Because of the company's policy of low profit margins on tractors and its failure to market a large tractor, both Coventry and Detroit had to make profits on implements. Detroit engineering developed a large number of these and it began work on a side-mounted hay baler, forage harvester, and combine. To speed its entry into the combine harvester field, it attempted to purchase self-propelled clipper combines from Massey-Harris, but the latter feared this would harm its own dealers and so rejected the proposal.

While the Ferguson Company was recovering quickly as a going commercial concern its attention and efforts were increasingly being devoted to the lawsuit with Ford. Harry Ferguson Incorporated filed a suit against Ford in New York on January 8, 1948. It was known as Civil Action no. 44-482 of the United States District Court, Southern District of New York. It was filed only after Ford had begun to market its tractor with patented Ferguson-System devices, and after Ferguson management had experienced what they regarded as deliberate interference with their efforts to become re-established. The complaint that was filed involved three claims for relief – the first involved the charge that Ford had virtually rendered the business of the Ferguson company unprofitable and valueless resulting in injuries and damages amounting to $80 million; and that under provisions of anti-trust laws Harry Ferguson Incorporated was entitled to recover treble damages or a total of $240 million. The second involved a charge that certain patents had been infringed, and damages for a total of $11.1 million were asked for by Harry Ferguson Incorporated. A supplementary complaint by Ferguson raised this total of $251 million to $342 million in March 1951. The third claim for relief was for damages arising from injuries suffered which violated the general principles of law and equity; for this no additional monetary relief was demanded. In this way began a trial that generated words amounting, it has been estimated, to the number that would be found in 2,000 or 3,000 full-length novels.

A chronology of the highlights in the litigation runs as follows:

January	1948:	Ferguson files suit against Ford in New York City
April	1948:	United States District Judge S. J. Ryan denies Ford motion to transfer suit from Southern District of New York to United States District Court in Detroit
June	1948:	Horace D'Angelo begins deposition
September	1948:	Harry Ferguson begins deposition in the United States
December	1948:	Harry Ferguson concludes deposition in the United States
July	1949:	Ford files counter-claims against Ferguson
July	1949:	Ferguson files reply terming counter-claims "wholly without merit"
October	1949:	Pre-trial examination of Henry Ford II begins

January	1950:	Judge Ryan denies second Ford motion to transfer suit from New York to Detroit
January	1950:	Pre-trial examination of Ernest R. Breech, Executive Vice-President, Ford Motor Company, begins
March	1950:	Ernest R. Breech completes pre-trial examination
March	1950:	Henry Ford II completes pre-trial examination
May	1950:	Circuit Court of Appeals denies third Ford attempt to transfer suit to Detroit
August	1950:	Harry Ferguson begins deposition in England
September	1950:	Harry Ferguson completes deposition in England
October	1950:	United States Supreme Court declines review of decisions of lower courts in connection with Ford motions to transfer suit to Detroit
December	1950:	Horace D'Angelo completes deposition
March	1951:	Harry Ferguson arrives from England for trial
March	1951:	Trial begins
April	1951:	Horace D'Angelo takes stand
June	1951:	Trial recessed for summer
October	1951:	Trial resumes
November	1951:	Horace D'Angelo completes his testimony
April	1952:	Judge Gregory F. Noonan issues consent judgment bringing settlement of case

As the lengthy litigation imposed severe strains on both Ford and Ferguson, it is not surprising that attempts were made from time to time to bring about an out-of-court settlement. But Harry Ferguson rejected this with a determination that appeared to many to border irrationality, arguing that he was fighting for all the small business men of the country and for the Ferguson System, knowing full well also that the publicity value of the suit was, at least to some extent, offsetting its mounting and burdensome expense. It is still believed by some people who were close to the events surrounding the litigation that had Ferguson chosen an opportune moment he would have been able to settle for two or three times the amount that he finally received. Still, when it is remembered that a determined effort had been made to impugn Ferguson's life work and monumental contribution to mechanized agriculture, when it is remembered that the basis of compromise suggested to him had frequently involved his losing control of the distribution company, and when it is also remembered that Ferguson management was genuinely convinced that Ford, with assets of over one billion dollars, had tried to do serious damage to Harry Ferguson Incorporated with assets of around twelve million, his obstinacy can be understood.

Then suddenly in early 1952 a curious change occurred on the Ferguson side. There seemed to be a strong desire to end the litigation as quickly as possible. What had brought about the change? The most likely explanation we have encountered is as unusual as it is simple. Ferguson counsel apparently began to fear, probably quite mistakenly, that circumstances surrounding the divorce action of a member of the Ferguson management might soon become public knowledge and might destroy, or at least tarnish, the aura of high morality with which the Ferguson case had been surrounded. Be that as it may, Ferguson was encouraged to agree to a consent judgment, which he did.

The consent judgment is much more significant for the history of Ferguson than might appear at first sight. It dismissed charges against Ford of conspiracy to injure or destroy Harry Ferguson Incorporated and of monopoly practices, but it awarded Ferguson $9,250,000 in payment of royalties on patented devices used by Ford after the break with Ferguson, devices consisting in large part the Ferguson System.

Perhaps the most convincing evidence presented to counter the Harry Ferguson Incorporated claim for damages under anti-trust legislation was its own financial recovery after the Detroit plant was brought into production. It became somewhat difficult to argue, as Ferguson counsel had to do, that severe damage had been done when business, in fact, was good. However, the following year, when the Korean War boom was gone and the weaknesses of a hastily developed distribution system were revealed, provided a better perspective on the permanent damage suffered by Harry Ferguson Incorporated in losing its source of supply and distribution system to Ford in 1947. This, of course, could not be known in 1952, but it does suggest that the Harry Ferguson Incorporated claim for damages under anti-trust laws was not unrealistic in principle even though the huge amount claimed seemed to weaken the Ferguson case.

But this was not the important issue in the controversy between Ford and Ferguson, at least not from a long-run point of view. During the trial many efforts were made to discredit Harry Ferguson's contribution in the development of the Ferguson System, and it must be said that Harry Ferguson was an easy target for attack. Views that he held with great tenacity and argued with vehemence frequently made him appear to be a crank. He found it difficult to think in terms of compromise. He had frequent periods of irrationality, and moods of elation and mental depression, which increased in his later years. His difficulty in working in harmony with people had amply been illustrated. Lack of formal training in engineering made it easy for critics to argue that his engineers, not he, were responsible for achievements attributed to him; and it is true, as it always must be in similar arrangements, that the group he brought together and held together was one that generated many ideas. He was also a difficult man. Many were the members of his management team who have walked away from a discussion with him in utter despair of his apparent ingratitude, his dogmatic views, his interference in personal matters. By the time of the merger in 1953 it almost seemed that the future of the Ferguson company would be brighter without Ferguson than with him.

But when all this is said there remains Harry Ferguson who had an idea, who had the vision and the tenacity to pursue it over several decades, who sought out people to pursue it with him, who inspired them to make contributions almost beyond their own talents, and who in many cases won from them an unusual and uncompromising loyalty. In the end, his concept of the Ferguson System and of a light tractor incorporating it, and his conviction that it should and could be produced and sold in mass quantities, changed the industry and in many areas of the world greatly hastened the process of farm mechanization. Look at tractors and farms at random – you will see everywhere what he accomplished.

Had the consent judgment detracted from that accomplishment it would have been a tragic injustice. This it did not do. In the area that really mattered, it fully vindicated Harry Ferguson for it upheld his patents on converging links, on the pump that hydraulically operated the draft control system and the power-take-off shaft, on the inlet valve, and on the coupling pin. Ford had in future to redesign portions of its tractor where patent infringement was involved.

The financial rewards from the suit were much less than the $9,250,000 suggests. Legal fees took about $4 million, and many other expenses had been incurred. Harry Ferguson Incorporated had also, as part of the settlement, to pay the Ohio Tractor Company and Berry Brothers $500,000 in settlement of their law suits against Ford. The award also included $1 million in payment by Ford for a licence on one of the patents and this Harry Ferguson Incorporated had to pay over to Ferguson. Ferguson actually wanted Harry Ferguson Incorporated to pay him another $1 million out of the award, for heartache and personal injury, but he finally gave way on this. It was a great victory – but much more a moral than a financial one.

With the law suit out of the way after April 1952, the future of the Ferguson company appeared bright. The general feeling was that Ferguson, the underdog, the "little man," had been victorious. Sales were growing and with an end to litigation expenses, profits were likely to increase. Then in mid-summer came a steel strike and, even worse, a strike at the Borg-Warner Corporation.

The Detroit plant soon lay idle, for without supplies of transmissions from Borg-Warner's Kercheval plant, the tractor assembly line could not operate. Borg-Warner had experienced recurring labour trouble, and a quick settlement was not in sight. Harry Ferguson Incorporated did not have the financial resources to lie idle for long. It was then that the Director of Procurement of Harry Ferguson Incorporated, A. A. Thornbrough, began to investigate a solution to the problem that was, in a sense, the beginning of a significant development – he thought of importing components from the United Kingdom for assembly in Detroit. On October 3, 1952, he sent the following cable to Harry Ferguson Limited, Coventry:

Borg-Warner strike settlement highly indefinite. Can resume tractor production if Harry Ferguson lease and operate one Borg-Warner plant provided Standard Motors can furnish parts per tractor as follows:

> (1) TE-7017 Matched (1) TE-7113; (1) TE-7110 Matched (1) TE-7121; (1) TE-7102 Matched (1) TE-7123; (1) TE-7119 Matched (1) TE-7118 and (1) TE-7111-A; (2) Matched Sets TE-7128-A; (1) TE-7061; (1) TE-7146; (1) TE-7141; (1) TE-7145; (1) TE-7140; (1) TE-716; (1) TE-717; (1) TE-618-A.

Request quotation of prices from Standard Motors of 5,000 separately packaged sets as listed above protected from corrosion C.I.F. eastern seabord United States and how quickly initial 5,000 could be delivered. Botwood informed of this request. If program for above parts is effected, we believe we should place our purchase orders directly with Standard Motors to obtain manufacturer prices. Letter follows. Appreciate prompt reply. Thornbrough / Farming / Detroit.

The Standard Motor Company sent a cable on October 7 saying it could send the 5,000 sets, the first within two weeks, provided Detroit could give them immediate instructions. But settlement of the strike came much sooner than had been expected and on October 16 Thornbrough had to send a cable placing the emergency programme in abeyance. By that time, however, a study had been undertaken at Detroit on the degree of interchangeability that existed between the components of the TE-20 tractor produced at Coventry, and the TO-20 tractor produced at Detroit, with emphasis on transmission components. It revealed that since just after the war when the two had started off as almost identical tractors, and so with completely interchangeable components, differences had crept in, but also that with careful matching and assembling the gears and shafts could be interchanged. There was no doubt that with care in the engineering stage, a high degree of interchangeability would be possible. Harry Ferguson stressed the importance of it.

At that time the Engineering Division at Detroit was working on a much improved model, the TO-35, and maximum interchangeability between it and the equivalent one to be built in Coventry some time in future was planned. This laid the basis for what, in future years, became an important part of Massey-Ferguson's engineering and manufacturing policy – the interchangeability of a maximum number of major components so as to increase the company's international flexibility in sourcing supplies and to reduce its unit manufacturing costs by sourcing them from centres of lowest cost. But that came somewhat later. For the present, Harry Ferguson Incorporated, with great relief resumed tractor production, only to find that the market had begun to shift.

The post-Korean boom was over. Harry Ferguson Incorporated felt this immediately. Soon it faced conditions that prompted it to view a merger with another organization much more favourably than it would have done before. As is always the case, these emerging conditions of declining sales revealed more clearly than previously other difficulties within the organization.

Harry Ferguson seldom visited Detroit and left the actual operation of the company to the Executive Vice President and the Board of Directors. The Executive Vice President, Horace D'Angelo, was for personal reasons absent from the company for prolonged periods. This double absence of leadership invited discord among management and the *esprit de corps* that lifted the corporation in 1948 began to vanish. It became increasingly difficult to deal with Harry Ferguson whose behaviour became more and more erratic. He regarded a minor quality difficulty as a major crisis warranting voluminous correspondence. Frequently, after misinterpreting a sentence in a letter sent to him by the management of the United States company, he would reply at great length in a highly emotional, even vitriolic way. Unpredictably he would range from the heights of enthusiasm to the depths of despair. By the time the negotiations for the merger had commenced Harry Ferguson was convinced that the United States operation was in serious trouble.

All this is unmistakably reflected in a letter Harry Ferguson wrote to Herman Klemm, Director of Engineering of Harry Ferguson Incorporated, on July 23, 1953:

My dear Herman,

FIRST THINGS FIRST

Curry wrote me a long letter dated July 8th. I have never received a worse shock than the state of affairs which that letter disclosed. We have reached the stage where something drastic must be done to save the situation so please forgive me if I write you in strength and frankness. . . .

The first and most vital thing to do is to put our present tractor right if all the troubles mentioned by Curry have not yet been put right. . . . This trouble is not the primary cause of our present situation in H. F. Inc. but it is one cause and it is the most urgent to be put right. . . .

Our Directors would appear to be temporising with everybody and he who temporises is lost. *What we need immediately is leadership.*

MY ATTITUDE TO YOU ALL

In what I said to you yesterday, and in what I am now saying, I have not one thought other than what is best for you all. I am quite willing to sink my own interests to help you and if you believe you can run the company better without me then buy me out at a reasonable price and we will all shake hands over it.

H.F. Inc. is going to pieces before our eyes. The home sales have slumped and I was told yesterday that last year in Canada we did nearly 80% of British tractor sales there and now we are doing only about 34%. In Brazil we were doing about 41% of the business last year and now we are doing no business there. If Horace comes here soon I will go into all this with him with the utmost seriousness and explain what must be done to save the situation. However, although I feel ill and deadtired over all this, I am writing these letters to you to stem the tide before it gets too strong for you.

YOUR LETTER OF JULY 10

... (Put) out of your head that we are going to change the colour of our tractor or its outline. Such talk is disastrous. *What is needed is to get our sales organization on the right basis immediately.*

The colour of our tractor is right and that is the way it will stay because it is right. ... We did that job with the most extraordinary care so that the machine could run for ten or twenty years without changes.

THE SOLUTION OF OUR SERIOUS PROBLEMS

Horace wrote me recently to express his confidence in the future regardless of associating ourselves with any other company. I utterly fail to see on what he bases his confidence. As I see it at this moment I feel we ought to make the best terms possible with some company and preferably M-H. Even if we do so we must reorganize our sales department and get a real drive behind selling. ...

Herman Klemm, deeply hurt, answered Ferguson's letter with a twenty-three-page reply. It showed that the quality problems had greatly been exaggerated, but also that Ferguson's worries about the United States company were justified. Klemm reported:

Our sales for the first six months in the United States have decreased 44.08 per cent over 1952; our Canadian sales have decreased by 29.24 per cent; our Latin-American sales have decreased 35.6 per cent, and our Eastern Hemisphere sales have decreased 6.9 per cent, or a total decrease in sales for the first six months of 42.32 per cent over 1952.

Our plant has been closed from July 1st and won't reopen until August 10th, and we are shipping tractors ... at the rate of approximately 60 a day. I realize the full impact of this because expenses are continuing, but since our Sales Division has been unable to ship more tractors it was best to close the plant and not to pile up a large inventory. ...

I know our situation is critical, but not hopeless.

So while Ferguson usually wrote with feeling, on this occasion it was at least partially justified. Klemm could also have mentioned that Harry Ferguson Incorporated was again experiencing a net loss on its operations which made it difficult for the company to cut prices in order to increase sales. The fear that Ford would do just that was a constant worry.

The declines in Ferguson sales and profits were worrying enough by themselves but they were accompanied by other problems; disharmony among management, lack of leadership within the United States company, a relatively young and inexperienced distribution organization, erratic and frequently disrupting direction from Ferguson whose absentee control was proving to be unequal to the task, and an inadequate product line, particularly the absence of a larger tractor suitable for the United States market and of harvesting equipment. It is questionable whether the United States operation could, by itself, have survived the difficult years that lay ahead.

It was a company with these problems that the President of Massey-

Harris came to inspect in July 1953. He too came from a company with worrying difficulties in its United States operation, as we have already seen. Both organizations badly needed new strength. Each seemed to have something to offer to the other.

CHAPTER 7

The merger

On August 17, 1953, the newspapers of a number of countries reported that the world-wide Massey-Harris and Ferguson organizations were going to be merged to form Massey-Harris-Ferguson Limited. The official press release, of course, made no reference to the mounting difficulties both companies were facing. It read:

> It has been jointly announced by James S. Duncan, CMG, President of Massey-Harris Company Limited, on behalf of its Board of Directors, and by Harry Ferguson, Chairman of the Harry Ferguson Companies, that the two world-wide organizations which they represent have agreed to amalgamate and to operate in the future under the name of Massey-Harris-Ferguson Limited.
>
> The uniting of the Harry Ferguson interests, whose tractors equipped with the revolutionary Ferguson System and Mounted Implements, have blazed a new trail throughout the world, and the 105-year-old Massey-Harris Company, which pioneered the self-propelled combine in every country where wheat is grown, and whose development and progress has been so spectacular in recent years, is probably the most important news in the farm equipment industry in the present century.
>
> It will not only bring together two progressive organizations, each of which has made an unprecedented contribution to agriculture and has set a pattern which the entire industry is seeking to emulate, but will pool their organizations, which are spread wide over the earth, including manufacturing facilities in the United States, where they have five plants, in Canada where they have four, and also in England, Scotland, South Africa, France and Germany.

While the merger was effected after a relatively short period of intensive negotiations, high-level contact between Ferguson and Massey-Harris first occurred in late 1947. After Roger Kyes had departed from Harry Ferguson Incorporated in October 1947, that company, in addi-

J. D. Barn

An MF tractor and baler work a hayfield in the south of France.

tion to being without a source of supply for its tractor, was also without a President. Harry Ferguson came personally to Detroit to deal with the crisis. On December 15, 1947, he met with J. S. Duncan, President of Massey-Harris, to try and interest him in joining his organization and to enquire whether Massey-Harris would undertake the manufacture of the Ferguson tractor in North America. Both proposals were declined. The latter proposal was declined because it was felt that Ferguson had been too severely weakened by Ford to be a good risk for capital expenditures and because Massey-Harris was preoccupied with expanding its own organization.

Then in the summer of 1948, G. T. M. Bevan, Vice President Engineering, Massey-Harris Company Limited, met Alan Botwood, Managing Director of Harry Ferguson Limited, Coventry. Botwood mentioned that he was thinking of obtaining a small pull-behind combine with auxiliary engine to be sold in conjunction with the Ferguson tractor, and wondered whether Massey-Harris might co-operate with them on the project. These ideas were further pursued in a meeting between the two in February 1949, and the President of Massey-Harris also expressed interest in the project.

On March 30, 1949, when Bevan for the third time visited Botwood, Sir John Black, Managing Director of the Standard Motor Company, happened to come into Botwood's office. Since Sir John could assure them of the adequacy of the Standard engine, the combine project seemed feasible. But Botwood was reported as having had to point out that:

> ... he wanted to regard these preliminary negotiations as exploratory and that he did not intend to put the matter before his Board of Directors at present, nor to inform Mr. Harry Ferguson as both he and Sir John said that that would probably "gum up the works." I gathered they had just had a Board Meeting and that Mr. Ferguson had, as usual, been very difficult.

Massey-Harris continued to consider the combine project but in the end, with their #744D tractor production and their new #726 combine production getting under way at Kilmarnock, they decided against it. When in August 1949 Botwood suggested that Massey-Harris might want to fit the Standard engine to their new small #730 combine, Massey-Harris rejected this suggestion as well.

Then in August 1951 Duncan went on an inspection tour through Standard's tractor plant at Coventry, and on September 10, 1951, wrote to Harry Ferguson:

> ... I was rather hoping that I would have had the pleasure of seeing you in England, but my trip was too hurried. I should like to have an opportunity, however, of talking things over with you, and if you would let me know when you are coming to the States, I could perhaps arrange to meet you in New York, or some other convenient location. If this does not work out, then perhaps I will be able to see you in England on one of my forthcoming trips during the spring of the year.

Harry Ferguson returned a short reply indicating that they might arrange to meet in the United States in October or November. But he then wrote immediately to Horace D'Angelo, President of Harry Ferguson Incorporated at Detroit, quoted Duncan's letter and went on to say, " . . . My impression is that they want to amalgamate with us but I feel about them just as I do about Oliver. Talk to the others and drop me a line on the subject."

D'Angelo discussed it with his senior executives and they agreed that they would not be interested in an offer of amalgamation should one come along. It did not come, and Ferguson and Duncan did not even see each other that year.

In the summer of 1952, Duncan requested Bevan to obtain a distillate or diesel engine, or even both if possible, for the baler and the new #750 combine to be manufactured at Kilmarnock. Bevan went to Botwood who at first was agreeable to supplying the engine used by Ferguson; but he had to decline shortly thereafter when Harry Ferguson objected to it.

In the meantime Massey-Harris was also beginning to experience severe difficulties in its Kilmarnock plant because of reductions in production and the increase in labour redundancy. Management wondered whether opportunities for manufacturing machinery for Ferguson still existed. Following a cable request by Duncan for information regarding the progress of the Ferguson company, Lionel Harper, A. L. Weeks, and E. G. Burgess, all of Massey-Harris, visited the Coventry factory of Standard Motors on January 5 where they met Trevor Knox and Alan Botwood of Harry Ferguson Limited. They came to the conclusion that the new large Ferguson tractor would be a threat to Massey-Harris, and they learned that Ferguson expected to be able to arrange for the manufacture of its existing tractor in France – and so pose a threat to the sale of the Massey-Harris Pony tractor. Very gently the Massey-Harris visitors broached the possibility of manufacturing some machines for Ferguson. They were told candidly that Ferguson sales had fallen off so sharply that the company was having difficulties with suppliers over cancelled orders and cancellation charges. Burgess had to conclude that there would not be much chance of manufacturing anything for the Ferguson company.

In order to make progress in obtaining an engine for the Kilmarnock baler and combine, Bevan went again to see Botwood in early 1953. Botwood told him that a clause in the contract with Standard prevented the sale of the Standard engine for agricultural purposes, but agreed to go back to Harry Ferguson and Sir John Black about it. In the end the latter two agreed to the proposal.

This agreement was followed by an invitation to lunch from Sir John Black to Massey-Harris management, and particularly to Bevan and Harper. They met in April and learned a great deal about the Ferguson organization: about the large tractor, about Ferguson's high hopes for that tractor in the United States market, about negotiations regarding tractor manufacturing in France and volumes expected, about output in Coventry, and finally about Ferguson's progress with a "wrapped around" or "mounted" combine. The Massey-Harris executives suggested to Black that they might be interested in building that combine for Ferguson, and Black regarded it as a good idea. After that meeting Bevan made an appointment with Botwood to talk to him about the small combine, and other matters. In a letter to Duncan dated April 29, 1953, Bevan expressed the view that "This get-together on engines might easily lead to further contacts which can be advantageous to us."

While these negotiations and discussions relating to the engine and mounted combine were developing, the two companies were being drawn together elsewhere as well. Jack Bean, who supplied certain items to Standard for the Ferguson tractor, and who was well known to both Duncan and Bevan, told the Massey-Harris executives that the Ferguson company in the United States had large inventories and that not too long ago even the United Kingdom company had found itself short of cash. He thought that the time might be propitious for Massey-Harris to buy the United States company. Bean was apparently encouraged by Massey-Harris to speak generally about such a possibility, and in a letter to Bevan dated April 17, 1953, Duncan remarked:

> I hope that Jack Bean will in his enquiries, or particularly in his conversation with Ferguson, not go beyond my own thoughts which were to ascertain whether the Ferguson people had ever given any serious thought to sell out their interests in the United States. We have gone no further in our thinking than the general conclusion that if the Ferguson outfit in the United States could be bought separately, it might be less expensive than to develop a tractor of our own and the equipment which goes with it. On the other hand, as I have frequently stated to you we fully recognize that whereas Ferguson in the United States has sold a very large number of tractors, they have never been able to make any money and the proposition as such, therefore, is not a particularly interesting one unless it was on the bargain counter.

Jack Bean telephoned Alan Botwood, and Botwood, in reporting this conversation to Horace D'Angelo, of Harry Ferguson Incorporated of Detroit, said:

> In order to find out a bit more about it, I treated the matter as being rather amusing. It then transpired that Duncan had without doubt been in touch with Bean and it seemed to me that a definite approach had been made. I naturally brought this to the attention of Ferguson, who, of course, did not spend much time on the matter because obviously we should not be interested. I therefore telephoned Jack Bean just merely to say that such an offer would be of no interest to us. I did not even invite him to come over and talk to Ferguson about it.
>
> I would like you to know that I have met Duncan on one or two occasions and that he is coming over to England in a couple of months or so and that I have suggested that we might meet when it might be that we could find out what he had at the back of his mind.

In spite of Ferguson's rejection, through Botwood, of Bean's approach,

Botwood wrote to Duncan on May 28 inviting him again to stay at his place when next in England, going on to say:

> There are many interesting subjects about which I am sure we could talk. . . . You will know I have had some talks with our mutual friend Jack Bean. . . . I shall be obliged if you would let me know when you are coming over and I am sure that we could quickly fix a time suitable to us.

Writing to Harry Ferguson Incorporated on June 4, Harry Ferguson told them of this letter to Duncan and, significantly, he pointed out that the letter had been carefully thought out by himself and Botwood. Obviously Ferguson was very much interested.

In June Duncan went to England and on June 22, 1953, he, W. Lattman, and G. T. M. Bevan met Botwood and visited Harry Ferguson Limited and Standard Motors. Ferguson was vacationing in Ireland at the time. Duncan visited Botwood's home and discussed the Massey-Harris offer to assist in designing the Ferguson mounted combine and in manufacturing it at Kilmarnock, as well as the matter of the supply of Standard Motors engines for the Massey-Harris combines produced at Kilmarnock. They seemed to agree that there were too many firms in the farm equipment business and that for Massey-Harris and Ferguson to co-operate in some way or another seemed mutually desirable. The meeting between the two went smoothly and they arranged to meet again on June 30 to sign agreements relating to the production by Massey-Harris of the Ferguson combine.

Then a curious thing happened. Bevan, Lattman, and Duncan returned to Coventry as arranged. They were met by a very upset and embarrassed Botwood, who had to explain that Ferguson had called him from Ireland, where he was convalescing after an operation, and told him to cancel immediately all the negotiations with Massey-Harris relating to the manufacture of the combine. Botwood went on to say that Ferguson had also told him that he was going to send his Rolls Royce that morning to Coventry in the hope that Duncan – and Duncan only – would lunch with him at Abbotswood, his country home, and discuss matters directly with him. Duncan went, leaving Bevan, Lattman, and Botwood behind and met Harry Ferguson for the first time since December 1947.

It may be wondered why Ferguson suddenly began to show positive interest in having discussions with Duncan. One important reason was that he was becoming increasingly concerned with his company in the United States. Indeed he had taken a most unusual procedure for obtaining information about that company. In late May he sent a cable to A. A. Thornbrough who, it must be remembered, was not the most senior executive of the United States company, asking him to come over to discuss the future manufacture of the TO-60 tractor. Thornbrough did not even have travel documents, but he and Mrs. Thornbrough went as quickly as they could.

Landing at London airport, they were chauffeured at great speed to

Abbotswood so as to arrive in time for sherry and lunch – an hour that never varied in the Ferguson home. After lunch Ferguson asked to be excused because of some important matter – later discovered to be a hair-cut. On his return he invited Thornbrough into his garden and there, in that strikingly beautiful and tranquil setting, he embarked on a most intensive and unblushingly indiscreet (since Thornbrough was not the senior executive) enquiry into the affairs of the United States company and its management. It was not the TO-60 he had wanted to discuss.

Thornbrough could not be optimistic. Harry Ferguson Incorporated he felt was severely under-capitalized. Market conditions were adverse. Ferguson distributors and dealers were not sufficiently strong to cope with a long period of low sales and intense competition. He recommended that Ferguson should either realize on his investment or make arrangements to obtain additional capital by associating with another company or by a securities issue. Additional capital would enable the company to reduce costs by moving further into manufacturing.

Ferguson had never been anxious to become deeply involved in manufacturing, so he began to enquire about the various companies in the industry. In doing so he revealed a particular interest in Massey-Harris, and he had in his hand a letter from Massey-Harris about mounted combines and engines. Which company would be the best to be associated with? Thornbrough went through the list. With International Harvester Company and Deere and Company there might be anti-trust implications. With Ford, there were obvious reasons against an association. The J. I. Case company was essentially a United States company, and others had similar disadvantages. Massey-Harris, on the other hand, had a rather complementary line of equipment in its combines and larger tractors, it was also a British Commonwealth company, and it was already operating in international fields. Since Massey-Harris was one of the smaller companies in the United States, the anti-trust implications were less than with the others and a merger might be permitted.

There is no doubt that Harry Ferguson then became increasingly worried over the management of the United States company and its sales prospects. A meeting with Duncan might lead to the solution of this potentially very dangerous problem.

The June 30 private meeting between Duncan and Ferguson was a most important one, if only because that exceedingly complex and frequently difficult and unpredictable man decided during it that he liked James Duncan. Mr. and Mrs. Ferguson met Duncan on their front step as he arrived and they went in to an excellent lunch. After lunch, Ferguson suggested that they walk out to the little tea house in the corner of his garden. There, as Duncan recalls it, Ferguson made him a most unexpected offer. He asked Duncan to leave Massey-Harris and join him as an equal partner. Duncan declined. Ferguson then began to raise the matter of a possible merger. It was clear then that he was interested in much more important things than the matter of Massey-Harris manufacturing combines for Ferguson. Ferguson's recorded views on his reasons for having the meeting and his interpretation of matters discussed appeared in a letter he wrote to the executives of his United States company on July 1, 1953:

In view of the pleasant nature of that meeting (Duncan and Botwood) and the world outlook generally, and yours in particular in the United States, and also in view of your letter to Alan and my talk to Al., I thought it advisable that I should see Mr. Duncan and have a frank talk with him, taking things a good deal further than Alan did.

In view of everything, I believed we should discuss the world outlook, and arranged for Mr. Duncan to come here yesterday morning, stay for lunch and tea, and have a nice chat. When Mr. Duncan arrived we started work immediately out in the shelter in the sunshine, and it was all very pleasant. I liked Mr. Duncan immediately, so did Tony, and so did John Turner when he arrived later in the afternoon and we all had tea together. . . .

I think you all know that I have far too much work to do, and that if I am to succeed with the new agricultural inventions and the new chassis and the new transmission, I absolutely must be freed from a great amount of the work which I am now doing. Indeed, my health would not stand up to doing more than my own job of product development. You can all, therefore, count upon it that I can do no more to help you build up the business in the United States other than help to provide you with the greatest products on earth.

Mr. Duncan said that all manufacturers in your hemisphere were concerned about business at present, and he believed Harry Ferguson Incorporated had a big job on its hands to win through against our imitators.

I then asked him to state what he believed his company and ours would gain by a close working together of some sort, either in the western hemisphere or throughout the world, if that could be developed. He mentioned a number of things. He said that if we would place orders with his company here for equipment, that would start our co-operation and we might develop from there. I then said that I did not see any advantage to Ferguson in anything he had put forward, and explained that from our point of view it would not be good business to give one of our competitors orders for, for example, our new harvesting machinery. He saw my point.

I then said that it was possible that the one great thing he could give us, he had not mentioned at all – and that was the strength of their business, the great many dealers they have and all the other good things attached to them.

I pointed out that on our part we had no interest in his line of tractors and equipment, and that I believed they were out of date and that if they did not get modern tractors of the type of our small and large tractors, their business would suffer seriously before long. . . .

I then said that if they could put up proposals to us which would strengthen our whole business throughout the world and leave me free to do my job, there might be some basis for getting together. Mr. Duncan says they have in the United States some 2500 dealers. It is for you to judge whether by adding these dealers to our strength we could make ourselves big enough to get a safe and profitable production figure. . . .

There arose the question of how we could start getting together. . . . I said that I felt the greatest benefit might come to us if we made a start in your hemisphere, and that that could be done by a purchase of some of my stock in Harry Ferguson Incorporated. He said that was simple, but if we were going to work together he would like to make it a world development. By this time, he was really interested in our world plan, our strategy, our product and, I think, all that is Ferguson.

Then arose the question of the amount of stock, and control, and all the rest of it. I told him that I could take the matter no further until he had a good talk with you nice fellows in Harry Ferguson Incorporated.

It is interesting to compare that report of the June 30 meeting with one made by Duncan, in the form of notes for an Executive Committee meeting. They suggest that Ferguson took the matter of an association with Massey-Harris rather further than his letter to the United States company had suggested. Duncan's notes read:

He (Ferguson) mentioned . . . his desire to be associated with me personally and gave his reasons for it. He went on to say that, in addition to this, our Company, with its world-wide outlook, progressiveness and reputation for honourable dealings, would fit into his picture very well. He believed that we could bring strength and experience to his organization, which, he felt, was not quite strong enough to handle the problems, which it would face with the introduction of the 3/4 plow tractor.

He asked me whether I would be prepared to live in England. I answered in the negative. He then suggested that the following procedure might be adopted:

1) That we would pool our interests in the U.S.A., I becoming President of the organization;
2) That we would look upon this as a first step and leave our participation in the U.K. Company to later.

This was discussed, and I advised him that we would not be interested in American participation, as our interests were not local but global, and further, that we would hesitate to invest a large sum in his company without being assured of eventual control.

He stated –

a) That he would be prepared to sell us an important share of his 90% interest at book value;
b) That he would agree to our purchasing control at then existing book value in a given number of years or at his death;
c) That he suggests that –
    1) We discuss amalgamation with D'Angelo, the President of their U.S. Corporation.
    2) If this looks satisfactory, that I write him a letter, indicating our interest in joining forces with him and outlining roughly how this might be done.
    3) He would then arrange for a meeting with me at his experimental farm, and he would show me and anyone else whom I would bring along, his 3/4 plow tractor; its line of implements and the combine.
    4) If, after seeing these demonstrated, we were satisfied that his claims were correct as to their advanced design, that we would then agree between him and me on the general lines of a merger, the details to be worked out by the officials of both companies subsequently.

> He stated, in leaving, that his desire was to spend the rest of his life designing machines for agriculture, that he was not interested in the commercial end of the business and would be very happy to see the end of it handled by our joint organization under my direction.

So Duncan left Ferguson having agreed to go discuss matters with the executives of his United States company, Harry Ferguson Incorporated, at Detroit.

After the June 30 meeting with Ferguson, Duncan went away and reflected on the advantages and disadvantages of a broad association between Harry Ferguson and Massey-Harris, putting these in note form. In the United States he felt that their very weak position in tractors would be strengthened, and if the new tractor were to work out well it might even give them the kind of advantage they had enjoyed in combines. The plants would benefit from manufacturing Ferguson implements that were now being manufactured by subcontractors. By manufacturing the new large Ferguson tractor at Racine the volume there would soon exceed, he felt, the production of all present models. Their immediate problem would become one of production rather than capital and sales. The amalgamation with Ferguson or their absorption of the Ferguson United States interests would add greatly to the company's prestige, and he felt that with their self-propelled combine on the one hand, and the Ferguson line on the other hand, their problem with respect to dealers in the United States would merely be one of selection.

True there were some difficulties. Ferguson patents were running out and Ford had the advantage of large-scale production and a vast network of dealers. Substantial capital might be required to buy out or buy into Ferguson, and to finance the expansion necessary for introducing the large Ferguson tractor. It might also be that the Ferguson company had substantial inventories that required financing.

As far as the United Kingdom was concerned the situation, he thought, was entirely different. In that country Ferguson was obtaining a huge share of the market – around 60 per cent. But Ferguson appeared anxious to retain control of the United Kingdom company for some time, and the problem of amalgamating distribution without control seemed very difficult. Still, the extent to which a joint enterprise would dominate the combine and tractor field made it a most tempting proposition. He thought that if Massey-Harris could finance the deal their best interests would be served by buying control of the entire Ferguson organization. If they controlled the United States company only, with merely an interest in the United Kingdom company, there was a possibility of friction developing between the two organizations which could preclude future absorption of the business. That friction he thought might develop because of Ferguson's temperamental character and dictatorial powers within his own organization. He reflected that it might just be possible that a deal could be worked out whereby Ferguson would retain his prerogatives in design engineering and be given a position of distinction on the Massey-Harris Board.

These views of James Duncan, written on July 3, 1953, in London, were remarkably prophetic because in the end the merged company did control operations in both the United States and United Kingdom, Harry Ferguson was content with prerogatives regarding engineering, he was given a position of distinction on the Board, and he was difficult to deal with. But all this came later.

Duncan then flew back to Toronto and began to prepare for the meeting with the management of Harry Ferguson Incorporated. He had his staff obtain as much information as possible on the Ferguson operation, and on July 10 reported to the Executive Committee of the Board. In his report he sketched in the background of Ferguson's success and pointed out that over a million Ferguson tractors had been sold, that Ford and Ferguson together were selling an enviable 24 per cent of all tractors sold in the United States and one-half of the tractors in the two-plow class. In the United Kingdom Ferguson was selling 61 per cent of the tractors, and all manufacturers including, belatedly, Massey-Harris itself were attempting to copy Ferguson tractors and implements. He also noted that the new dealers would have access to the full Massey-Harris line and that there would be no radical changes in the line except that the Ferguson tractor would replace the Massey-Harris #22, the Mustang and the Colt. Massey-Harris factories would undertake the manufacture of Ferguson implements manufactured by subcontractors, but the small Ferguson tractor, the TO-30, would continue to be assembled at Detroit. The new 3/4 plow Ferguson tractor could be produced at Racine which might eventually replace the #33, #44, and perhaps #55 Massey-Harris tractors.

Duncan would become Chairman of the new United States company, H. H. Bloom would become President, and senior ranking positions would be given to officials of Harry Ferguson Incorporated, including Herman Klemm in particular. However, Duncan said he would favour this proposal only if at the same time Massey-Harris would obtain a substantial interest in the Ferguson United Kingdom operation, including an agreement whereby Massey-Harris would obtain control of the United Kingdom operation after a given number of years, or at the time of Ferguson's death.

In addition he told the Executive Committee that Massey-Harris should obtain first choice in the manufacture of all Ferguson implements in France, the United Kingdom and Germany, and the purchase of Ferguson patents and Massey-Harris rights to them would have to be clarified. World-wide sales policy of the two organizations would be blended wherever this could be done to mutual advantage. Care would have to be taken to ensure that the new company would retain its source of supply of tractors from Standard.

Duncan also reported to the committee that he was going to go to Detroit on July 13 and 14, 1953, accompanied by H. H. Bloom, M. F. Verity and

R. H. Metcalfe, as arranged between himself and Harry Ferguson. But Duncan was convinced before he left, and after reviewing all the complications, that the only desirable course as far as the United States operation was concerned would be outright purchase of Harry Ferguson Incorporated.

The Massey-Harris representatives met with Horace D'Angelo and Herman Klemm at Detroit. The purpose of the meeting, they felt, was not to discuss or settle the form that a closer association between the two organizations might take but rather to concentrate on the advantages and disadvantages of closer association. To form a view on this, the Massey-Harris representatives wished to obtain more detailed information on the position of the Ferguson Company. They were told of the improvements to the TO-30 tractor that were being considered and which would overcome the Ford advantage of an independent power-take-off and constant running hydraulics. The major components for the proposed large tractor were to be imported from England. They were shown implements that were soon to be released which included the one-way disc, the heavy non-reversible disc plow, the two-row middle buster planter, a subsoiler, side-mounted mower, side-mounted forager, side-mounted baler, and a three-ton trailer. They did not see the new side-mounted combine or the two-row mounted corn picker, and the one-row side-mounted corn picker. But they did see implements suitable for the large tractor, which included the 3/4 furrow mounted plow, four-row mounted planter, four-row mounted cultivator, four-furrow mounted mouldboard plow. They did not see the large combine soon to be semi-mounted on the large tractor, but they had seen enough to be greatly impressed.

They were given details of the Ferguson distribution organization, of the Ferguson training programme, of the suppliers of components important to Ferguson – particularly Borg-Warner – and statistical information on costs and profit margins. They were told that Ferguson manufacturing costs in the United States were actually higher than the cost of landed tractors made at Coventry, but that it was necessary to manufacture in the United States because of the strong customer reaction there to buying British-made Ferguson tractors.

On July 15, after the meetings with D'Angelo and Klemm, Massey-Harris management representatives went to their room and discussed what they had seen and heard. On balance they concluded that a closer association would be desirable, but only if this took the form of a complete fusion of the two United States organizations. Duncan then enumerated in detail the advantages that he saw of an association with Ferguson: prestige of Massey-Harris would be enhanced as a result of acquiring the Ferguson line; it would bring into the organization the most advanced engineering practices known to the industry and a new philosophy of implement and tractor designing; they would inherit at relatively low cost the spare parts business arising from the 200,000 Ferguson tractors and many

Ferguson implements on farms in the United States; it would increase their dealers to more than 4,000; they would acquire the original Ferguson products which in some form had been copied by all manufacturers; they would receive the benefits of the inventiveness of Harry Ferguson and the services of Herman Klemm; their strength in the export field would be increased although this would depend on their arrangements in the United Kingdom; they would have a better chance to compete with Ford than they would with the #22 or Mustang, or even with the disappointingly long-delayed Pitt tractor; they would be able to compete more effectively with the orthodox tractors in the market; they would perhaps be able to take a lead in the 3/4 plow tractor class; their manufacturing programmes for implements in the United States and United Kingdom would be increased; they would be able to offer a more desirable dealership franchise than that offered even by Ford because of the combined appeal of the Ferguson tractors, the Massey-Harris lugging tractors and the Massey-Harris combines, as well as Ferguson and Massey-Harris implements. It was an impressive array of advantages.

Duncan went on to say:

> I place great stress on the engineering quality of the Ferguson designed implements. This all stems from Harry Ferguson's genius and may never be duplicated by Ford, just as after 18 years it has not been successfully duplicated by anyone in the orthodox implement industry. . . .
>
> Basically we must ask ourselves if the trend is towards the Ferguson concept or the old line implement company concept of engineering.
>
> If it is the former – and I believe it is – then the value of an association with its originator might be very great and might well place us in a position of leadership in the industry and should, within the next few years, add at least another 100 million dollars to our turn-over.

At the same meeting Morley Verity pointed out, among other things, that Massey-Harris did not have a tractor which embodied the basic Ferguson design, and that a fusion of the two organizations would enable them immediately to enjoy the benefits of such machines without the risk of mechanical defects of unproven designs. He also noted that the Ferguson combines might make inroads on the Massey-Harris combines in future, and if the Ferguson machine manufactured at Coventry would later be made available to them their exports would greatly be improved. He did not think that Massey-Harris could meet international competition from Ford and Ferguson in the small tractor class. But Verity noted that the profit margin on Ferguson tractors was low and Ford competition was keen, particularly since the Ferguson tractor was priced higher than the Ford tractor. He, too, emphasized that a closer association with Ferguson would be desirable only if it included a substantial and immediate investment in the United Kingdom company and assurance of ownership later on. But this, he felt, would mean a momentous change in the Massey-Harris organization in the United States, Canada, and Latin America which, along with the interest that would have to be taken in the United

Kingdom business would "... put a very great additional strain on top management and on the organization as a whole, which would only be partially offset by the additional personnel which would come to us from Ferguson. The man behind the vast success of Ferguson is Harry Ferguson himself, and he is 69 years old. How much has he left behind him in skills and knowledge? This would require to be investigated."

Since the Detroit meeting resulted in the conclusion that a fusion of the United States organizations would be desirable if it were combined with some sort of association with the entire Ferguson organization, James Duncan wrote to Harry Ferguson on July 16 and suggested that they arrange to meet, pointing out that August 4 would be a suitable date. He said there was some urgency in their meeting early because of the imminent release at Detroit of new implements which might best be placed in Massey-Harris factories if a closer association were to develop between Ferguson and Massey-Harris.

He also explained his view on distribution. This is significant because it so clearly indicated that Duncan's judgment regarding this matter, and which he was later unfortunately to revise, was correct.

> We examined in some detail the feasibility of operating both units as separate entities, but, at the same time, co-operating and encouraging our dealers to handle both lines. This situation was discarded as being impracticable. It was felt that it might lead to misunderstandings and would not effectively decrease the expense or increase the turnover of either company.
>
> By the afternoon of the first day's meeting, we had arrived at the joint conclusion that a fusion of our operations was the only practical solution, and that such action would undoubtedly widen the distribution of both companies' products.
>
> It would enable us to offer the dealers a most attractive package, namely,
>
> a) Your TO-30 tractor and its mounted equipment.
> b) Our large lugging tractors and pull-behind implements.
> c) Your 3/4 plow tractor and its mounted equipment.
> d) Our self-propelled combines.
>
> This proposal would prove to be so attractive, in our opinion, that not only would we retain the great majority of your dealers and ours, but we would attract many of our competitors'.

Duncan's letter to Ferguson continued:

> You will gather from what I have said that as a result of the 3 days which I spent in Detroit, I have every confidence that we could advantageously co-ordinate our efforts in North America. Providing you are still of the opinion expressed when I had the pleasure of meeting you at Abbotswood, namely, that we look upon the North American proposition as a first step towards a still closer union with your organization as a whole, I would be most happy to come over to England, bringing with me a small group of my associates at your earliest convenience. The next step would be to attend the demonstration of the 3/4 plow tractor and its mounted equipment, including the combine, as you suggested.

Harry Ferguson did invite Duncan, indicating that he was interested in something more than just the purchase by Massey-Harris of the United States Ferguson operation. His letter also reveals him to be the superb salesman that he was:

I hope you will make arrangements to stay with us at least on the night of August 4th. Horace will stay with us also so we will be able to have a pleasant chat. If we have a satisfactory talk then we will proceed at once with your technical investigation and the necessary demonstrations. The only qualm is that we may not be able to get conditions sufficiently difficult to show the merits of our new big tractor, in particular because the harvest will not then be gathered and all the land is occupied. However, we will do our best.

I note you have some commitments on the continent. I am sure, however, that you would not think of leaving any technical investigations to others without being present yourself. It is these technical investigations which will create that necessary confidence between us which could not be created in any other way.

A company can be no greater than its products and I know that we have something of immense value to put before you for the whole future of your company.

Thank you so much for your very nice letter of July 3rd. You are a nice gentleman and I do look forward to seeing you again.

On July 17 Duncan reported to the Board of Directors on his visit to Detroit and he outlined all his enthusiastic conclusions. But he also observed that Ferguson was facing a number of difficulties in the United States including loss of sales to Ford through the latter bringing out a new tractor, and he noted that Harry Ferguson Incorporated had made adequate, although not unusually large profits, when related to capital invested. He went over all the advantages that he saw emanating from a fusion of the two organizations in the United States as well as the various disadvantages. He repeated the warning earlier referred to by Verity about the burdens that such a merger would place on the organization.

We would . . . be faced with a very major reorganization which would affect our factories, engineering, sales and dealers' organization, and this will throw an additional burden on an organization which is already over taxed and will create confusion in our sales forces at a time when they badly require a period of normalcy to get solidly entrenched.

Unfortunately, as things later turned out, great confusion of the kind referred to developed not because of major reorganizations, but because of the failure to introduce them.

Massey-Harris management turned their attention to the various forms of association that might be considered. The one that was preferred above all the others was an exchange of stock between both companies at net book value so that the Ferguson organization would become fused with Massey-Harris, the new organization being called "Massey-Harris-Ferguson Company." But management thought that Ferguson would not agree to a complete merger, in which case Massey-Harris could suggest taking over all the assets of Harry Ferguson in the United States against Massey-Harris parent company stock, together with an arrangement for control at some future date of the rest of the Ferguson organization. In the event of a complete merger, they thought that it might be desirable to offer Harry Ferguson the title of Honorary Chairman, or even Chairman for a given number of years. Ferguson would have to be permitted to control the development of his products, and he would have to be given a substantial salary to compensate him for his future contributions. They saw a number of defects of co-operation without complete integration, including that of the two organizations competing abroad in areas where their product lines conflicted.

Before his meeting with Ferguson on August 4, Duncan was also reminded by W. B. Wedd who had long experience with the Massey-Harris company in France, that the success of the Pony tractor in France was, in part, because the Ferguson tractor was not being manufactured there, and that since Ferguson was planning to do so Massey-Harris might find it difficult to maintain its volume of Pony sales. The probability was that there would be a sharp drop in Pony sales, and a merger with Ferguson might, in that event, appear to be almost necessary as far as France was concerned.

At a meeting of the Executive Committee of the Board just before Duncan was off to see Harry Ferguson the Committee generally supported the line of thinking that had developed. It was agreed that if common ground could be reached, Duncan would cable heads of agreement to Lieutenant-Colonel W. E. Phillips, Chairman of the Executive Committee of the Board, and would send for the company's counsel to investigate legal ramifications, particularly concerning Standard and other suppliers. The Board agreed that Ferguson could be offered a high position in the new organization and that he could remain responsible for the design of Ferguson products. They also agreed that the name "Ferguson" could be incorporated in the name of the new organization although they objected to the name "Massey-Ferguson" or "Ferguson-Massey-Harris."

Duncan arrived at Ferguson's home on August 4, went into a meeting with Harry Ferguson, attended also by R. H. Metcalfe, G. T. M. Bevan, and Horace D'Angelo, and was greeted with a most pleasant surprise. Ferguson had always given the impression that while he might dispose of his United States operation he certainly would not consider a complete merger of his entire organization. Now he announced that he had given the matter of an association with Massey-Harris considerable thought, and that he was prepared to permit Massey-Harris to acquire complete control of all his interests provided that certain factors could be agreed upon. All the alternative arrangements that Massey-Harris management had considered suddenly were irrelevant, for Ferguson was offering the very arrangement that Massey-Harris management had desired from the start. Why this sudden change?

The answer is not clear but as we saw in the preceding chapter, in July

Ferguson had become very depressed about the state of the United States company and its management, and he also had misgivings about management at Coventry. Relations between himself and Alan Botwood, Managing Director of Harry Ferguson Limited, Coventry, were no longer very close and he was convinced that Harry Ferguson Incorporated at Detroit lacked leadership. This, in the backdrop of weak sales, must have made it tempting for Ferguson to relieve himself of the operational worries ahead and to devote much of his time to the development of the new land vehicle on which he had been working.

The conditions that Ferguson felt had to be met to make a merger possible were that the prestige of both companies had to be preserved, which meant that he wished to be given a position of honour on the Board. The Executive Committee had already expressed to Duncan their willingness on this point. Ferguson also said that he would not want to attend meetings other than about one a year, and wanted no responsibility in running the company: he was prepared to relinquish his control if he could avoid such responsibilities and if he was convinced that the company was in capable hands. This also was easily arranged, for Duncan indicated that he would relinquish the chairmanship of the company for a period of five years in favour of Ferguson, with the understanding that the presidency would then be responsible for all policy making and for the active direction of the organization.

Harry Ferguson also wanted two of his senior directors to join the Board of the new organization, and this too posed no particular difficulties. Ferguson then expressed his desire to perpetuate the Ferguson name, and he thought that this might be done by leaving intact the Ferguson organization as a Ferguson division within the over-all organization. Duncan could not accept this suggestion, and Ferguson finally agreed to the proposal that the name of the parent company should become "Massey-Harris-Ferguson." Again this was something that the Executive Committee had agreed to in advance. When discussion turned to the manner of effecting the merger, Ferguson expressed his faith in the Massey-Harris-Ferguson company, and his desire to obtain payment in stock. It was also agreed that the exchange of stock between the two organizations should be on the basis of book value.

Then arose the point that was in succeeding months to cause so much difficulty in the new organization. Ferguson indicated that he wished to remain in charge of the design of Ferguson equipment and to exercise certain control over its application. Here, too, the Executive Committee had given its approval in advance, and so this created no difficulty during the meeting with Ferguson. Other minor points were quickly clarified. Ferguson then declared himself satisfied with the general basis of the arrangement and offered to demonstrate his machines the following day. He requested Duncan to draw up heads of agreement indicating that he thought the deal could be completed in short order.

Duncan said he would telephone Toronto and ask some of the directors and the company's counsel to come over so that they could see the machines in operation and assist in the preparation of necessary documents. By seven in the evening, the unvarying dinner hour in the Ferguson home, all this had been arranged. The only thing that had not firmly been decided was the crucial issue of the basis of the exchange of stock between the two organizations.

Duncan telephoned Phillips that evening and was advised next morning that Phillips, Tory, McDougald, and McCutcheon – all directors of the company – would be in London on August 7. Meantime, on August 6, Harry Ferguson demonstrated his full line of implements, including those on the secret list, to Duncan, Bloom, Pitt and Adams, and this demonstration was repeated for the Massey-Harris directors on Saturday, August 8.

On Sunday, August 9, heads of agreement were gone over with Harry Ferguson and a Ferguson executive. They were soon agreed to, subject only to the exact number of Massey-Harris shares to be involved in the exchange. On Tuesday morning, August 11, W. E. Phillips, Harry Ferguson, John Turner, J. S. D. Tory and R. H. Metcalfe met finally to decide on the exact nature of the exchange of shares. The salient financial facts were these: the Ferguson assets to be acquired were valued net at $16 million as of December 31, 1952, and approximately at $16.7 million as of June 30, 1953. With Massey-Harris shares at just under $9 each, Massey-Harris proposed to give Harry Ferguson 1,805,055 Massey-Harris shares, in exchange for all the shares of the operating Ferguson companies. Ferguson for his part suggested that the book value to be used should be $17 million instead of $16 million, and that therefore he should be given 1,930,000 shares of Massey-Harris stock.

It was at this point, and while Harry Ferguson, W. E. Phillips, James Duncan and W. A. McDougald were together on a drive to a demonstration, that the "haggling" reached its climax. This it did when Harry Ferguson finally suggested that they should toss a coin for the million dollars. Phillips immediately agreed. The car was stopped in front of the Lincoln Arms in the village of Broadway. Harry Ferguson produced a half-crown, and the toss was made. Massey-Harris won. It was a rather curious way to do business. But the fact was that a figure of $17 million for book value could be as easily justified as $16 million. Phillips later indicated to Duncan that he would have been prepared to go up to two million shares to make the deal. And what about the famous half-crown? When Phillips made a motion to return it to Ferguson, McDougald intervened and made an alternative suggestion. He subsequently took it to Asprey and Company Limited on Bond Street, London, had it mounted on a silver cigar humidor, and had the box engraved with kind words for Ferguson as well as with the signatures of those who had participated in the negotiations. It was in this form that the "million dollar" half-crown was returned to Harry Ferguson.

The only thing that now remained was approval for the merger from the Bank of England and the United Kingdom Government Board of Trade. Duncan and Tory went to see the relevant officials, many of whom Duncan knew personally. Approval was granted on Friday evening, August 14, after the officials were satisfied that United Kingdom exports would not suffer as a result of the merger. A press release announcing the merger was prepared on August 16, 1953, and was made public the following morning. That was it. It was a historic occasion for Massey-Harris and for Ferguson.

As an immediate consequence of the merger there now was a company called "Massey-Harris-Ferguson" with the second highest sales volume of any farm machinery company in the world, exceeding even that of Deere and Company. Paradoxically, for the next few years the relative position of the company deteriorated and some of the important advantages of the merger were almost lost. These were to be years in which the new organization faced serious difficulties.

There is one question about the merger that should be clarified at this point: was it a merger? Legally a new organization was not formed. Rather, Massey-Harris Company Limited issued shares to Harry Ferguson and merely changed its corporate name to Massey-Harris-Ferguson Limited. Most of the influential positions in the parent company were filled with Massey-Harris personnel. Within a year even Harry Ferguson had left the company, and had sold his stock to the Argus group, the shareholders who had been close to Massey-Harris for many years. Consequently it is not incorrect to say that Massey-Harris Company Limited acquired Harry Ferguson Limited, and this was also the way the President of Massey-Harris-Ferguson Limited privately referred to the event. Officially it was always referred to as a "merger" and it cannot be denied that that word carries much warmer connotation than does the term "take-over" or "acquisition."

CHAPTER 8

McKay of Australia

One of the many interesting consequences of the merger between Harry Ferguson and Massey-Harris was that it eventually led to Massey-Ferguson acquiring a major manufacturing operation in Australia, including a factory of 1.8 million square feet – second largest among the company's many factories. The company acquired it by purchasing H. V. McKay–Massey Harris Proprietary Limited with its Sunshine Harvester Works just outside Melbourne. It will be remembered that in 1930, when faced with prohibitive trade restrictions, Massey-Harris sold all its assets in Australia to the H. V. McKay company in exchange for a minority interest in the merged company and that the new company became the distributor for Massey-Harris products in Australia. Massey-Harris had enjoyed a strong market for its machines in Australia from the turn of the century.

When Massey-Ferguson acquired the H. V. McKay company it once again inherited an organization with its own long history for the McKay name had been a familiar one in Australia for a great many years. It is a name that will always occupy an honourable place in the history of the development of harvesting machinery.

The development of farm mechanization in Australia involved principally a simplification of breaking and cultivating new lands, and a reduction in the labour of harvesting cereals. Before the turn of the century land was cleared in Australia by cutting or pulling down timber at ground level, or smashing down light scrub with a heavy roller, and using the debris, when it dried, to burn off the stumps above ground. No attempt was made to remove the roots and stumps left beneath the ground. A unique plow was developed which was hinged so that the share would slide over a root, stone, or stump, and plow the roughly cleared ground without sustaining damage. It was called a "stump-jumper." This principle was soon applied to disc plows, as well as mouldboard plows, and to cultivators and drills.

The subsequent crop could not be cut by any form of ground level reaper because the ground was littered with stumps, branches, and the suckers thrown up by the persistent eucalyptus stumps. Consequently a new type of harvesting machine, the stripper, was conceived as early as 1845 by John Ridley at Adelaide. It stripped the heads of the standing crop with a comb, threshed them with revolving beaters, and delivered the threshed grain and chaff into a container. This machine had clearance to pass over the stumps and debris without interference. Periodically the mixture of grain and chaff was emptied into a tarpaulin, and the grain was then winnowed from the chaff with the use of a stationary winnowing machine.

These harvesting machines were later copied by Massey-Harris and sold in large numbers in South America as well as Australia. While representing a considerable advance in harvesting over scything, binding, stacking, and stationary threshing, the winnowing after stripping still required more labour than could easily be found. This led Hugh Victor McKay to experiment with combining the winnower with the stripper, in a single mobile machine, and in 1884 he had ready for testing a "stripper harvester" that did indeed combine the two. The stump-jump implements, strippers, and bulk harvesters, enabled new land to be put under wheat and Australia rapidly became an exporter of cheap wheat and flour.

The 1884 McKay machine was not immediately perfected, and until 1894 McKay devoted much of his time to converting existing strippers to stripper harvesters. After financial failure following the end of a period of land speculation and a wave of bank failures, McKay in 1895 finally began manufacturing his own model of the stripper harvester at Ballarat; he gave it the trade name "Sunshine," a name inspired by the theme of an address by the travelling evangelist Dr. Talmage. By 1905 almost two thousand machines were produced. A sharp contraction of the local market because of drought motivated McKay, in 1902, to start selling his machines in the Argentine – yet another country with large farms and a labour shortage. By the First World War over 10,000 Sunshine harvesters had been sold overseas.

Expansion of sales and the growing demand for other farm implements induced McKay in 1904 to purchase the Braybrook Implement Company Limited of Braybrook Junction, seven and a half miles out of Melbourne – where the plant is still located – and in 1906 the move to the new facilities was begun. McKay virtually built a new community, providing houses, roads, water, with street and house lighting. In August 1907 even the name of the community was changed from Braybrook Junction to Sunshine – one of the few instances where a community has adopted the trade name of a product.

The plant, always referred to as the Sunshine Harvester Works, was soon producing a wide range of implements, including grain and fertilizer drills and mouldboard and disc plows. In 1916-17 McKay introduced three machines of considerable importance: the "Sundercut," a combination stump-jump cultivator, the name of which later was recognized as the generic designation of that type of machine; the "Suntyne," a combined grain and fertilizer drill and cultivator; and later, the "Sunderseeder," a combination of the Sundercut and the seed box of the Suntyne. With no more than a Sunderseeder and a harvester, a farmer could break up, seed and harvest his newly cleared land, and this enabled rapid expansion of wheat growing in the vast semi-arid inland plains.

In 1913 Headlie S. Taylor developed the Taylor header, which, while retaining the comb front similar to that of the harvester, and with the latter's high ground clearance, added a knife to sever the heads from the straw and a meshed spiral conveyor for delivering them to the elevator. A threshing cylinder, straw walkers, and winnower were also incorporated. McKay soon saw the value of this invention which, with the aid of crop-lifters, could pick up and harvest a lodged crop, a feat beyond the capacity of the harvester. Taylor joined McKay in 1915 and McKay began to manufacture the Taylor header in 1916. The commercial introduction of this machine coincided with a season in which most of the crop was lodged, and its success was assured.

A contemporary development was the first Massey-Harris reaper thresher, which combined the stripping principle with a cutting knife and utilized a reel, canvas feeder, iron conveyer and threshing cylinder. This was designed in Australia for Massey-Harris by South Australians Charlton, Chapman, and East, but it was made in Toronto. The reaper thresher enjoyed wide sales in South America, where it was sold in competition with McKay machines. Many were sold in Australia but the reaper thresher was not quite so well suited to local conditions as the Taylor header.

In 1924 Taylor developed the Sunshine auto header, a self-propelled combine. It cut a twelve-foot swath, and was propelled by a single driving wheel (see illustration on page 409). This machine was the first self-propelled combine to be manufactured on a regular basis. Many were sold and they gave encouragement to the future development of the self-propelled headers and combines, although even in Australia they were for many years superseded by power-driven tractor-pulled headers.

In 1916 McKay began to manufacture tractors with imported motors but the venture did not succeed, and so McKay regretfully deferred adopting a machine that in two decades was to become vitally important to implement manufacturers. Five years before H. V. McKay died in 1926, he turned his business into a proprietary company, H. V. McKay Proprietary Limited. Upon his death the McKay family continued to own and operate the Sunshine Harvester Works through that company. They had inherited a company that enjoyed perhaps half of the local market for farm machinery and a valuable trade name – but a company without a line of tractors. So when Massey-Harris management came to Australia in 1930 to discuss amalgamation they too had something to offer to McKay – the Wallis tractor which could help fill a widening gap in the Sunshine product line.

During the negotiations in 1930 Massey-Harris management was given details of the intention of Sunshine to begin manufacturing the auto header in North America. Sunshine had been selling the Australian-made auto header in North America, but decided that local manufacturing would improve its position in the market. A new company, the Sunshine Waterloo Company, was formed and it acquired the assets of a separator company at Waterloo, Ontario. When Massey-Harris management heard of this they expressed regret that this had happened since they themselves had ample manufacturing space available for such a project. But it was too late, and negotiations were confined to the amalgamation in Australia.

After the 1930 merger Massey-Harris and Sunshine products were sold separately. There were Massey-Harris and Sunshine sections at each branch which controlled separate agencies throughout the various states. This arrangement inevitably proved unworkable and integrated distribution was finally effected (similar to the Massey-Harris and Ferguson integration of 1958). The reorganization proved profitable and in only two years of the Depression, and never after, did the company miss paying a dividend.

Expansion of the Sunshine Harvester Works continued until, by the Second World War, it covered just under two million square feet – all on one floor. The Sunshine Waterloo Company in Canada, however, did not succeed in its original purpose. At first it imported all its components for the auto header from Australia and assembled the machine at Waterloo for sale in Canada and the United States. Toolage was then sent from Australia and local manufacturing content was increased. But the Depression wiped out the market for headers before the company could determine whether that machine could successfully be marketed in North America. The company also cancelled its plans to tool up for the Sundercut, which probably was an unfortunate decision, for that machine might well have been a success in North America.

Consideration was then given by McKay to selling the Waterloo plant or locking it up. But the local management felt that the company should instead venture into other lines of manufacturing, which it did. First, parts for motor cars were manufactured, then tricycles and baby carriages. Profits were made and this permitted the company to tool up for more products. Household electrical appliances were manufactured and marketed, as were lockers and shelving. The most important Sunshine product eventually to be manufactured was office equipment, a venture that began in 1949 and soon placed the company among the few large manufacturers of office equipment in Canada. This continues to be its primary activity.

During the war the Sunshine factories in Australia produced armoured vehicles and a wide range of war equipment, but farm machinery continued even then to be its most important lines. It supplied several thousands of binders, drills and disc harrows to the United Kingdom, which were serviced by Massey-Harris in that country.

As soon as the war was over the company reverted to its traditional activity of engineering, manufacturing, importing and selling farm machinery. But to an outsider it seems that Sunshine approached that activity with a degree of caution and conservatism that had not really been part of its tradition. The self-propelled header was dropped during the war and was not replaced, and the company marketed pull-type headers only. Not until 1952 did it import Massey-Harris self-propelled combines and so offer farmers a machine of a type that it had itself pioneered in earlier years.

TABLE 9

Sales of Major Machines in Australia (1948–1955) (Massey-Harris, Sunshine, Ferguson)

	TRACTORS			COMBINE HARVESTERS		
	Massey-Harris (Imported)[1]	Ferguson (Imported)[2]	Total	Massey-Harris S.P. (Imported)[1]	H. V. McKay P.T. (Local Prod.)[1]	Total
1948	674	1482	2156		1639	1639
1949	772	3768	4540		1567	1567
1950	655	5852	6507		1349	1349
1951	1493	8213	9706		1485	1485
1952	1347	7851	9198	299	1579	1878
1953	1304	7257	8561	200	2022	2222
1954	792	6445	7237	101	2291	2392
1955	315	6538	6853	102	2642	2744

1. Sold by H. V. McKay–Massey Harris Proprietary Limited.
2. Sold by Ferguson Distributors.

Improved models of the pull-type machines were developed and new models of the Suntyne drills were marketed – another unique Australian machine combining the function of the cultivator with the seed drill. The company imported all its tractors from Massey-Harris, most of them from the latter's Kilmarnock plant. Sales of Massey-Harris tractors did not expand greatly until 1951 and, as almost everywhere else, sales of Ferguson tractors by the Ferguson distributors greatly out-distanced them. For example, in 1951 Ferguson sold 8,213 tractors in Australia, while only 1,493 Massey-Harris tractors were sold (see Table 9). In part, this unsatisfactory sales performance reflected the technologically non-competitive nature of the Massey-Harris tractors.

The conservatism of the company was most clearly evident in its manufacturing activity. A period of excess demand existed in Australia right after the war, and a company could sell what it could produce. Unlike some of its competitors, Sunshine made little effort to expand or modernize its facilities during this period. The value of its fixed assets, less depreciation, was only 13 per cent higher in 1950 than in 1931, reflecting the relatively small capital expenditures during and after the war; and this was also suggested by the relatively high depreciation reserves in relation to capital expenditures.

The national shortage of labour, local labour unrest, and shortages of steel were regarded by the company as major factors limiting the volume of its production. The company seems to have felt that it was unethical to purchase steel in the "grey market" from companies that were not using their full quota, and it appears to have been reluctant to increase its labour force by paying a few shillings extra for it. Not all companies took the same view, and McKay's share of the market declined when its output remained relatively constant in the early post-war years. For example, from 1948 to 1952 the number of combines or headers manufactured and sold by the H. V. McKay company did not increase at all (see Table 9). The value of the company's sales did not begin to rise significantly until 1951 (see Table 10).

The gains that were made by other companies, particularly by Ferguson and the International Harvester company, were greater than those of McKay. By 1952 McKay's share of the market seems to have declined to about 16 per cent – from over 40 per cent of former years, although these figures can only be approximate. Undoubtedly, the company had lost its position of dominance. But in the years 1952 to 1954 both sales and profits (Table 10) increased markedly, reaching a peak at about the time when Massey-Ferguson began negotiating for the acquisition of the McKay company.

There was no doubt that even before 1952 the company was very profitable in relation to its sales volume, as merely a glance at Table 10 will show. But its future position in the industry and its effectiveness in selling Massey-Harris products did worry Massey-Harris management. They studied the Sunshine business closely in 1951 and even before this, in

TABLE 10

H. V. McKay–Massey Harris Proprietary Limited
Sales and Profits (1945–1955)

	Total sales (Millions £A)	Profits (£A Net after taxes)
1945	3.1	270,340
1946	3.1	316,455
1947	2.4	179,832
1948	2.9	214,321
1949	3.4	231,999
1950	3.8	326,191
1951	5.5	430,955
1952	6.7	394,931
1953	9.1	674,614
1954	9.1	994,850
1955	9.1	845,496

1948, J. S. Duncan, R. H. Metcalfe, and E. G. Burgess had inspected the Sunshine works and left with doubts about its efficiency. The resulting report spoke of the dire need for new and up-to-date machine tools, for a review of methods, for young men to replace a labour force with an unusually high average age, and particularly for a more progressive approach to problems. (Somewhat similar comments could have been made about some aspects of Massey-Harris operations in North America at the same time.)

Around 1950, Massey-Harris management began to peruse the agreement of 1930, which was due to expire on August 30, 1955. They wondered whether what they read in the agreement was really true, and referred it to legal counsel. Their suspicions were confirmed. In October 1951 Australian solicitors advised them that, among other things, they had permanently denied themselves the right to use the name "Massey-Harris" in Australia.

Various plans were conceived by Massey-Harris in 1951 to anticipate the difficulties that would arise in 1955. But nothing came of them before the merger of Massey-Harris and Ferguson had been effected and had made the former's legal position in Australia infinitely more difficult. It was also, commercially, a much more valuable position because of the market strength enjoyed by Ferguson tractors there. From 1948 to 1953 McKay sold 6,245 Massey-Harris tractors while Ferguson distributors sold 34,423 of the TE-20 tractors.

With the merger in 1953 Massey-Harris acquired Harry Ferguson of Australia Limited, a company formed in 1952 to supervise subcontracting of Ferguson implements and to oversee the operations of the five independent Ferguson distributors. The agreements between the Ferguson company and its local distributors could be terminated within a year but this Massey-Ferguson executives did not want to do. They were convinced that the Australian Ferguson distributors and their dealers represented a much stronger marketing system – particularly for the technically novel Ferguson tractor – than did the Sunshine branches and their agents (not dealers). For this reason Massey-Ferguson not only honoured the Ferguson agreements but took no action in issuing termination notices.

Cecil N. McKay, Chairman and Managing Director of Sunshine, objected almost immediately, and for this he had good legal grounds. He argued that under the 1930 agreement he had acquired not only the right to distribute all the Ferguson machines in Australia, but also to manufacture the Ferguson machines that were just beginning to be manufactured by other subcontractors in Australia.

Massey-Ferguson once more consulted legal counsel, this time with particular reference to points raised by the 1953 merger. The difficulty of its position was confirmed. There was no doubt that H. V. McKay-Massey Harris Proprietary Limited was sole agent not only for the sale of Massey-Harris products in Australia and Papua/New Guinea until August 30, 1955, but also for the products of all the subsidiaries of Massey-Harris.

This would include Harry Ferguson Limited. It was also indisputable that when Massey-Harris acquired the Ferguson interests it assumed Ferguson's commitments to the Australian Ferguson distributors. These included exclusive distribution rights for Ferguson products in certain areas of Australia subject to cancellation upon receiving twelve months' notice. Certainly in a technical sense Massey-Harris could be viewed as having breached its contract by continuing to distribute Ferguson machines in Australia through Ferguson distributors after 1953. At the same time if it had begun after that date to ship Ferguson products to H. V. McKay it would have been breaking its contracts with the Ferguson distributors. Massey-Harris was faced with a dilemma which could not easily be resolved.

But there was another complication. In 1930 Massey-Harris, in addition to selling its name and goodwill in Australia, had sold to Sunshine exclusive right to manufacture Massey-Harris products in Australia for twenty-five years and, apparently, a non-exclusive right to do so thereafter. Massey-Harris had even committed itself to furnishing Sunshine with drawings, jigs, fixtures, and other items necessary to begin production. But the Standard Motor Company had the legal right to manufacture Ferguson tractors in Australia, a right that could now also be claimed by McKay. The same applied to Ferguson implements made by subcontractors in Australia. Again the legal dilemma faced by Massey-Ferguson was complete and again no obvious resolution of it seemed to exist.

Counsel once more pointed out that after August 30, 1955, McKay's exclusive distribution rights would no longer exist, nor would its exclusive rights to manufacture, but it would still have the right to manufacture those implements, and it did have exclusive right to use the name "Massey-Harris" in Australia forever. Massey-Ferguson could undoubtedly operate under the name "Harry Ferguson" but it would probably not be permitted to use the name "Massey-Harris-Ferguson." One final complication, even after August 30, 1955, Massey-Ferguson would not be permitted to dispose of its shares in H. V. McKay–Massey Harris Proprietary Limited without first offering them to that company.

Meanwhile the number of possible solutions to these several dilemmas was restricted because Massey-Ferguson did not wish to transfer distribution of the Ferguson machines to Sunshine. It feared that this would reduce the sale of Ferguson machines and also that if Sunshine were given the Ferguson distributorship it would be only a matter of time before that company would manufacture a complete line of implements themselves. This would reduce the Massey-Ferguson return on sales in Australia to the dividends it received on stock in the H. V. McKay–Massey Harris Company.

J. S. Duncan invited Cecil McKay to come personally to Toronto to negotiate all the various points in dispute. Prior to his arrival Massey-Ferguson management believed that the best approach would be to attempt to convince him that a renewal of his agency agreement for a definite number of years should be concluded, but that it should exclude Ferguson products. A new contract was waiting for him when he arrived.

McKay came to Toronto in March 1954 and Massey-Ferguson management had their suggestion ready: that the McKay agency should be extended for ten years, excluding the Ferguson lines, and that the two-line concept should be perpetuated in Australia with Ferguson distributors continuing to handle the Ferguson products. Massey-Ferguson argued that competition against the Ferguson tractor would probably increase in future, that McKay was unreasonable in believing that the 1930 agreement included the Ferguson business, that while McKay did have the right after August 1955 to manufacture any lines then in production the agreement did not require Massey-Ferguson to supply "know-how." Finally, it was contended that any breach of contract on the part of Massey-Ferguson had resulted because there had been no alternative, and that Massey-Ferguson could not put any pressure on the Australian distributors to persuade them to join the McKay Company. Duncan then suggested to McKay that another way out of the dilemma would be for them to acquire control of H. V. McKay–Massey Harris Proprietary Limited.

McKay did not agree with Massey-Ferguson that after June 30, 1955, there would be any limitation at all on the Massey-Ferguson goods they were entitled to manufacture, and insisted that they were entitled to have the Ferguson products irrespective of any obligations that Ferguson might have had in Australia at the time of the merger. McKay insisted that the proper approach was to bring the Ferguson lines within the operations of the McKay company. The agreement that Massey-Ferguson had so carefully prepared remained unsigned. At the time Massey-Ferguson management did not realize that this was just as well. If McKay had accepted it, it would have made subsequent improvement of the company's position in the Australian market most difficult before expiry of the new agency agreement, particularly when the policy of single-line products and single-line distribution came to be adopted world-wide.

The arguments of both sides were clearly articulated in a letter written by McKay to Duncan on May 31, 1954. In response to the Massey-Ferguson suggestion of acquiring sole ownership of the McKay company it read:

> We had some discussion on your suggestion that you might acquire sole ownership of our Company, but in the absence of any concrete proposal, we did not get very far. As I told you in Toronto, none of us had been thinking on these lines and at first sight, my Board is not inclined to be interested. Perhaps the general view was that this suggestion should be put aside for the time being, but if all other solutions failed, then you might feel inclined to put up some definite proposal which we could consider.

Actually when McKay arrived in Toronto and immediate settlement seemed impossible, Massey-Ferguson management were rather anxious to delay further negotiations because their time was occupied with another and even more difficult negotiation – that of purchasing the Massey-Ferguson shares held by Harry Ferguson.

Therefore, in part to delay negotiations, it was arranged that Massey-Ferguson management would go to Australia in October. During the summer the company again sought legal advice and again received substantially the same opinions. It could not use the name "Massey-Harris" after August 1955 in Australia; it would breach the provisions relating to the distribution of Ferguson products in Australia; and McKay would enjoy at least some manufacturing rights after the expiration of the agency agreement. Again it was clear that no problems would definitely be solved by merely permitting the agency agreement to expire, and indeed it might well invite a suit for damages since McKay would have been denied the right to distribute Ferguson products from the time of the merger to August 1955. The latter possibility had been mentioned by McKay when he was in Toronto.

Massey-Ferguson management did go to Australia and negotiations with McKay extended from October 16 to December 14. For Massey-Ferguson there were: J. S. Duncan, J. A. McDougald, M. F. Verity, and C. N. Appleton; for McKay: C. N. McKay, H. V. McKay, and J. A. Forrest (trustees of the estate of the late H. V. McKay), G. M. Taylor, and A. D. J. Forster. Massey-Ferguson was also at various times represented by legal counsel.

In the initial talks, Duncan told McKay that they intended to distribute Ferguson goods through Ferguson distributors, and emphasized that the new line of Massey-Ferguson goods that would soon appear would offer great advantages to the H. V. McKay company. McKay was not moved by any argument of advantages to be derived, and he persisted with the view that the Ferguson line should be added to his Massey-Harris line. His view was that Standard could be induced to link up with Sunshine for the distribution of Ferguson machinery.

Massey-Ferguson management said they would explore the possibility of an amalgamation with the major Ferguson distributor, Standard Motor Products Limited and that company's subsidiary, British Farm Equipment Proprietary Limited. However, Standard Motor Products Limited soon advised them that they could not separate their motor car business from the tractor business and that therefore they were unable to consider any offer of amalgamation or sale of shares. When it was realized that the combined net worth of Ferguson distributors would be in the neighbourhood of A£6,000,000 there was no further discussion of amalgamation between the Ferguson distributors and the McKay company.

With Massey-Ferguson management insisting that the Ferguson line would not be added to the Sunshine Massey-Harris line, and since McKay could not acquire the assets of the Ferguson distributors and apparently did not wish to re-purchase shares held by Massey-Ferguson at the price he was offered for his shares, the conclusion seemed obvious: Massey-Ferguson would have to acquire ownership of the McKay company. Both sides appear to have accepted this logic even though "bargaining" continued for many weeks.

It was agreed that both companies would appoint independent auditors to give an opinion of the value of the McKay company shares. The Massey-Ferguson auditors valued them in a number of different ways and found that on the basis of audited accounts, as of June 30, the shares of the McKay company could be valued at A£3.0.4. and after adding undisclosed reserves this would rise to A£4.0.10. If average earnings were capitalized on a 12 per cent basis they would amount to A£3.6.10 and if the estimate was based on an average dividend and a yield of 7 per cent per year, the amount came out at A£3.14.8. Consequently the auditors decided that a fair value of the one pound shares would be A£3.10.0. The McKay auditors on one basis placed a valuation of A£5.2.6 on the shares of the McKay company, but on another basis it was discovered that the auditors had also arrived at the figure of A£4.3.4. Massey-Ferguson offered A£3.15.0 for the shares and, alternatively, also offered to sell their own holdings in the company to McKay for that amount. McKay did not agree either to buying the shares held by Massey-Ferguson or to selling his shares at A£3.15.0 and at that point it appeared that the negotiations were deadlocked. Indeed several days elapsed before further discussion took place.

When they met again price continued to separate the two sides. Massey-Ferguson made a final offer of A£3.18.6. This offer was not immediately accepted. J. A. McDougald felt that the time had come when a determined effort should be made to reach a settlement. He did so by outlining somewhat forcefully what to him appeared to be the major difficulties that the McKay firm would face in the future.

These difficulties were real enough. From August 1955 onward, the McKay company would lose its important source of tractors, its source of self-propelled combines, and its source of new technology as far as Massey-Harris machines were concerned. Competition from International Harvester was likely to be intense. Taxation complications over past inventory valuation practices might arise. In addition, McKay had no one to succeed him in maintaining the family tradition in the approaching period of stiff competition in price and technology. Many family companies in many countries have faced broadly similar circumstances in the swift race towards increasingly large and complex industrial organizations.

After all points were raised and weighed, agreement was finally reached. The settlement seems to have been a generous one. Massey-Ferguson paid A£3.18.6 for each of the 1,785,128 shares, or a total of A£7,006,627, and it did so in a year when the post-war boom was just beginning to subside and competition was becoming increasingly intense.

To help finance the acquisition, the local operating company, H. V. McKay–Massey Harris Proprietary Limited sold A£2,500,000 debentures in Australia, and a local holding company formed for acquiring the McKay ordinary shares – H. V. McKay–Massey Harris Holdings Limited – sold A£1,500,000 redeemable preference shares.

After the new owners took possession they discovered that the inventory valuation had been too high by about three-quarters of a million pounds, but that against this, other assets acquired proved to be more valuable than expected, so that the two offset each other. What really mattered was not the exact price paid but rather the success of the company in developing the new business acquired. In this respect the immediate experience was one of disappointment because both sales and profits in 1955, which at the time of the negotiations appeared to be strong, turned out to be somewhat weaker than had been expected. What the local organization required was not the injection of outside capital but the injection of new technology and new managerial and control concepts.

Effective February 1955, Massey-Ferguson found itself represented in Australia by its new subsidiary, and L. T. Ritchie was sent out to take charge of it. It was, however, also represented by Harry Ferguson of Australia Limited. Whereas the former company was responsible for selling the Sunshine and Massey-Harris implements, the latter continued to oversee the sales operations of the five independent Ferguson distributors. A dual-line policy was thereby perpetuated.

It was seen at the time of the acquisition of the McKay company that the local Ferguson operation would within a year be made responsible to the local H. V. McKay–Massey Harris Company, but in the meantime the Ferguson Company would answer to the Managing Director of the Eastern Hemisphere Division in London, and the other to a Group Vice President of the parent company in Toronto. The arrangement was changed as planned on July 1, 1956. But this still left Massey-Ferguson in Australia with a Ferguson organization that operated through independent distributors, and the McKay organization with its own retail agents.

When the President of Massey-Ferguson reported to the Executive Committee of the Board on the state of the newly acquired company, he referred to its seriously run-down condition – it was a very old plant. Some important management gaps had also to be filled. But because of the goodwill of the half-century old McKay business and name, the acquisition really represented a challenge to the new owners – a challenge to reorganize and modernize the Australian company so that it could survive and prosper in the period of increased competition and technological change that was emerging.

Unfortunately, Massey-Ferguson itself had not at that time clarified its own organizational and management concepts and so improvement in its operations in Australia took much longer than expected. It eventually involved combining the Ferguson and McKay distribution systems by eliminating the independent Ferguson distributors and placing the McKay agents on a dealership basis (who no longer received goods on consignment); it involved introducing a new organizational structure and re-organizing each of the functional divisions; it involved changes in management; and it involved expenditures for modernization. Not until the early 1960s had these changes substantially been adopted.

CHAPTER 9

Aftermath of the merger

1953/1956

The merger of the Massey-Harris and Ferguson organizations had been discussed in an atmosphere of friendship and goodwill and had finally legally been effected in a spirit of good humour. On August 12, 1953, the day of the official signing, Massey-Harris executives gave Harry Ferguson the silver cigar box on which was mounted the famous "million dollar" coin commemorating the occasion that made it famous. But seldom has a honeymoon been over more quickly. Soon it became only too obvious that making the merger work in practice would be a most difficult matter, and that working with Harry Ferguson would be impossible. It is no exaggeration to say that the merger did not work at all successfully during the three years after it was effected – a period that finally ended with heavy losses for the company in its North American operations and a change in management.

In retrospect it seems that this aftermath was difficult because of problems created by the merger itself, because of the internal problems that each organization had been suffering from even before the merger, and because of depressed economic conditions in North American agriculture. All these groups of problems suddenly came together after the merger, and it is difficult to avoid the conclusion that management failed to cope satisfactorily with them.

The difficulties created by the merger itself were monumental. Any merger is more painful than is suggested by that innocent and even pleasant-sounding term. Old loyalties, well-developed lines of personal communication, familiarity with routine and with company policy, and the sense of security that all these have helped to create are usually shattered. Some individuals can never attach themselves to new policies and new loyalties, and they remain as an irritant to the smooth operation of the organization. Others, suddenly faced with an enlargement of their area of responsibility, find themselves out of their depth; while still others, whose responsibilities have declined – through elimination of management duplication – may interpret this as a gross injustice, or as a mark of personal failure. The potential disruption to management and labour, the sudden confusion in distribution and manufacturing arrangements, the uncertainty felt by everyone over future policy, all make it essential that a merger be preceded, or if that is impossible, immediately succeeded by a detailed organizational, operational and personnel plan for the new organization.

It is a curious, even incredible, fact that management of Massey-Ferguson did not seem to realize that a fundamentally new organization had to be created; they apparently did not think it necessary to develop a detailed plan for the structure and the operation of the new organization. There is no evidence that any members of senior management felt that such a plan was necessary; there were only sporadic protests from time to time over specific actions taken. Policy emerged piece by piece as emergencies were encountered one by one.

From the point of view of personnel relations the merger started badly and remained so. Massey-Harris executives filled most of the senior positions in the parent company. For example, of the sixteen positions shown on Organization Chart 2, dated March 1954, only two were filled by former Ferguson executives. This was largely because the Ferguson organization had few senior executives available for such positions. But difficulties arose when Massey-Harris executives showed insufficient sensitivity in dealing with Ferguson executives, in spite of the President's instructions to the contrary. An invisible line began to appear between the "red" (Massey-Harris) and "grey" (Ferguson) personnel. In the United Kingdom, visiting Massey-Harris executives were referred to as the "Mau-Mau." Ferguson personnel for their part, particularly those in the United Kingdom, complicated matters by their suspicions about all things North American. While these conditions did not exist everywhere in the company they were sufficiently important to contribute to the company's difficulties. These problems did not disappear even though, right after the merger, J. S. Duncan had emphasized the need for utilizing Ferguson talent and ideas.

A number of the difficulties that the company experienced arose from the failure of management to cope satisfactorily with the product and organizational problems created by the merger. There was of course the inevitable and immediate matter of duplication of distribution facilities. But it was perhaps more complex in this merger than in most others. Ferguson had always operated through independent distributors, which meant that in the United States, Canada, United Kingdom, France, Germany, South Africa, Denmark, Belgium, New Zealand, Argentina, Brazil, and Uruguay, Massey-Harris branches faced independent Ferguson distributors and local Massey-Harris dealers faced the dealers of the

ORGANIZATION CHART 2

Massey-Harris-Ferguson Limited
March 1954

FIRST VICE PRESIDENT
H. H. Bloom

CO-ORDINATING AND POLICY DIVISIONS

VICE PRESIDENT AND SECRETARY
C. N. Appleton

Corporate affairs (relations with shareholders and Board)
Legal matters
Banking arrangements
Donations
Insurance (broad supervision)
Pensions (general policy)
Aircraft operations
Executive salary administration

VICE PRESIDENT MANUFACTURING
E. G. Burgess

Factory operations
Cost research
Defence production

VICE PRESIDENT ENGINEERING
H. G. Klemm

Engineering – North Ameri
Liaison with Eastern Hemisphere Division

DIRECTOR OF PERSONNEL AND INDUSTRIAL RELATIONS
C. B. C. Scott

Personnel relations
Union negotiations
Group insurance
Wage and salary administration (other than executives)

DIRECTOR OF RESEARCH
A. Pitt

Engineering research
Process research

OPERATING DIVISIONS

UNITED STATES DIVISION
H. H. Bloom

First Vice President
President, The Massey-Harris Co.

CANADIAN DIVISION
L. T. Ritchie

Vice President
General Manager Canadian Division

HEAD OFFICE CO-ORDINATING COMMITTEE: *Chairman*, R. H. Metcalfe; *Members*, Abaroa, Appleton, Burgess, Klemm, Lattman, Verity; *Ex-officio members:* Duncan, Bloom, Ritchie, Young.

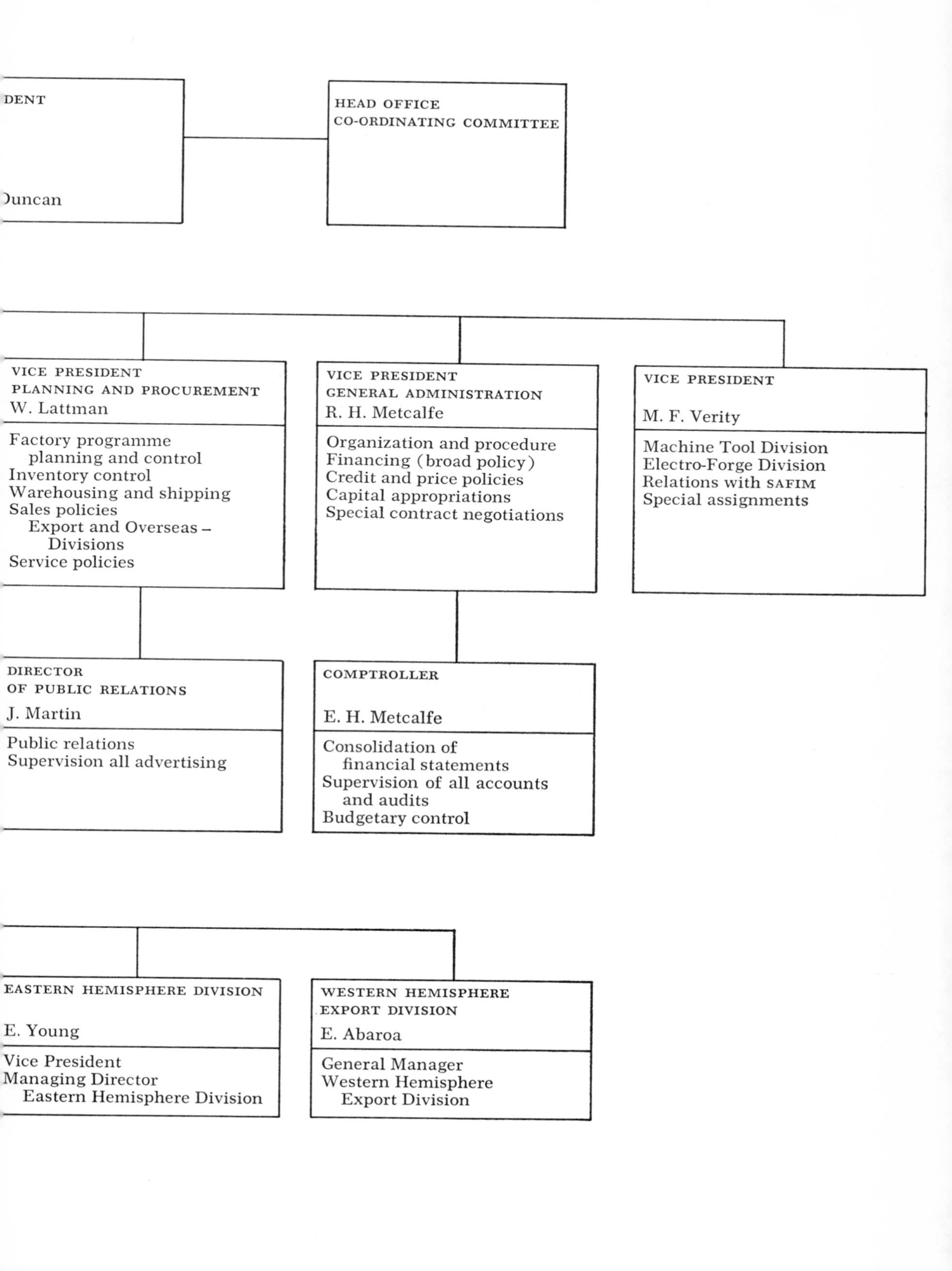

DENT
Duncan
HEAD OFFICE
CO-ORDINATING COMMITTEE
VICE PRESIDENT
PLANNING AND PROCUREMENT
W. Lattman
Factory programme
planning and control
Inventory control
Warehousing and shipping
Sales policies
Export and Overseas –
Divisions
Service policies
VICE PRESIDENT
GENERAL ADMINISTRATION
R. H. Metcalfe
Organization and procedure
Financing (broad policy)
Credit and price policies
Capital appropriations
Special contract negotiations
VICE PRESIDENT
M. F. Verity
Machine Tool Division
Electro-Forge Division
Relations with SAFIM
Special assignments
DIRECTOR
OF PUBLIC RELATIONS
J. Martin
Public relations
Supervision all advertising
COMPTROLLER
E. H. Metcalfe
Consolidation of
financial statements
Supervision of all accounts
and audits
Budgetary control
EASTERN HEMISPHERE DIVISION
E. Young
Vice President
Managing Director
Eastern Hemisphere Division
WESTERN HEMISPHERE
EXPORT DIVISION
E. Abaroa
General Manager
Western Hemisphere
Export Division

independent Ferguson distributors. In a number of export territories, independent Massey-Harris distributors and independent Ferguson distributors found themselves operating in competition with each other as before, but now they were also expected to further the interests of the same company – not an easy task.

Since the merger gave the organization two product lines, there was immediate competition between Ferguson tractors and Massey-Harris tractors, and between a number of implements manufactured by both firms. And while no conflict existed in combines this could develop if the Ferguson mounted and semi-mounted combines were to be put into production. The new organization also inherited the quality difficulties that had begun to appear in Massey-Harris implements, as well as the difficulties arising from the delay in introducing the large Ferguson tractor and the small TO-35 Ferguson tractor to replace the TO-30.

Difficulties in manufacturing created by the merger were not really ones of duplication of facilities, for Ferguson owned only the Detroit assembly plant. Standard Motors produced Ferguson tractors at Coventry; in co-operation with Hotchkiss that company had also just begun to produce them in France in 1953, and it was planning to assemble a small number of them from completely knocked down packs (C.K.D.) in India. But the manufacturing problems that the merger did present to the new organization were those associated with the increased degree of subcontracting and reduced control over supply that it involved.

Harry Ferguson had subcontracted the manufacture of all his implements and all his tractors except those produced at Detroit, and even these were largely assembled from components supplied by other companies. Before the merger the Ferguson company frequently expressed dissatisfaction with manufacturing costs at Standard Motors, and its United States costs were higher than those of its major competitors; while Massey-Harris, too, had its cost problems particularly in the United States. The new organization therefore depended much more heavily on subcontracting than the old Massey-Harris company had done, and it inherited cost problems from the manufacturing and assembly operations of both organizations. In consequence, cost of goods sold suddenly rose to much higher levels. For example, whereas the ratios of cost of goods sold to net sales of Massey-Harris in North America in 1950, 1951, 1952, and 1953 were 68, 72, 73, and 75 per cent, for the Massey-Ferguson organization in 1954 and 1955 they were 81 and 82 per cent.

Problems of distribution and costs were augmented by disturbing market conditions. Just before the merger, on July 29, 1953, the President outlined for his executive group, his impression of economic conditions, and he emphasized the need to reduce costs:

On July 10th, I submitted to our Executive Committee a tentative 1954 forecast for North America; figures for other units were not yet completed. This forecast was based upon the assumption that we would finish up the present year with inventories of finished machines comparable to 1952, both in our own stock and dealers' hands, and that wherever costs were increasing, it might not be possible to increase selling prices.

Since that date, the position has worsened somewhat. Crops have deteriorated, and there is evidence on every hand that our competitors are cutting prices, either directly or by the subterfuge of trade-in allowance or stock liquidation concessions. . . . Both the gross margin and the manufacturing programme contained in our July 10th estimate appears, therefore, to be too high.

The results forecast, even with the very important contribution of war work, were discouraging. Without war work, we estimated that we would fall short of earning our dividend. Fundamentally, these poor results are the consequence of lower volume, higher costs and considerably higher expense ratio to sales.

We should take a fresh look at our whole set-up and ruthlessly cut back any expenses which are not absolutely essential, even down to minor details, such as postage, telephone calls, cables, etc. Travelling expenses will have to be cut, the number of people authorized to travel carefully scrutinized. Older employees of pensionable age retired and replaced by younger men, institutional work decreased. Staff will require to be substantially reduced. We should examine the pension position, the production tooling, the Research Department and donation budget.

Emphasis on the need to cut expenses, which had been a recurring theme within the company even before the merger and which persisted for several years after, was not accompanied by any detailed, formal, investigation of why the costs of goods sold by Massey-Harris and Ferguson were higher than those of competing companies. Consequently the emphasis was simply on cost reduction in general rather than on devising plans for structural changes in the company that would reduce costs through increased efficiency in operation and organization.

The new organization began its life with other difficulties. We have already seen that the two companies went into the merger at a time when they both were beginning to experience disturbing sales and profit results in their North American operations. Since both companies were in difficulties there, the merger could not be relied upon automatically to solve the problem of the new organization in North America. Nor did it do so. One immediate problem was the depressed state of industry sales. An estimate of the North American sales of six North American implement companies (International, Deere, Case, Oliver, Minneapolis-Moline, and Massey-Ferguson) shows that their 1951 sales level was not exceeded until 1957. A glance at Chart 3 below shows that from 1953 to 1957 North American sales of Massey-Ferguson stagnated.

The effect of all these difficulties on the company's profits in its North American operations was disastrous (see Chart 6, p. 169). But the truly worrying aspect of these developments was that they were distinctly more severe in Massey-Ferguson than in its major competitors. Massey-Ferguson profits on its North American operations (including defence contracts) amounted to $6.6 million in 1953 and $1.4 million in 1955, a decline of almost 80 per cent; over the same period the profits of Deere and Company – which at that time were derived essentially from operations in North America – actually increased on balance. In the year 1955, taken by it-

CHART 3
Massey-Ferguson sales 1953-1966

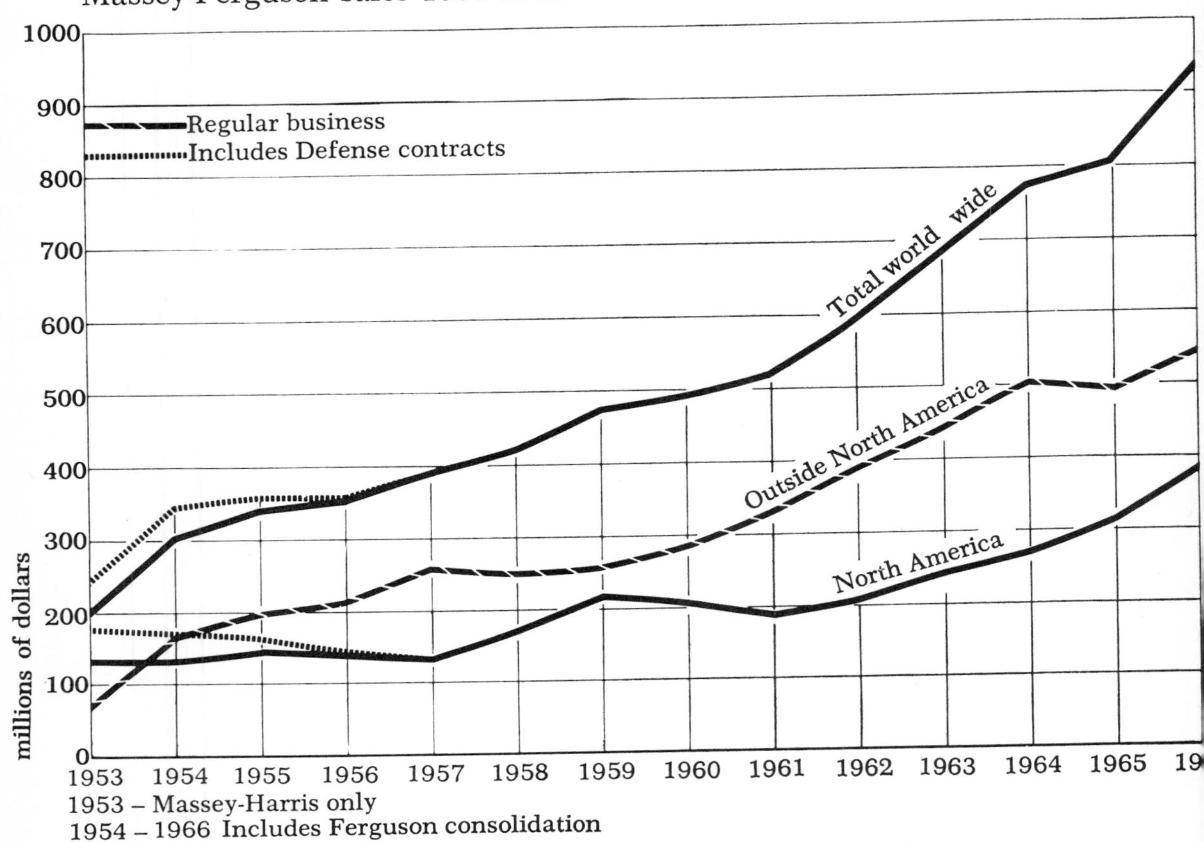

self, sales of regular goods by Deere increased by 18 per cent (which were essentially sales in North America) and total profits after taxes increased by 37 per cent; Massey-Ferguson sales of regular goods in North America increased by 10 per cent but total profits in North American operations decreased by 45 per cent.

The deterioration of the company's North American business and particularly its United States business was offset to a substantial degree by profits in its operations abroad. In 1954 profits of the new organization totalled $8.9 million, but this hid a sharp decline in the profits of the United States operation, which fell to $.7 million from $2.4 million; and if defence contracts are excluded, a loss of $1.5 million was experienced in that operation. In 1955 the loss in the United States operation (excluding defence contracts) was $.9 million, but strength elsewhere gave the company an over-all profit of $12.2 million. As 1956 began, United States losses accelerated which, together with large inventory write-offs initiated by new management in the later part of the year, produced the frightening loss of $6.5 million; and this was now joined by a loss of $1.1 million in Canada. World-wide profits consequently declined to $3.2 million. Finally in 1957 losses in the United States reached a staggering $14.4

million, losses in Canada were $1.1 million, and world-wide losses were $4.7 million. Obviously the deterioration in the company's United States operations had not come quickly, but rather for some time had been hidden by profits arising from business outside North America. At the same time these profits from outside North America were the principal reason why the company could not be regarded as standing on the brink of bankruptcy in 1956, as some articles suggested.

It needs to be emphasized that the company's difficulties did not arise because of any fundamental unsoundness about the merger. Subsequent events showed that it was sound, indeed, that without it the company in all likelihood would never have enjoyed the position of prominence in the industry that it came to enjoy. Massey-Harris executives had expected that their company would acquire urgently needed strength in tractors from Ferguson, and this it did. Charts 4 and 5 show the amount in dollars and percentages of Massey-Harris sales of tractors, combines, and all other items for the year 1953, and the same for the merged company after 1953. It is easily seen that the merger immediately created a balance between tractor sales and total sales that changed little thereafter – tractor sales accounting for from 45 per cent to 49 per cent of total sales in the post-

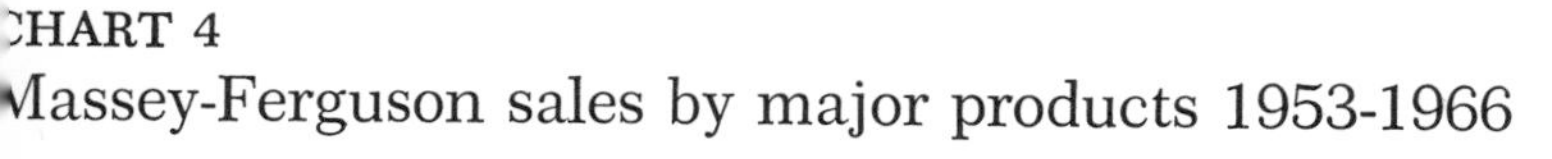
CHART 4
Massey-Ferguson sales by major products 1953-1966

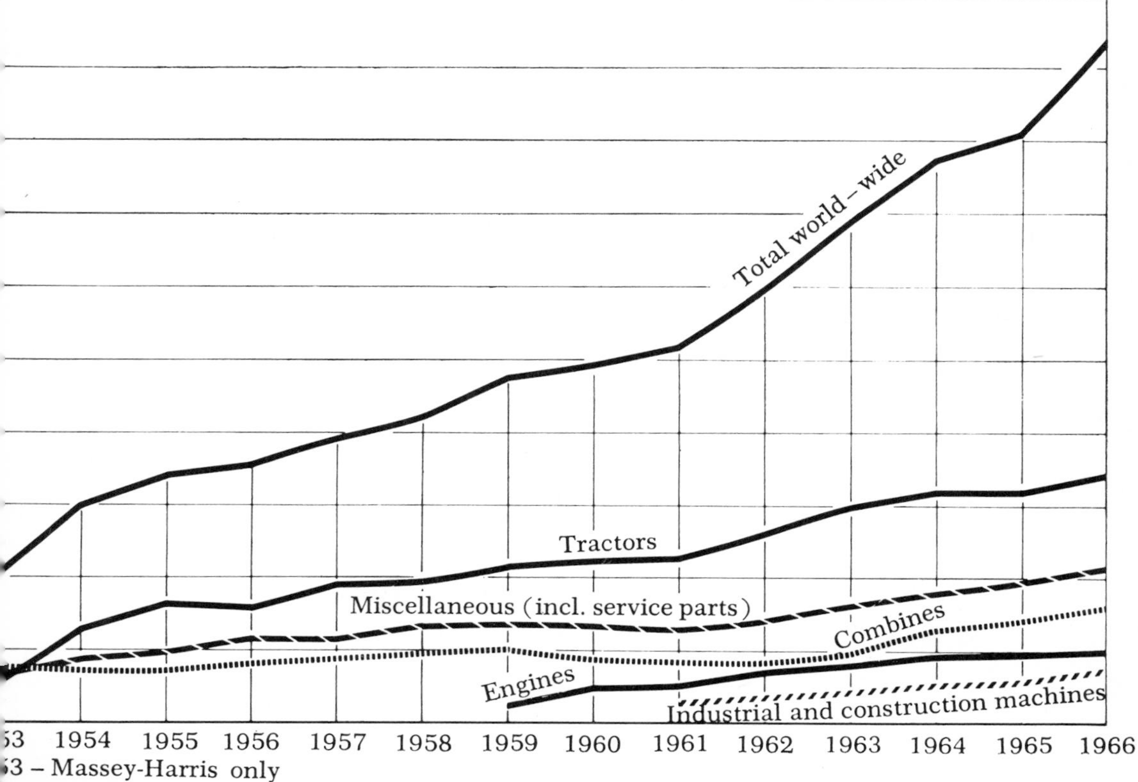

53 – Massey-Harris only
ustrial and construction machinery included in "miscellaneous" prior to 1961

CHART 5
Massey-Ferguson sales 1953-1966
Percentage of distribution by major products

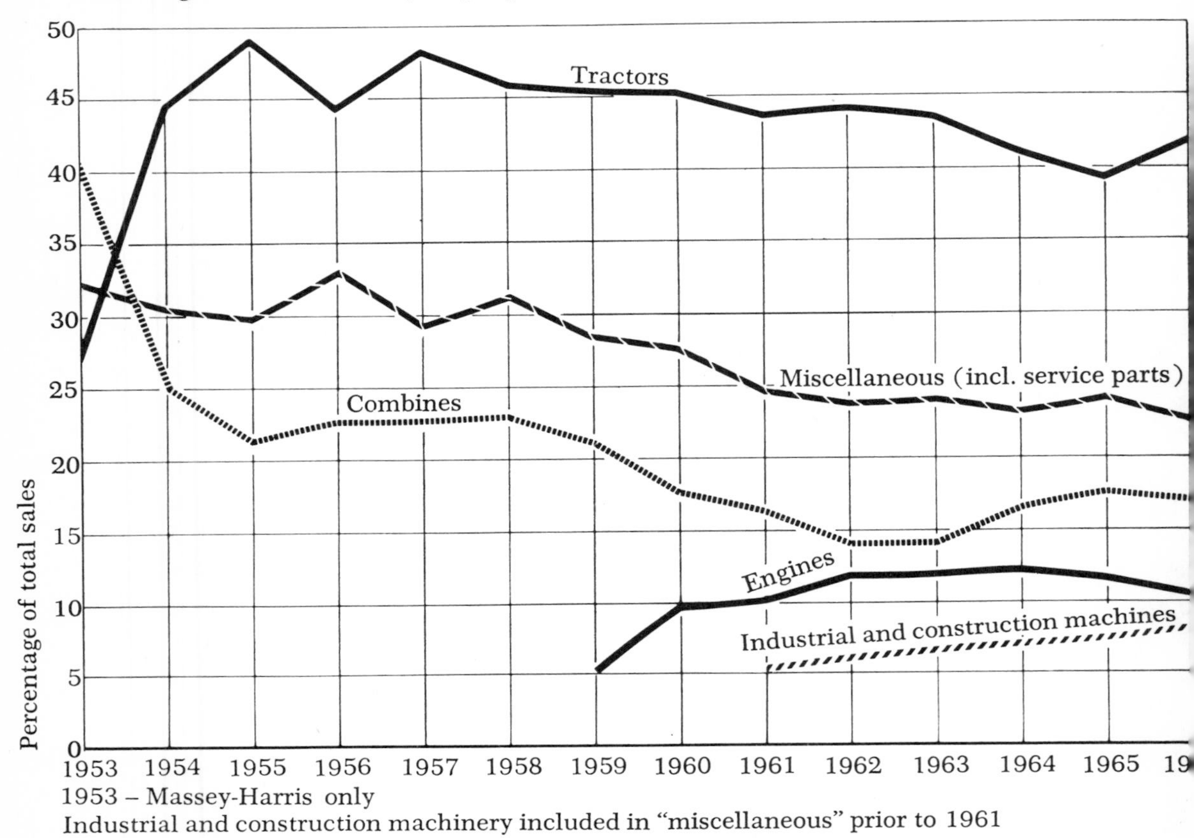

1953 – Massey-Harris only
Industrial and construction machinery included in "miscellaneous" prior to 1961

merger years. The relative importance of sales of combines declined – although the decline after 1958 was not a consequence of the merger. Without doubt the merger created a more balanced product mix, considering the sales structure of full-line implement companies.

It was thought at the time of the merger that the Engineering Department of the Massey-Harris organization would be strengthened by the contribution of the Ferguson engineers. This proved to be only a partially accurate appraisal. It was entirely accurate as far as tractor engineering was concerned; it was partially accurate in the area of implement engineering – although Massey-Harris executives at the time of the merger almost certainly over-emphasized the importance of the new implements that Ferguson had been developing in Detroit; and in combines it can be argued that the merger reduced design activity for a period to the detriment of the company.

The merger had other effects. It permanently increased the over-all size of the organization and it permanently increased the relative importance of the sales of the company outside North America. Chart 3 is based on Massey-Harris sales for 1953 and Massey-Ferguson sales for the years thereafter. Obviously, Massey-Harris sales outside North America in 1953 were substantially smaller than sales in North America, a pattern that was permanently reversed by the merger. Massey-Harris sales outside North America were 28 per cent of total sales in 1953, while the figure for Massey-Ferguson in 1954 was 49 per cent. This trend continued and reached its peak in 1957 with 66 per cent. A period of instability followed but it was still at 59 per cent in 1966.

This over-all view of the merger in its first few years provides a necessary background but hides most of the details of what actually happened. To these we now turn, and we begin with operations in North America, leaving until later a discussion of the developments abroad that provided such timely support to the company. For it was in its North American operations that the company's problems were magnified to larger-than-life proportion and where certain misguided efforts to solve them were first introduced.

No policy decision did more harm to the new organization in its North American operations than the one of perpetuating and extending the two lines of machinery and of perpetuating the two separate distribution systems inherited from the merger. How this happened, and what the consequences of it were, is an important part of the company's experience from 1953 to 1956.

It will be remembered that the President of the United States Massey-Harris company, H. H. Bloom, and the Vice-President of the United States Harry Ferguson company, Horace D'Angelo, had participated in the dis-

CHART 6
Massey-Ferguson profits 1953-1966
(Net after taxes)

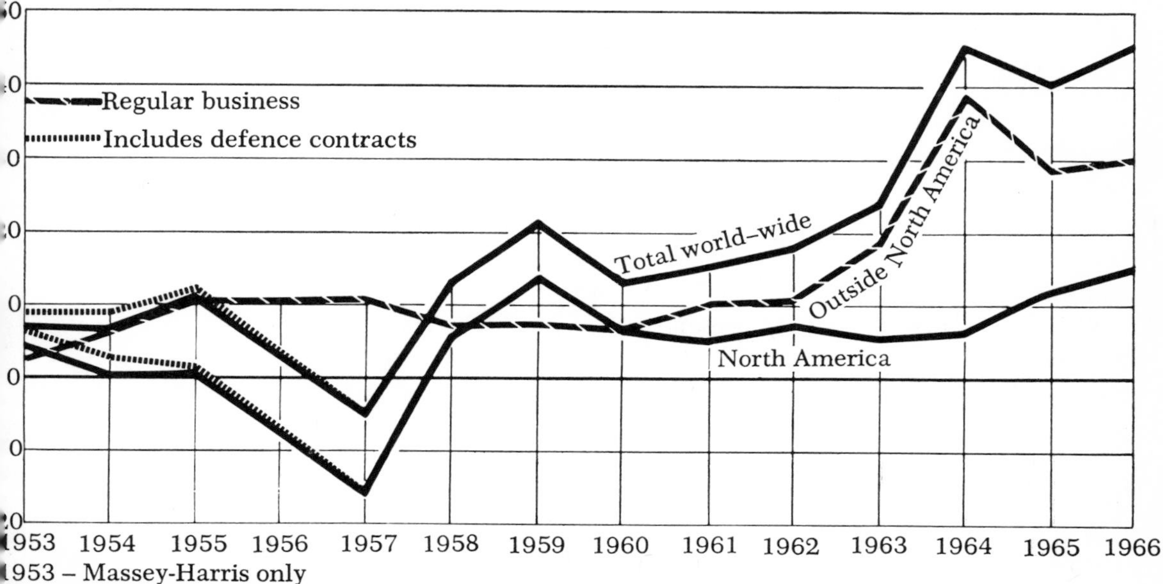

cussions in England that led to the merger. When they returned to the United States they discovered that the United States Ferguson distributors had stopped ordering tractors and implements. The distributors were uncertain as to whether they would be permitted to continue as distributors and so wished to avoid building up inventories. At the same time Ferguson dealers stopped ordering equipment from distributors because they too were uncertain over their status and also because they felt that the more favourable Massey-Harris credit terms and discounts would soon be made applicable to the Ferguson line. Bloom and D'Angelo, together with their sales managers, had a meeting in August to discuss the problem. At that meeting they devised a plan that involved keeping all good dealers in the business, on the assumption that they would need all the Ferguson dealers and all the Massey-Harris dealers to sell the maximum volume of merchandise.

A letter sent to Ferguson dealers on August 16, and another one on August 27, 1953, outlined the new distribution policy in this way:

> There will continue to be Ferguson tractors and Ferguson System implements bearing the Ferguson name; and there will continue to be Massey-Harris tractors and equipment bearing the Massey-Harris name. There will be a separate franchise covering each line . . . the Ferguson line will continue to be sold under present Ferguson terms and discounts and the Massey-Harris line under their present basis of terms and discounts.
>
> Every effort will be made to give you the opportunity of expanding your business by adding the Massey-Harris franchise, on the basis of the above provisions, if you are so located that it would be advisable. For example, if there is a Massey-Harris dealer currently and adequately covering your territory, naturally the Massey-Harris dealer will continue to handle that line.

It was also decided that nine of the twenty-seven Ferguson distributors whose volume was small and who, for the most part were located in poor territories, would be terminated and their territories would be covered by Massey-Harris branches. This meant that the company would have eighteen Ferguson distributors with their system of dealers and sixteen Massey-Harris branches (eight of them with a separate Ferguson division) with another net-work of dealers.

These initial moves made outside of the parent company appeared innocent enough at the time, but actually they marked the beginning of what was later known as the company's "two-line policy." In a memorandum dated October 2, 1953, addressed to E. G. Burgess, Vice President and General Manager of the Canadian Division, James Duncan supported the arrangement whereby there would be exclusive Massey-Harris dealers, exclusive Ferguson dealers, and some Massey-Harris-Ferguson dealers. He explained that:

> The tendency will be towards a gradual development of dealers handling both lines, but this process should be allowed to grow naturally and should not be hastened. The important factor in this whole situation is that no dealer should be given a dual agency, unless he can prove his ability to handle both lines.

This policy, it was felt, would give Ferguson machines better coverage immediately, and it would also strengthen the competitive position of Massey-Harris dealers selected for dual agencies. It was recognized by the President that difficulties were involved in such a programme. For example, an advertising campaign directed at the Ferguson line would weaken the position of Massey-Harris dealers. But it was thought that if the Ferguson product was definitely better everything would be taken care of by the gradual weakening of the competitive position of the Massey-Harris competitor in the same area. The President went on to say:

> In the course of the next twelve months or so, we will become very much better acquainted with the pros and cons of this sales organizational picture and we may be able to see much more clearly what road should be pursued, that is to say, whether we should carry on a mixed bag of dealers, as above suggested, whether we should work towards having nothing but dual dealers, or whether, on the contrary, we should work towards having two separate sales organizations, handling the Ferguson line or the Massey-Harris line.

Duncan also indicated that where dual dealerships had been arranged they would be the direct responsibility of the Massey-Harris district managers, but that the Ferguson supervisor should also visit those dealers as an inspector to see that they were being handled suitably; and in the event that he discovered that this was not the case, he would report immediately to the Massey-Harris district manager who would take the necessary action. This divided authority and responsibility was to lead to great difficulties. It affected every field man in the company.

The difficulties that would be involved in having two sets of representatives visit the same dealer extolling the virtues to him of two lines of equipment, and in having individual dealers selling competing lines to farmers, were not discussed. What is significant is that by October 2, the President had supported a policy of marketing two lines, through two distribution systems, in the sense that he opposed a swift move away from that system and had an open mind as to whether it should eventually be changed or not. At the same time there was a slight degree of integration in that some dealers were permitted to market both lines.

In the export territories the company's efforts right after the merger were directed toward moving to one distribution system. This was strongly favoured by the President. We have seen that, before the merger, the President had favoured integration. In France integration of the two distribution systems went forward right after the merger, although not with the unqualified support of parent company management. There is no doubt that the drift into a two-line concept occurred in operations in the United States.

By early November it was plain that the distribution problem in North American operations had not been settled by policy pronouncements. Ferguson dealers began to complain about the lack of wholesale and retail financing, about increased feeling among farmers that the TO-30 tractor would soon be obsolete as a result of the introduction of new models, and about the interminable delay in obtaining the large tractor. To increase the strength of the Ferguson dealers the company decided in November to transfer to them the Massey-Harris Pony tractor and Pony components. It was also decided that these would be painted grey to blend with the rest of the Ferguson line. This was the first significant action taken to bolster the policy of maintaining two groups of dealers, and it did not take long before the folly of such a policy became evident.

Massey-Harris branch managers within a month began to point out that such a policy would result in reduced sales. They argued, convincingly, that it involved Ferguson dealers selling a style of tractor which for years they had ridiculed, and at the same time it involved taking from a number of Massey-Harris dealers a machine upon which they had depended for their sales volume. The branch managers also feared the consequences of the next management decision, that of giving the Massey-Harris #15 manure spreader and the Goble disc harrow to the Ferguson dealers. Rumours were rife that other Massey-Harris implements would also be transferred to Ferguson dealers, and that this would result in a sharp deterioration in the competitive position of Massey-Harris dealers. These rumours, and the suspicions and fears that they generated, were inevitable from the moment that the company was faced with two lines and two systems of dealers. When the rumours began to fly, the morale of the dealers began to drop.

The company, in spite of those difficulties, soon confirmed its intention of pursuing a "two-line" policy. On November 30 the anti-monopoly bureau of the Federal Trade Commission of the United States Government wrote a letter to Harry Ferguson Incorporated, Detroit, and also to the Massey-Harris Company, Racine, requesting information relating to the proposed conduct of the businesses of these two companies. A memorandum in reply to this letter, dated February 15, 1954, said in effect that it was the intention of the two organizations to merge in view of the substantially complementary and not competitive nature of their lines; that the merger was necessary for survival in the highly competitive farm machinery industry; that it was proposed to continue merchandising two lines of equipment, namely, a Massey-Harris line and a Ferguson line; and that it was proposed to distribute the former through Massey-Harris dealers and the latter through dealers of the independent Ferguson distributors. This essentially reflected what the company had already been doing. It did not involve a change in policy.

Such a policy had important implications for the engineering programmes of the company for it required that both groups of dealers would have to be provided with equally attractive products to sell. These implications were outlined by Herman Klemm, Vice-President Engineering, in January 1954. The policy also raised organizational difficulties. Who, for example, would be responsible for the various engineering programmes that would be required? None of these problems was either minor or transitory. So it was decided to call a conference of top management from both eastern and western hemisphere operations to discuss them – the San Antonio, Texas, conference of March 6 to 11, 1954. In some vital respects it was a fateful conference for the company.

Discussion of the company's two-line policy was bound to occur at San Antonio, and it was raised by the President of the United States company, H. H. Bloom. His view was that to disband any of the distributors and dealers would mean a serious loss of sales volume. Reference was also made to the possibility of legal difficulties in disbanding the two-line policy, although there was no evidence that such difficulties were serious, and the subsequent decision of the Federal Trade Commission proved that they were non-existent. It was emphasized that more dealers, offering more goods, would lead to larger turn-over, and it was also felt that the merged line was too big to receive concentrated attention from individual dealers.

The Massey-Harris representatives felt strongly that it was necessary for Massey-Harris dealers in the United States to have a tractor incorporating the Ferguson draft control system. It will be remembered that Massey-Harris attempts to develop such a tractor had failed. And by the time of the conference there was little doubt that tractor engineering in all companies was going to move away from the old "lugging" type of machines. In other words Massey-Harris dealers wanted a tractor that would make them more competitive with Ferguson dealers. Moreover, Massey-Harris personnel at the conference at times reminded Ferguson personnel of who had acquired whom, and the "red" and the "grey" split remained wide.

But there was really no opposition to the two-line concept as such at the San Antonio conference. A few weeks earlier Walter Lattman had read with horror Klemm's letter which had considered the co-existence of two sets of dealers and two product lines. He wrote to the President arguing in favour of one line of tractors, pointing out that the company had not the time or resources to create a line for Massey-Harris dealers. In his notes he also wrote: " . . . To administer 2 lines we would have to ask our Management & our Sales organization to 'prostitute' themselves when talking to the 'other' Dealer or Distributor!"

His letter had no visible effect. It was decided at the conference that in the United States the only advisable policy, not from a legal or anti-trust viewpoint but for obtaining maximum sales volume, was to maintain two separate distribution organizations. While there would be one general sales manager there would be a sales division for each line. This decision was not modified by any suggestion that the two distribution systems would, in future, gradually be merged, and in that respect it went further towards the two-line concept than had been implied in the President's previous statements.

The two distribution systems, to be competitive with each other, would have to be supplied with competing and not merely complementary products. Therefore, the conference turned its attention to an engineering programme that would provide required strength for both lines. Engineering and manufacturing representatives argued that the differences between the two lines should be minimized, but marketing considerations led to the view that outward appearance at least should not be similar.

To provide Ferguson dealers in the United States with a tractor that would be competitive it was decided to speed the introduction of the TO-35 to replace the TO-30 tractor. Herman Klemm had supervised the design of that tractor and, after unceasing opposition from Harry Ferguson including withdrawal of approval of it on one occasion, he had obtained the latter's final consent for it to go into production. That tractor was undoubtedly Klemm's greatest contribution to the company. Appearing first in January 1955 it was made available to Ferguson dealers about a year earlier than had at one time seemed possible.

It was thought, however, that introduction of the TO-35 would give Ferguson dealers an advantage over Massey-Harris dealers, and thus it was decided that a tractor comparable to the TO-35 should be developed for the Massey-Harris line, a decision subject only to advice on the legal implications of such a move. A favourable legal opinion was soon received, and such a tractor, differing from the TO-35 in its sheet metal, colour scheme, engine specifications, and in its longer wheel base and capacity to accommodate mid-mounted implements, was introduced as the MH-50 in December of 1955. It came in four models – the standard four-wheel models, the high clearance four-wheel models, the single front wheel tricycle models, and the dual front wheel tricycle models. Being distinctly superior to the TO-35 for row-crop work, this tractor was soon to cause a crisis in the company's relations with its distributors.

The conference discussed the introduction of the TO-60 tractor, that long-awaited large Ferguson tractor. There were misgivings – particularly over the fact that it was not a tricycle tractor and so could not accommodate a mid-mounted four-row cultivator or a mounted two-row corn picker. There was also some reservation over its likely success in the plains region where tractors were used in many instances for lugging purposes only and where the Ferguson weight-transfer principle was not applicable. Final decision on the TO-60 tractor was not taken at the conference but it was decided that the company should develop a Massey-Harris tractor equivalent to the TO-60 to be made available to Massey-Harris dealers.

Because of these decisions, the company embarked on a costly engineering programme designed to develop not one, but two, complete lines of farm machinery, rather like the automobile companies were doing. The assumption presumably was that what was good for General Motors was good for Massey-Ferguson.

To ensure co-ordination in engineering between the Massey-Harris and Ferguson lines of machinery it was announced, at the San Antonio meeting, that Herman Klemm would be chief engineer for North America covering both lines. The Racine Engineering Department would progressively be moved to Detroit and in addition there would, of course, be the

Engineering Department in Toronto.

While these engineering and distribution decisions were later to prove disastrous, the decisions taken relating to manufacturing were quite different. We have already noted that a policy of supplying the Detroit tractor plant with United Kingdom components appeared in embryonic form in the Harry Ferguson organization. The earliest engineering plans for the TO-35 tractor included plans for interchangeability of major components with a similar tractor to be manufactured later at Coventry. This interest in international sourcing of tractor components was carried into the merged organization. In January 1954, a group headed by A. A. Thornbrough was established to study the economies of importing Ferguson tractors and components into North America from England. In February Thornbrough sent a memorandum on the subject to H. H. Bloom, President of the United States subsidiary company and First Vice President of the parent company. It showed that the tractor manufactured in Coventry could be delivered to the east coast of the United States at a saving of $262 over the Detroit-produced tractor and at a saving of $296 and $208 at the west coast and at Detroit. Furthermore the memorandum estimated that the four major components – rear axle, transmission, lift cover assembly and hydraulic pump – could be landed at Detroit at a saving of $96 per tractor. It was this study that provided the background for the decision taken at Detroit to obtain the main assemblies for the new TO-35 – transmission and centre housing and the rear axle – from the Standard Motor Company plant at Coventry. The TO-35, which was to replace the TO-30 Detroit tractor and the TE-20 Coventry tractor, not only eventually used major components manufactured in Coventry but also was the first tractor permitting full interchangeability of United States and United Kingdom electrical systems and other accessories.

In effect, therefore, the San Antonio conference adopted the policy of international sourcing of major tractor components, and with it, the policy of interchangeability of major tractor components. This policy was later to be extended to other machines, and it became increasingly sophisticated.

Actually the conference ignored important implications and problems of the decision to source TO-35 components in England. It did not take into account the risk of a suit for damages if the company suddenly ceased ordering tractor components from Borg-Warner of Detroit, with whom it had a supply contract; there was also the questionable ability of the Standard Motor Company to supply the TO-35 components on time. As it happened, major components for the TO-35 had for a period to be obtained solely in the United States. Eventually, agreements with Borg-Warner were terminated and Standard did begin to ship components.

Opposition to the decisions taken at San Antonio was not long in coming, and it came from Harry Ferguson. Even before the conference it had become obvious that working with Harry Ferguson would be most difficult.

Just after the merger he expressed the view that he should perhaps be made Chairman of the Board of the English company. This was contrary to the agreement with him, and Duncan had John Turner of Armitage and Norton, auditor of the Harry Ferguson organization, convey this view to Ferguson. Duncan also suggested to Turner that Ferguson was unacquainted with the limitations of his responsibilities as Chairman of the Board of the parent company. He explained to John Turner that " . . . there was a great difference between the functions of a British Chairman of a Board and that of an American or Canadian Chairman of the Board whose functions, as in our case, are usually confined to presiding over Board Meetings but have nothing to do with the policy or management of the business."

Ferguson accepted Duncan's view on the matter and for several months Harry Ferguson wrote letters of almost exaggerated cheerfulness to James Duncan. True there was a note of things to come in his letters of October 12 and October 16, in which he argued that the design of the small Massey-Harris combine that was being developed in Germany was wrong – a design used by all Massey-Harris self-propelled combines with spectacular success – and that his own quite different approach involving mounted and semi-mounted machines was right. But Ferguson did not persist in his opposition, and his primary interest centred on the development of a chassis for the revolutionary automobile that he hoped to perfect.

Then his health deteriorated and he left for a two months' holiday, only to return without having recovered from his illness. His major ailment seems to have been of a nervous character which manifested itself in severe mental depressions, and he was required to spend some time in a nursing home. All his outside activities had to be suspended, and he was not permitted to see anyone outside his family nor to read any letters. When Duncan wrote to him on March 9, 1954, during the San Antonio conference, requesting his approval of certain industrial equipment, Ferguson's son-in-law had to answer that while Mr. Ferguson had come out of the nursing home he still could not be consulted about the company's business affairs. He recommended therefore that Duncan should consider that Clause 6 of the agreement with Ferguson, the clause relating to Ferguson's prerogatives in engineering, be suspended owing to illness – a suspension that was subsequently approved by Ferguson for a period of six months. However, the suspension did not formally exist from the end of December 1953 to March 1954, that is, at the time when important decisions were taken at San Antonio.

Two weeks later Duncan wrote Ferguson a letter regarding the San Antonio conference. He informed him of the fears of the sales organization regarding the unsuitability of the TO-60 for the North American market; of the appointment of Klemm as Vice President in charge of engineering; and of his concern over competition from a new Ford tractor to be built in the United Kingdom. Ferguson replied on March 30:

> Your letters are sound, practical and breathe that spirit of good fighters that does a man's heart good to read. I do wish to thank you all, most warmly.

> It is a real joy to be associated with all you grand people and it certainly was a good day for both you and us when we got together. We certainly needed you and you certainly needed what we can contribute in tractors and implements. . . .

In a letter to Herman Klemm of March 25 he said:

> I have had some lovely letters from Toronto, at least Tony and my secretary say so, but I have not been allowed to deal with them yet.
>
> I shake hands with myself twice every day for the fact that we have joined up with these wonderful Massey-Harris people.

Up to this point Ferguson obviously did not understand the implications of the decisions taken at San Antonio, but he did begin to understand them a few days later. On April 1, he wrote to Duncan saying that he was deeply concerned about the talk he had heard that, as a result of the San Antonio conference, former Ferguson engineers were to be diverted to developing a new design of the Massey-Harris tractor to take the place of the old MH-44 and MH-33. What Ferguson recommended was that the Massey-Harris tractors should be declared obsolete as quickly as possible and that the company should concentrate its energies on bringing the new big Ferguson tractor and the new TO-35 into production. He said:

> Ed. Burgess, who is with me just now, says that at San Antonio it was decided to sell two separate lines of tractors, one red and one grey, and keep them apart. Jimmy, you don't need me to tell you that you cannot sell as many grey tractors in that way as you would by selling grey tractors only. Now, if you do anything to reduce the sale of the grey tractors you are going to sell a far less number of tractors of all kinds because your price will be higher.
>
> It is just as clear to me as noon-day that, with our new big tractor and the TO-35 we have machines of the future that will lead the world. In all conscience they cover agricultural conditions to such an extremely wide extent that we would be 100% justified commercially in concentrating on them. . . .

This letter made it plain that Ferguson had not begun to appreciate the character of the United States demand for tractors, something that had eluded him even in the days prior to the merger. His fears over the dispersion of engineering energies were, however, fully justified.

From the time of that meeting with Burgess, Harry Ferguson began to write numerous, very long letters to James Duncan which became ever more biting, and even unfair in their criticism. On April 7, he expressed his opposition to the concept of a North American three-wheel tractor in this way:

> Now, Jimmy, this is a serious business. Any 3-wheel tractor is a mechanical monstrosity and we want to keep away from monstrosities if we are to lead the industry. It seems I will be asked to approve the designs of a 3-wheel machine. I find all our staff here much concerned about this, and if I approved it I might be doing harm to our future. . . .

His opposition to mid-mounted implements was equally direct. The letter also revealed what, in fact, was happening at that time, and even prior to the merger: a sort of estrangement between the engineering people in Coventry and those in Detroit. This, perhaps, was inevitable as long as Harry Ferguson was located at Coventry and was involved in matters of engineering.

On April 8, Duncan wrote Ferguson a somewhat general letter about problems of distribution in the United States and their engineering implications. He also said that there should be closer relations between the engineering department in Detroit and the one in Coventry through frequent visits by senior personnel. This was treading on sensitive ground, and he made it even more sensitive by saying " . . . if this had been done when the TO-60 was being developed, we would have had a tractor which would have not only embodied all the great advantages which we all know of, but would have met the requirements of the North American continent. We will have to see that in the future we avoid working in water-tight compartments. . . ." Duncan had misjudged his Harry Ferguson, as he soon discovered.

In his first reply to that letter, dated April 13, Ferguson was friendly and he may perhaps not have read Duncan's letter. He said that he did not reply in detail because he thought that Duncan had not yet received all the information that he had sent him. Still, he did go on to reiterate his view that the Massey-Harris tractor should be dropped. He was also pleased that the engineering activity necessary for replacing the MH-44 and MH-55 tractors would be located at Racine for this meant, he thought, that the company's skilled staff – by which he meant Ferguson engineers – would not be occupied with red tractors at a time when competition required that the two Ferguson tractors be brought forward. He also said that he assumed that the company was not going to ruin the design of the big Ferguson tractor.

The detailed reply that he had referred to came on April 20, and with it the dam broke. He spoke of Duncan's " . . . wild and irresponsible words . . . " and when referring to Duncan's letter of April 8, he described it as " . . . the rudest I have ever received. . . . " In that letter he also said that the company should return to farmers a £20 price decrease that Standard Motors had granted them which, of course, was in accord with the view that he had held for many years, that prices on tractors should be reduced even if this meant very low profit margins. Ferguson was not about to change convictions of a lifetime. Duncan had decided to visit Ferguson even before this letter arrived. Its arrival made a visit essential.

Arriving in London, Duncan found a long letter from Ferguson waiting for him. In it Ferguson said that if they could not resolve their differences he would sell his stock, and even if they did find a solution he would still want to sell some of it. When Duncan arrived in Coventry he found that a meeting had been arranged in the engineering department and that it had been carefully rehearsed by Ferguson. The object of that meeting, it seemed, was to impress Duncan with the importance of the Ferguson influence on the company's engineering departments at Coventry and Detroit; to emphasize his view that the San Antonio decisions were wrong;

and to hint at the possibility that, if his views were not accepted, he and his top engineers would resign, leaving the company in a difficult position. The possibility of seeing Massey-Ferguson stock depressed by the sale of Ferguson's holdings, or of it getting into "unfriendly" hands, and of losing engineering talent that had made the merger so attractive, was not pleasant to contemplate.

In the conference that followed, Ferguson's well-known views, including his misconceptions, were repeated as they were several times over the next few days. Ferguson left the impression that he was determined to be responsible for engineering in the United States and in the United Kingdom, and that this should include educational work, field demonstration and sales presentation. Increasingly his comments related to matters clearly outside his area of responsibility. By agreement he enjoyed the right to approve engineering designs, but not the right to manage engineering activity, let alone activity outside of engineering. Ferguson did not accept that interpretation of the agreement. The impasse remained. This situation was not simply irksome for the company, it was untenable. Complete disagreement between Chairman and President was bad enough, but since it could involve delaying the approval of newly designed machines it was also dangerous.

Was it simply a matter of a difficult temperament, one that had already been revealed in Ferguson's associations with David Brown, Roger Kyes, Ford Motor Company officials, Sir John Black and others? Probably it was, although this was complicated further by ill heath. However, in almost every instance, including his disagreement with Massey-Ferguson management, there was an element of truth in his views. He was wrong about the suitability of the TO-60 tractor for the United States market, but he was right about the dangers involved in engineering two lines of tractors and about the need to develop even the large tractors along Ferguson-system principles.

By the time Duncan reported to the Executive Committee on May 27, 1954, there was no doubt that Harry Ferguson, having in mind his opposition to company policy and his difficult temperament, as well as his precarious health, would become a serious obstacle to the future hopes of the new organization. The danger that he would develop a sort of Ferguson group among the company's engineers, that he would interfere with the company's relationship with suppliers such as Standard Motors, and that a growing feeling of unrest would be created in the organization could not be ignored. Duncan felt that " . . . he has sold his business, but wants to continue to manage it. Under these circumstances, my recommendation is that we should accept his resignation and his sale of our stock. . . . "

There was, however, the danger that Ferguson would agree to do both those things but would insist on retaining his authority to approve designs. A special meeting of the Board of Directors was called for June 4, 1954. Harry Ferguson wrote saying that he regretted that it would not be possible for him to attend. It was then agreed that J. S. Duncan, W. E.

Phillips, J. A. McDougald, and J. S. D. Tory, would go to Coventry to attempt to deal directly with him. They arrived in London on June 13. Surprisingly enough, excellent progress was made in the first meetings with Ferguson, and the company's counsel took copious notes of the areas upon which agreement was being reached. By late afternoon, when Massey-Ferguson representatives left for London, agreement had been reached on every contentious point except the price to be paid for Ferguson's shares. Massey-Ferguson executives arranged to meet in London to draft necessary agreements with the help of the company's counsel. But strangely enough, the company's lawyer did not arrive, and notes of the conference with Ferguson could not be found. McDougald had to arrange for new counsel and got Sir Sam Brown of Linklaters and Paines. Sir Hartley Shawcross was retained in case a law suit developed. It seemed in subsequent encounters with Ferguson that he sensed the predicament of the Massey-Ferguson representatives, and despite many meetings and much discussion it was not even possible to agree on what had been decided in the first meeting.

Some of the Massey-Ferguson directors had finally to attend to other business and McDougald and Duncan were left behind to attempt to finish the work. Their immediate success was no greater than that of previous efforts. It seemed that if settlement was to be reached, it would be reached only through Massey-Ferguson negotiators establishing an improved bargaining position and exploiting it successfully.

McDougald began to think that Harry Ferguson might soon be needing money. It was known that he was negotiating with an Italian firm for a transmission for his experimental car, and also that he was actively interested in buying a famous and very expensive Constable painting. Curiously enough, it was Ferguson's interest in that painting that was finally to encourage McDougald to press negotiations to a successful conclusion.

Briefly what happened was this. McDougald had been arranging with Mr. Oscar Johnson of Leggatt Brothers, a well-known firm of art dealers in London, to have Sir Alfred Munnings paint a picture of a pair of his horses. He wished Mr. Johnson to make an appointment with Sir Alfred for a certain Thursday, but Mr. Johnson told him that he could not meet him on that day since he had a very important appointment with a Mr. Harry Ferguson, of Stow-on-the-Wold. That appointment, of course, related to the Constable painting in which Ferguson was interested. Being frustrated over the interminable delay, and having received this added bit of information, McDougald decided to "play a little poker. . . . " He contacted John Turner, Ferguson's auditor, and told him that he was weary to the point of distraction of the "on-again off-again" negotiations with Ferguson, and that he was leaving for Canada the next day. However, he said to Turner, if Ferguson was interested in one last try to reach a settlement he would be prepared to defer his departure and go to Stow-on-the-Wold on the Thursday. By choosing Thursday he would force Ferguson to choose between seeing McDougald and seeing Mr. Johnson about the Constable.

It may be wondered why this would be regarded as a difficult choice for Ferguson. However, Harry Ferguson was deeply sensitive to things of beauty – the interior of his house was invariably furnished in excellent taste, and his immaculate country estate was a joy to see – so that having to choose between negotiating for a Constable painting and seeing McDougald about selling his stock was not as easy as it might appear to be. McDougald reasoned that if Ferguson chose to see him instead of Mr. Johnston, with the understanding that it would be a last attempt at reaching a settlement, he would know then that Ferguson really did want to sell his stock. What Ferguson's choice was became known when Mr. Johnson by chance, and in all innocence, informed Mrs. McDougald that his appointment had been cancelled because Mr. Ferguson had to have a meeting with a very important gentleman from North America. That gentleman, of course, was McDougald.

McDougald, Duncan, and a few others went to Ferguson's home on the appointed day. McDougald opened the discussion by saying that if agreement could not be reached that day, he and his colleagues would return to Toronto immediately and that the negotiations would be terminated. Negotiations began at 10:00 a.m. and Ferguson, as always, was an exceedingly gracious host. He missed few bargaining manoeuvres. One intriguing one was that whenever discussion became particularly heated he arranged for his butler to pass around cigars out of the silver cigar humidor that the Massey-Harris negotiators had given to him at the time of the merger; on it were inscribed comments highly complimentary to Ferguson and the signatures of the very people that were now negotiating with him.

In the early evening a strange thing happened. Ferguson remarked that he was tired and suggested that everyone, except himself, should go to Burford for dinner at the Cotswold Gate, and then return for further negotiations later in the evening. He said his daughter and son-in-law would be their hosts. McDougald stayed behind to telephone his Canadian banker to arrange immediate payment in London if the deal went through. His presence there was unknown to Ferguson. Suddenly McDougald heard Ferguson remark heatedly to his butler that since everyone had deserted him they could wait until tomorrow for further discussion. McDougald came out of the "Blue Room" and reminded Ferguson that the deal would be closed that day or not at all and that if he wanted to work on it he would immediately recall his colleagues. McDougald went to the Cotswald Gate, interrupted his colleagues who were just having their soup, and asked them to return. Negotiations were resumed and the deal then was finally closed. Duncan and McDougald had their dinner at 3:00 a.m. in Leamington Spa on the way to London. The agreement terminating Harry Ferguson's association with Massey-Harris-Ferguson Limited was dated July 6, 1954. His stock went to the Argus group.

With the departure of Harry Ferguson from the company, management could again devote most of its attention to implementing decisions taken at San Antonio although, as we saw in the preceding chapter, it was during this period that disagreement with Cecil McKay also demanded attention. Before tracing the developments that led finally to a change in management, two other aspects of the company's experience from 1953 to 1956 warrant attention: the approach to organization and control taken after the merger, and the nature of the company's performance outside North America from 1953 to 1956, a performance that provided a welcome balance to its North American experiences.

The organizational structure which was introduced and which remained relatively unchanged until the 1956 change in management is shown on Organization Chart 2. Even before the end of 1953, it had been decided that there would be an Eastern Hemisphere Division which would be a successor to the former Massey-Harris European Division. The managing director of the Eastern Hemisphere Division was to be E. W. Young, a former Harry Ferguson Limited executive, with offices in London. That division would be responsible for manufacturing and marketing operations of the company in the United Kingdom, France and Germany, as well as for a limited part of the company's total export activity.

United States and Canadian divisions introduced into the organizational structure in 1950 were perpetuated, although another division was established called the Western Hemisphere Export Division which was responsible for exports to territories not covered by the Eastern Hemisphere Division. In addition to these four operating divisions – all of which, incidentally, except the Western Hemisphere Export Division, were headed by a vice president of the parent company – there were established within the parent company a number of co-ordinating and policy divisions and also a co-ordinating committee. The Co-ordinating Committee, which was a sort of successor to the 1944 Operating Committee, was composed essentially of parent company vice presidents. These vice presidents were heads of the various parent company co-ordinating and policy divisions, divisions that included Manufacturing, Engineering, Planning and Procurement, General Administration, Secretary's Office, and another division responsible for the company's machine tool division, the Electro Forge Division, and for relations with South African and later Australian operations. Finally, there was the usual group of service heads including the Director Personnel and Industrial Relations, Director Research, Director Public Relations, and Comptroller.

The position and role of the operating and co-ordinating divisions was explained in a memorandum dated December 28, 1953. It is worthwhile examining that memorandum for it indicates the confusion in matters of organization that existed at the time. The memorandum pointed out that each of the operating and co-ordinating and policy divisions was directly responsible to the President of the Parent company. Furthermore, it indicated that the Director and First Vice President of the parent company, H. H. Bloom, while being primarily responsible for the operations of the United States company had at all times to be consulted on major decisions and, in the absence of the President, had the deciding voice as to policy or action.

The memorandum said that the operating divisional heads carried full responsibility for all phases of day-to-day activities of their divisions and pointed out that the usual procedure was to clear problems through the co-ordinating and policy divisions as far as possible, so as to minimize direct contact with the President on other than matters of considerable importance. It also said that:

The functions of the Co-ordinating Vice-Presidents are to develop and co-ordinate the policies of the Company on a world-wide basis; to assist the operating divisions in their planning; to hold a watching brief over all the operations of the Operating Divisions; and to draw their attention to any action which appears to be contrary to the Company's policy or undesirable from any other point of view.

Normally, these problems will be cleared and settled directly between the two respective heads of the Co-ordinating and Policy Division and the Operating Division. In the event that a satisfactory solution cannot be reached, the question should be brought to the President for decision.

It must not be assumed, from what has been said above, that the functions of the Co-ordinating and Policy Divisions are purely advisory. On the contrary, the Vice-Presidents of such Divisions, who are to be fully conversant with the way in which the functions which come under their Divisions are being carried out by the operational units, will be held equally responsible by the President for the satisfactory operations of such functions.

This last sentence typifies the confusion that surrounded matters of organization at the time. By stating that responsibility for operations rested *both* with the parent company Vice Presidents *and* with management of the operating divisions, individual executives could not be held accountable for results. When everyone is accountable it usually means that no one is accountable. In a period of difficulty, such an arrangement frequently encourages arguments over who is responsible.

Another characteristic of the concept of organization perpetuated after the merger was its high degree of centralization. This is clear from the organization chart which shows that the United States, Canadian, and Eastern Hemisphere divisions were headed by Vice Presidents of the parent company, so that their management groups were in effect an extension of parent company management. But centralization, in practice, was also encouraged by the already-mentioned failure to define unambiguously the responsibilities of senior executives, for this led to the drift of detailed decision-making to dominant personalities – in this case particularly the President.

To control the operations of the company in France, Germany, and United Kingdom, the Eastern Hemisphere Division also introduced what it termed co-ordinating and policy departments similar to those of the parent company. They covered the functions of Engineering, Finance, Manufacturing, and Supply and Control. Furthermore that Division also had service departments including Public Relations, Education, and Market Research.

One senses that groping within the Eastern Hemisphere Division towards a traditional organizational structure was more firmly based than similar efforts in the parent company. The words "staff" and "line" occasionally appeared in the organizational memoranda of that division whereas they were foreign to the organizational memoranda of the parent company. However, the whole development of the "corporate" group within the Eastern Hemisphere Division meant that a further executive layer was being created between the operating heads of that division and the Chief Executive Officer of the parent company.

The decision to design and market two separate lines of machinery prompted the perpetuation of separate Ferguson and Massey-Harris engineering departments in the Eastern Hemisphere Division. Co-ordination between the engineering activity of the Eastern Hemisphere Division in general and the engineering activity in the rest of the organization was to be effected by an Eastern and Western Hemisphere Engineering Co-ordinating Committee.

While there were a number of detailed changes within the various divisions of the company after 1954, organization did not change fundamentally until the change in management in mid-1956. Defects of that organizational *structure* from a longer-run point of view were that it encouraged the creation of an executive layer between operating companies in Europe and the parent company, and that the parent company co-ordinating and policy divisions did not include all the functional divisions of the operating companies. But the main difficulty was the low priority that management placed on matters of organization, with the result that rational and comprehensive concepts of organization did not really emerge. There was no clear concept of "line" and "staff"; there was no serious attempt to define executive positions in terms of responsibilities. Both the organizational structure of the company and the apparent responsibilities attaching to individual executive positions shifted in response to shifts of individuals from one part of the company to another.

The approach to planning and control also remained unchanged after the merger. Essentially it involved salesmen making a projection of sales for the coming year, financial officers incorporating these sales figures into a projected balance sheet and income statement for the coming year, and periodic revisions of this "budget" as realized results deviated from it. Typically results were given in terms of absolute figures; not only was the approach narrowly financial, but even financial developments were difficult to interpret intelligently.

Perhaps the most significant point to be made is that general developments in the science of management pertaining to the organization and control of complex profit-maximizing corporations had gone well beyond the formal training and practical experience of Massey-Ferguson senior management. It is not often realized that the concept of technological unemployment can apply as harshly to management as it can to labour.

Outside North America, developments in the Massey-Ferguson organization from 1953 to 1956 carried considerable long-term significance. This was particularly so for developments in the French, German, and United Kingdom operations, all of which generated profits for the parent company at a time when North American operations became increasingly unprofitable.

By the time of the 1953 merger, the Massey-Harris company in Germany had begun to design its small #630 self-propelled combine, it had purchased a straw press manufacturing factory at Eschwege, and it was manufacturing roller chain and some spare parts at its old factory at Westhoven. Its distribution system was still rudimentary and the volume of its local production very small. Ferguson had been represented in Germany by an independent distributor of indifferent quality. A merger between the two organizations in Germany therefore encountered few obstacles and it was effected simply by transferring rights for the distribution of Ferguson tractors and implements from the distributor to the old Massey-Harris Company, but now changed to Massey-Harris-Ferguson G.m.b.H.

For the German company, the right to distribute the Ferguson machinery was an important step in its post-war reconstruction. Earlier it had begun to revive through the manufacture of roller chain, the acquisition of the Rausendorf straw press business (a principal adjunct to the European combine business), the distribution of imported combines, and particularly through its designing the small #630 self-propelled combine.

Thirty prototype machines of the #630 combine were produced in 1954 and their preliminary reception was enthusiastic. It was thought that with it, as well as with the introduction of other implements, the company in West Germany would become an important manufacturing centre. Plans were considered for expanding the German manufacturing facilities as well as for re-establishing a strong network of German retail dealers to sell the new combine and the Ferguson tractor. The location for this expansion, it was decided, should be Eschwege which, it was thought, was as well placed for distribution in West Germany as Westhoven; and in the event of reunification of Germany its location would be ideal. It was estimated that the somewhat higher transportation costs for exported machinery in comparison with Westhoven would be compensated for by lower labour rates, and also by lower finance charges since expansion at Eschwege could be financed by additional funds borrowed from the Land of Hesse. The latter wished to increase employment opportunities at Eschwege. A copper mine that had previously depended on a smelter in what is now East Germany had, after the partition, become uneconomic but had been receiving governmental assistance to avoid creating a local unemployment problem.

By early 1955 negotiations with the Land of Hesse for capital to finance expansion of manufacturing facilities at Eschwege had successfully been completed, and the Board of Directors at Toronto had approved the expansion. Under the agreement the Land of Hesse was to construct a new factory involving a building of approximately 200,000 square feet. It would also give the company a thirty-year option to buy the land and buildings. In addition, it would grant the company two long-term loans, at a very favourable interest rate, for machine tools, equipment, and working capital. The loan for working capital was not to exceed twice the capital of the German company. For this reason the company's capital was increased from $198,000 to $620,000, of which a loan from the parent company supplied $83,000 and a loan from the United Kingdom Massey-Harris-Ferguson Company supplied $183,000, the rest coming from surplus. In addition to the loan, Hesse agreed to pay for moving staff and machines from Westhoven and for training new workers, and it also assisted in providing housing accommodation. The agreement committed Massey-Ferguson to create eleven hundred jobs and it was this prospect that induced the Land of Hesse to provide assistance. The company on the other hand hopefully looked forward to a production of 500 #630 combines in 1955 and 3,200 in 1956.

The project went speedily ahead and 850 #630 machines were produced in 1955, 3,910 in 1956, and 5,185 in 1957. Sales of the German company soon reached record levels rising steadily until 1957, and profits by 1957 exceeded three-quarters of a million dollars – which curiously was more than Massey-Ferguson profits in either North America or South Africa in that year. And so in this way the German company, with very little financial assistance from the parent company, was successful in establishing itself as a source of supply for small combines in Europe and certain other export territories. The large Massey-Ferguson combines for the continental market came from both Marquette and Kilmarnock.

On December 5, 1955, the manager of the German company wrote to the Minister of Finance of the Land of Hesse pointing out with pride:

> We have great satisfaction in informing you that our self-propelled No. 630 combine employed for the first time on a larger scale during the last harvest has met with such approval that our Board of Directors at a recent meeting decided to increase our original Manufacturing Schedule – underlying our Contract with the Land Hesse of February 16th, 1955 – by appr. 40%. In this connection it may be of interest that 50% of next year's production are earmarked for export, whilst in 1957 the total of the additional planned output is also scheduled for export. . . .

In the same letter the company went on to state that its production programme for the period after 1957 would require additional building space of about 129,000 square feet, and that because of the long delay in the delivery of necessary raw materials, orders for them had soon to be placed. It therefore requested an additional loan on the same terms as the previous one to help finance expansion of the company's operations.

The Finance Minister of the Land of Hesse felt confident that the request would receive a good reception. But articles appeared in the German press expressing opposition to "a foreign competitor" receiving such assistance. The Land of Hesse itself suddenly experienced a certain degree of financial stringency because of new credit restrictions and, in any case, the general shortage of labour reduced the need for special assistance to create employment. A period of uncertainty ensued. Finally the company was informed that it would be granted a further credit for every job created but at a less advantageous interest rate. The relevant agreements were soon signed.

Unfortunately just about that time sales of combines in Germany weakened, factory employment had to be reduced, and public criticism reappeared. Official opening of the new plant at Eschwege had been scheduled for July 13, 1956, but it seemed prudent to delay it. Inventories of the German company, because of declining sales, began to rise in 1956. Special efforts had to be taken by mid-1956 to reduce them so as to improve the company's cash position. As this was happening, the change in the management of the parent company was under way.

The position of the Massey-Harris and Ferguson companies in France at the time of the merger was quite different from what it was in Germany. Massey-Harris was fully occupied manufacturing the Pony tractor and some implements (binders, plows, haying equipment, fertilizer spreaders, one-way discs) at Marquette, and it was also beginning to produce the large #890 self-propelled combine. By the time of the merger it had developed a dominant position in both tractors and combines in the French market. Harry Ferguson de France S.A. had been formed in 1952 with Ferguson holding 70 per cent of the equity and three French shareholders holding 30 per cent; the purpose of this company was to supervise subcontracting of Ferguson machines and to supervise the distribution of both tractors and implements. To begin production of the TE-20 in France the Standard Motor Company Limited of Coventry and Société Anonyme des Anciens Etablissements Hotchkiss (later Société Hotchkiss-Delahaye) of Paris had formed a new company, in May 1953, called Standard-Hotchkiss S.A. It was planned to produce that tractor at St. Denis, a suburb of Paris, the motor to be supplied by the Hotchkiss company. Therefore when Massey-Harris and Harry Ferguson merged, the new Massey-Ferguson organization acquired a 70 per cent interest in Harry Ferguson de France; it inherited the arrangement to purchase the small Ferguson tractor to be produced by Standard-Hotchkiss; and it also had the tractor, combine, and implement factory at Marquette, the Ferguson and Massey-Harris head office facilities in Paris, and two retail distribution systems.

It seemed that one of the first things that had to be done was to acquire the stock of the minority shareholders in Harry Ferguson de France. After most difficult negotiations this was achieved. Then there was the dual distribution system. Right after the merger it had been assumed by management in France that the two systems would be merged and merging at the dealer level was soon under way. But after the San Antonio conference of mid-March 1954 management was told that they would have to "unmerge" in France.

This was met with consternation by French management. On April 8,

1954, J. W. Beith, who was Directeur Général of the French company, wrote a long memorandum on distribution policy as it applied to France. The memorandum is worth quoting at length for it raised specifically a number of difficulties that were in the next few years to cause the downfall of the two-line distribution system in North America; and it constitutes the most complete criticism of that approach to distribution made at the time.

Today we have reached a point where merger at the dealer level is 20% completed with many other settlements nearing completion or in the course of being discussed with the parties concerned. If we wished to arrest this process now an immediate alternative would have to be found in order to avoid an anticlimax with its attendant loss of face for us and disillusionment amongst our dealers, particularly those on the M.H. side who have been promised a two-plow tractor for some time. Needless to say it would be impossible to undo what has been done already. As for those that have not been settled, it is feared that without the immediate promise of a complete line a good many of the best M.H. dealers would strengthen their ties with the competition for the supply of larger tractors than the Pony . . . Ferguson dealers would continue selling opposition Combines and Harvesting Machinery.

The only immediate alternative would be to create two lines immediately other than in horse drawn machines which must be counted out anyway in such an arrangement. Basically, it would mean turning out a red version of the grey Ferguson Tractor in its identical form and perhaps also a grey Pony. In the case of the Ferguson tractor production prospects are such for the next two years under the limitations imposed by the currency situation, that any red tractors would have to be built at the expense of grey tractors. It is questionable whether we could turn out sufficient red ones to satisfy the expectant Massey-Harris dealers and it is hardly likely that the Ferguson dealers would enjoy seeing an identical red tractor encroaching without any sales effort on their preserves. It is also inconceivable that Ferguson dealers could be compensated by the addition to their line of a grey Pony which they would have to sell in a more competitive atmosphere against red Ponys handled by more experienced colleagues. In self-propelled combines of the type now being manufactured it is difficult to visualize two identical types being sold concurrently.

If the principle of selling two lines in France could be accepted at all it would necessarily have to be on the basis of machines differing sufficiently to render them readily distinguishable. If the I.H.C. gave up their dual network it was in large measure because the sale of easily distinguishable Binders, Mowers, etc., was gradually being replaced by Tractors, Combines, etc., which could not be produced in different versions except at great expense. It must not be overlooked that France is a relatively small country very intensely worked and that dealers handling practically identical lines would be sitting pretty well one on top of the other.

Supposing we could afford to wait for two years in order to market two versions of the Ferguson tractor, it is questionable how far we could go in making them different and at the same time keep the cost down. Merely to change the styling would in itself be costly for the runs envisaged by Standard Hotchkiss. But would this be enough and would it not at least be necessary to have different engines? This might be practicable for the Diesel by introducing the Perkins

P3 but not for the Petrol where no obvious alternative exists for the Standard. To divide our effort in the engine line would not assist the suppliers in arriving at the lowest costs. . . .

For a grey version of the Pony, the problem of re-styling and the use of a different engine is still greater and less justifiable in view of the lesser quantities involved. In the case of the Combine, distinguishable versions can be justified even less.

Quite apart from all the foregoing it is seriously questioned whether in the long run any lasting advantage could be derived from marketing two lines in a country like France by the same organization even in the most favourable circumstances.

The . . . more numerous school holds the view that we should strive immediately for a single strong network of dealers rather than two weaker ones. The feeling is that France is too closely worked to allow for two parallel networks without ultimately giving rise to price cutting, friction and loss of good will. We might sell up to 20% more Ferguson type tractors altogether for a limited number of years but that situation would not last. It is questionable whether many more Ponys or Combines would be sold by these means. This same school also holds that we could not market two identical lines and that having to wait two years for two different lines would prove disastrous. The majority of our dealers on the M.H. side would not stand for it. . . .

It must also be borne in mind that if we ran two lines it would not be possible to distribute them both through the Branches under the same local management. It was awkward enough when we handled M.H. and Johnston this way in the days when we dealt mainly with blacksmiths and small country dealers. It would mean keeping a separate commercial organization in Paris for the grey line with all its attendant expense. This naturally in addition to a different set of travellers and servicemen and the separate handling of spares.

These comments could as readily have applied to North America as to France. At the time they were not endorsed by either the Managing Director of the Eastern Hemisphere Division, E. W. Young, or by the President, J. S. Duncan. The latter wrote on September 2 to Young that "... the fusion of the organizations is certainly a very debatable one. My own inclination for France was toward two competitive sales organizations along the U.S.A. pattern. . . . " The argument that there was a shortage of Ferguson tractors and that Massey-Harris versions of them would not soon be available held the day, and France moved towards a single distribution system. Personality clashes involved in doing so were not pleasant, but slowly progress was made.

Sales in France increased greatly from the time of the merger until 1957. Massey-Harris sales in 1953 had been $20.6 million whereas Massey-Ferguson sales in France in 1957 totalled $87.7 million. Profits naturally reflected this expansion totalling almost $3 million in 1956, about three times as large as the figure for Massey-Harris in France in 1953. From 1953 to 1957 production of the Pony tractor at Marquette increased from 8,133 to 13,847; and production of the Ferguson tractor at the Standard-Hotchkiss plant at St. Denis rose from 2,007 to 18,767; while output of the #890 self-propelled combine rose from 99 to 1,527.

There was at the same time similar expansion in the volume of production of a wide range of implements. Generally it was a period when sales were limited not by demand but by productive capacity, and when the company consolidated its position in tractors by the timely introduction of locally produced Ferguson tractors.

The company became hard-pressed for manufacturing facilities. It operated with three shifts and even so began more and more to rely on subcontractors. Thought naturally turned to expansion. Additional land was purchased at Marquette in 1955, and a new tractor assembly line was constructed there in 1956. Plans began to appear for construction of a new combine plant on the land purchased in 1955, and later Standard-Hotchkiss began to plan for a new tractor plant at Beauvais. But all this did not materialize until after the 1956 change in management at Toronto. Still, it may be noted that after 1956 and 1957 the French company began to experience a decline in its sales and profits, that is, just at about the time when it was embarking on a new round of capital expansion. The years after 1957 were to be very much more difficult than those before 1957.

The company's operations in the United Kingdom were permanently and materially altered by the 1953 merger in terms of sales and manufacturing activity. Ferguson sales in the United Kingdom totalled £9.0 million in 1953, whereas Massey-Harris sales had been £3.8 million. The merger made Massey-Ferguson a significant factor in the United Kingdom market in tractors and combines, and a similarly striking change occurred in the importance of the United Kingdom as a source of the company's products. The Massey-Harris Kilmarnock plant had produced tractors, combines, and balers essentially on an assembly basis, and implements were similarly produced at the Manchester factory, all of which accounted for a relatively small portion of the company's total output. Ferguson, on the other hand, supplied virtually all of its markets outside of the United States with tractors manufactured by the Standard Motor Company at Coventry. Consequently, from a manufacturing point of view, the attention of Massey-Ferguson management in the United Kingdom had to be more on the operations of the Standard Motor Company than on its own plants.

The San Antonio conference confirmed that the two distribution systems in the United Kingdom – Ferguson and Massey-Harris – should be perpetuated, and so they were for a number of years. The dominant position of the Ferguson tractor and the Massey-Harris combine in the United Kingdom reduced somewhat the difficulties of operating these two systems, but not for long. The Massey-Harris sales organization began to press hard for production in the United Kingdom of a Ferguson-type tractor for their dealers. A scheme was developed by the United Kingdom Massey-Harris personnel to assemble the proposed MH-50 at the Manchester plant. What this would do to relations with Standard, to costs of tractor production in the United Kingdom, and to the Ferguson dealers in the United Kingdom and export markets seems not to have been clearly appreciated by those proposing the scheme. But before it could be implemented management at Toronto had changed and the scheme was then immediately killed. Separate Ferguson and Massey-Harris sales organizations in the United Kingdom were accompanied by separate engineering departments, and so in important respects the merger there simply did not take place until the later 1950s.

As in France, operations in the United Kingdom from 1954 to 1957 gave strong and timely support to the financial position of the parent company. No large capital expenditures were made, and no significantly new products were introduced, but sales of both Massey-Harris and Ferguson products expanded. This was a pleasant surprise considering the worries Massey-Harris executives were having in 1953 about sales of machines produced in the United Kingdom. United Kingdom operations in 1954, 1955, 1956, and 1957 made profits before taxes of $4.4 million, $6.5 million, $5.8 million, and $8.9 million.

Of long-run significance was the experience of the company in its relations with Standard Motors. Ferguson personnel in the United Kingdom had long known that great care was required to maintain a satisfactory working relationship with Standard, particularly since Sir John Black, President of Standard, and Harry Ferguson did not get along. Massey-Harris personnel and some North American Ferguson personnel as well did not immediately realize that the relationship, to be satisfactory, required more than a show of authority on their part. Relations with Standard were immediately complicated. At the end of August 1953 Duncan called on Sir John Black and discovered that the latter had not even heard of the merger until he read about it in the newspaper. Continued production of Ferguson tractors after the conclusion of existing contracts was important to Standard. Duncan, in return for an assurance to that effect, negotiated a £20 reduction in the price they paid Standard for the Ferguson tractor. In spite of this a long period of disagreement followed and Massey-Ferguson became convinced that the price it was paying for its tractors was too high. Disagreement did not disappear until Massey-Ferguson purchased Standard's tractor manufacturing facilities in 1959. Behind that disagreement and eventual acquisition is a complex and fascinating story which we shall recount in chapter 13.

Why was it that these three Eastern Hemisphere division operations performed so much more profitably and smoothly after the merger than did the North American divisions? Obviously, of course, market conditions there were still buoyant whereas in North America the post-war expansion had ended. But there were other reasons. Conflict between Ferguson and Massey-Harris distribution systems was more successfully avoided. In Germany the Massey-Harris company assumed responsibility for marketing the Ferguson products, and France was the only country besides Canada where Massey-Harris and Ferguson dealers were permitted to become part of a single system. In the United Kingdom, Ferguson dealers were content to sell Ferguson tractors and Massey-Harris dealers, not having sold any significant quantities of Massey-Harris tractors, could not really object; at the same time Massey-Harris dealers continued to market the much-demanded self-propelled combine, to which Ferguson dealers could not object since they had never had their own combine. But conditions in North American operations were quite different, as we shall now see.

After the San Antonio conference it was vitally important to bring the TO-35 into production at Detroit, for without it the company would be seriously vulnerable in the tractor market. A. A. Thornbrough of the United States company was responsible for seeing it put into production. It will be recalled that in February he had indicated the substantial savings that would be involved in sourcing the major tractor components from Standard at Coventry, England, rather than from Borg-Warner at Detroit, and that in March, at the San Antonio conference, management endorsed such a sourcing policy. However, by about the time of that conference it had become uncertain whether Standard would be able to tool up for the TO-35 components in time for the 1955 spring season, and so Thornbrough began to cast around for an alternative solution to the sourcing problem. By mid-May it was clear that Standard could not supply the component parts until the fall of 1955. What this meant for the company and how Thornbrough thought the problem should be overcome is indicated in his memorandum headed "Interim Ferguson TO-35 Tractor Program" sent to Bloom on May 22, 1954. It said in part:

> I wish I could state that delay of the TO-35, although serious, would not greatly damage the Ferguson distribution. I cannot help but feel in view of the reports coming in from the field that by delay we face the strong probability of most seriously weakening the Ferguson distribution to the point where the TO-35 cannot be capitalized upon in late 1955 and early 1956. . . .
>
> Because of this probable sales organization deterioration, I wish most strongly to recommend the adoption of an interim program for TO-35 to be undertaken at Borg-Warner for approximately 25,000 TO-35 tractors if we can successfully negotiate such a program, which I strongly believe can be done. . . .

After Thornbrough made a detailed cost study of the interim tractor programme, it was decided to adopt it, that is, to continue for a period to obtain major tractor components for Detroit assembly from Borg-Warner. This decision, by the way, also avoided the real possibility that Borg-Warner might submit huge cancellation claims if components were purchased from Standard before expiration of Massey-Ferguson's contract with that company.

Most of the parts for the TO-35 were new and so posed complicated sourcing problems. Thornbrough almost lived at the Borg-Warner Kercheval plant at Detroit to keep the programme on schedule and he had with him the same group of people who had succeeded in getting the TO-20 out in record time in 1948. Pilot models of the new TO-35 tractor were actually available for the distributor conference of January 1955. It was this project and Thornbrough's performance in bringing the TO-35 forward that probably brought his abilities to the attention of the President. Later when some members of the Executive Committee insisted that the President appoint an Executive Vice President to assist him and further insisted that he should not appoint certain old-time Massey-Harris executives, the President appointed Thornbrough.

The implications of the San Antonio conference for engineering activity within the company were important. Should the controversial TO-60 Ferguson tractor, so long in the planning stage, finally be put into production? This the San Antonio conference had left unsettled, pending information regarding its durability when run at 2,000 r.p.m. – a speed necessary to give it sufficient lugging power. There were also continued misgivings as to its unsightly appearance when changed to a three-wheeled model for row crop cultivation and corn picking. On April 22, H. H. Bloom wrote to J. S. Duncan explaining that, on the matter of speed and its possible harm to the engine, there had developed a difference of opinion between the United Kingdom Ferguson engineers on the one hand, and the engineers of Continental Motors and of Massey-Harris on the other. The latter held that there was sufficient doubt about it that it would be a grave mistake to introduce the tractor without extensive tests.

In commenting on this memorandum to E. W. Young, Duncan again expressed his unhappiness over the TO-60 because it did not have the power range required in the western States or western Canada, it was not suitable for mid-mounted cultivation, it could not be sold for corn picking as a regular four-wheeled machine, and at least one competing tractor had a more favourable price. He ended by saying: "... All this adds up to small sales, and will be a grievous disappointment to the Ferguson distributors, because, following the U.S. Conference and Mr. Klemm's talk, they are expecting a mid-mounted and corn picking tractor, and also one which will be acceptable to the West. ..."

Herman Klemm was asked to reanalyse the TO-60 tractor and he sent his report to the President on June 10, 1954. The report was unfavourable to the TO-60 on many counts besides the one of engine speed. It recommended that the TO-60 project should be shelved, and that its place should be taken by another tractor, referred to as the MH-70; it would be more modern than the TO-60 which after all, by that time, was approximately six years old. And so the TO-60 was dropped, to the bitter disappointment of many former Ferguson people in the Coventry organization. What a waste of time, money, and talent it all had been!

There can be little doubt that the TO-60 was unsuitable for the United States market. It stands as a classic example of the consequences of engineering development being too far removed from market considerations and also of the consequences of delay in bringing new machines to market. Had the TO-60 been introduced in the early 1950s, following the phenomenal success of the TO-20, there is little doubt that in many parts of the world it would have been a commercial success.

Having discarded the TO-60, the company's tractor engineering resources were concentrated on developing, as quickly as possible, the already discussed TO-35 tractor for the Ferguson distributors and dealers and an equivalent tractor, subsequently called the MH-50 tractor, for the Massey-Harris dealers. But with the TO-60 dead, a larger tractor was badly needed and so it was decided to develop the MF-85 (actually called at the time the TO-70) and a Massey-Harris equivalent as well. It was recognized as early as August 1954 that when the MH-50 was made available to Massey-Harris dealers it might provide unduly hard competition for the TO-35 tractor retailed by Ferguson dealers, and that therefore work should begin on a successor to the TO-35, a tractor referred to as the TO-40. The principal differences between the Ferguson tractors and their Massey-Harris opposites were to be ones of sheet metal, colour scheme, and other inconsequential characteristics. Ferguson tractors would be painted grey, and Massey-Harris tractors would be painted red. As far as implements were concerned, certain Ferguson implements were adapted for incorporation into the Massey-Harris line, and certain Massey-Harris implements were adapted for the Ferguson line. It was in this way that two lines of competitive equipment were to be created, lines that were very substantially alike but which, for lack of engineering talent and capital could not be made available to both systems of dealers simultaneously at all times.

Complaints from Ferguson distributors and dealers, and to a lesser extent from Massey-Harris dealers, did not disappear with the introduction of a policy of juggling implements between the two lines. Cross-franchising of dealers was soon strongly objected to by dealers and distributors and on July 2, 1954, the company had to announce that after that date cross-franchising would be permitted only in unusual cases, and only if Massey-Harris branch management and Ferguson distributor management were in full agreement that the cross-franchise would be in the best interests of both of them. Even this move did not remove complaints and objections, particularly those emanating from the independent Ferguson distributors and their dealers. Consequently, on April 25, 1955, the company went a step further and completely prohibited further cross-franchising of dealers. Still complaints persisted. They reached crisis proportions when the Ferguson distributors began to consider the implications for them of the introduction of the new MH-50 tractor to Massey-Harris dealers. On April 4, 1955, J. L. Hooker, the General Sales Manager of the Ferguson line in the United States, wrote to H. H. Bloom stating that introduction of the new Massey-Harris tractor would so impair sales of the TO-35 that they might as well liquidate the selling effort represented by non-cross-franchised Ferguson dealers. His view was that they could not successfully carry through the sales year of 1956 with the TO-35 competing against the MH-50.

As the date for the introduction of the MH-50 neared, the company became increasingly aware of rumours among distributors and dealers that the Ferguson line of tractors and implements would soon be withdrawn

This MF combine equipped with an eight-row corn head can pick and shell 1,000 bushels of corn an hour.

from the market. The effects of these rumours on the morale of the Ferguson sales organization were devastating. Special letters were sent in November 1955 to distributors to dispel the rumours but this was unsuccessful. Difficulties also arose because Ferguson distributors thought that district managers and other members of the Massey-Harris branch organization were openly soliciting orders for the MH-50 tractor from cross-franchised dealers to the detriment of TO-35 sales, and on February 10, 1956, the company had to forbid this activity on the part of the Massey-Harris branch organization. More and more effort had to be expended in explaining what could or could not be done by members of the two sales organizations. The objective of maximizing sales and profits of the organization as a whole became increasingly blurred.

By February 1956, after the MH-50 had been in the market for two months, the crisis so long obvious within the company reached a critical stage. On February 22, fifteen out of the nineteen independent Ferguson distributors demanded by telegram that the company eliminate the competitive lines of merchandise from the Massey-Harris division, which in effect constituted a demand for withdrawal of the MH-50 tractor and its complement of implements, or for giving Ferguson distributors an exclusive franchise for them. In correspondence some distributors also mentioned a third alternative, that of the company buying out the independent Ferguson distributors. Massey-Ferguson could not comply with the demands placed upon it and even though a number of the signatory distributors subsequently said that they were not in accord with the telegram, there was no doubt at all that the company's two-line policy was in a shambles. It was after this revolt on the part of the Ferguson distributors that J. S. Duncan wrote a memorandum to A. A. Thornbrough (Executive Vice President since October 1, 1955) in which he said:

Like all of us, I have been giving a great deal of thought to the reasons back of the upheaval in the Ferguson organization.

There are many, and although I still feel that the consequences will be less serious than they look at present, the fact remains that the impact of the MH-50 and its similarity to the TO-35 has been greater than either we or our distributors expected, and the fact is not yet proven whether we can effectively sell the two tractors competitively over the years, although I still hope that we can if we prove ourselves sufficiently good merchandisers.

On analysis, the difficulty, of course, is that fundamentally there is not sufficient difference in the two tractors to provide our respective organizations with adequate selling points, so that competition is apt to degenerate into price cutting.

To have done the job properly, we would have required a different engine or transmission, plus greater appearance differences in the hydraulic controls.

Perhaps we have made a mistake in not facing up to this problem, and at the cost of delay in coming out with the MH-50 made sufficient fundamental changes to facilitate the sale of the product without causing unacceptable manufacturing disadvantages.

We have been talking about the successful way in which the automobile people have sold similar products competitively, but it looks as if we have gone so much further than they have in our similarity that we have, in some measure, defeated our own objective.

Some of the Ferguson distributors tell us that had we just gone a little further in appearance changes, the present trouble might have been avoided. I doubt that it would have been seriously diminished, but on the other hand, it would undoubtedly have helped matters somewhat if we had gone a little further than we have.

The important point in all this is that we should not fall into the same error twice – once is excusable, twice is not. Our present experience will probably cost us the sale of 5,000 tractors and a number of good dealers, plus an upheaval in our distributor organization. This should be enough of a lesson to all of us – and I say all of us because we were all in accord with the action which was taken and are therefore all equally responsible.

This brings me to the large tractor, and it is in this connection that I say a second error in judgment of the magnitude of the present one cannot be made.

It is, therefore, essential that Mr. Klemm's department should immediately prepare a dual project:

(a) Design changes in the 65 and 75 which are of a sufficiently important nature to give our sales forces real talking points, but which retain, to the greatest degree possible, basic component inter-changeability (a different engine could, however, be usefully considered).

(b) Design changes of a superficial nature, but going much further than in the TO-40 and MH-50. Here the recommendation would be how to modify the outward appearance of as many of the lesser components as possible without changing their functions or centres.

This was a surprising memorandum quite apart from the concept of "collective responsibility" that it espoused. It showed that parent company management had not come to the conclusion that the two-line policy was basically an unworkable one for the company. It favoured pursuing that policy even more vigorously. This view rested on no comprehensive study of the economics of such an approach, nor was such a study contemplated. It ignored, in a way a competent study could not have done, the cost of the engineering effort required to make it work. It ignored the possibility that multiple-line policies in the automobile industry were tenable because superficial product differentiation has much more appeal for purchasers of consumer-oriented products than for purchasers of capital goods; and it also ignored the very real possibility that the size of the automobile industry and the individual multiple-line companies in it were such that even individual lines could be produced at near optimum levels of output. Finally it suggested that fundamentally new approaches were not likely to come quickly from existing management.

As the company's distribution policy disintegrated, the serious consequences of that disintegration for sales volume became clear; and to make matters worse sales were further depressed by adverse economic conditions. We have seen that there were fundamental problems in company: absence of serious concern over matters of organization, control, and planning, and a high degree of centralization of authority; the high costs of manufacturing and the heavy reliance on sub-contracting; the failure really to merge, as indicated by the dual distribution system and two-line concept; the demoralization of the sales force; the failure to develop executives with professional business training, and the continued emphasis on the superiority of the "self-made man." But it was the increasing concern in 1956 over sales, profits, inventories and liquidity, and over the ability of management to deal effectively with them, particularly since the President was suffering from recurring illnesses, that induced the Board in early July to effect a change in management. It was also in this area of the company's performance that a fundamental difference of opinion with respect to policy developed between the President, J. S. Duncan, and his Executive Vice President, A. A. Thornbrough. Late in 1955 and early in 1956 mistrust and suspicion between the Board and management, and within senior management itself, seems to have emerged as a significant disturbing force.

Over the years relations between Duncan, who held the position not only of President but also Chairman of the Board, and W. E. Phillips, Chairman of the Executive Committee of the Board, seem to have been very cordial. However, Duncan seems to have been quite sensitive to any implied criticism, and he was upset because he thought that the Argus group were attempting to strengthen their position in management matters. This is how he seems to have interpreted the formation of the Executive Committee itself in 1948, the over-ruling of his views on dividend policy, and also the instruction given him by Phillips and another Director at a meeting in the York Club of Toronto in 1955 to appoint an Executive Vice President who would be senior to H. H. Bloom. However, it was Duncan himself who chose Thornbrough for that position, an appointment approved by the Executive Committee on October 5, 1955.

The chronology of events leading to the change in management can begin at about the time of that appointment. Duncan went on a trip to the U.S.S.R. in late October and returned at the end of November a sick man, suffering from phlebitis. He was frequently to be sick over the next six months which undoubtedly aggravated the difficulties of this organization – one that had for years been centralized on the dominant personality of the President. One also senses that Duncan did not realize the extent to which, because of illness, he became isolated from the details of the difficulties that the company was undergoing.

In mid-December Duncan asked Thornbrough to take charge of the 1956 budget that had to be presented to the Board in mid-January. When Thornbrough began this work he discovered that three senior executives needed for the task were away on a cruise without his knowledge or permission. It seemed to confirm his suspicions that he was being by-passed in his new position at Toronto. The budget presented to the Board on January 26 was based on the exceedingly optimistic sales forecast of 11 per cent over 1955 for the North American and United Kingdom operations, and it estimated that there would be an increase in profits in those operations. At that same meeting the President referred to the developing shortage of working capital because of increased turnover and needed capital expenditures: in March a $21 million issue of 4½ per cent debentures was sold to take care of that problem.

February figures began to show how over-optimistic the 11 per cent 1956 sales forecast had been; far from exceeding sales of the equivalent period of 1955, actual sales were 16 per cent below 1955 figures and 29 per cent below budgeted 1956 figures. Inventories in North American and United Kingdom operations in February were 33 per cent higher than the year before. This pattern of sales well short of budget and inventories substantially higher did not change over the next several months. At the end of June sales in North America and the United Kingdom operations continued to be far short of the original budget (revised budgets were introduced along the way) and were running 4 per cent below actual sales over the equivalent period 1955. Inventories in those operations were running 39 per cent higher than a year ago, and worse still they contained an abnormally high proportion of obsolete finished goods. Losses were being experienced in the Canadian and United States operations and profits in North America and United Kingdom combined, far from running higher than the year before, stood at $.4 million, compared with $4.4 million the year before. In brief, the financial position of the company in early 1956 was about as ominous as it was to be in June. Even inventories stood at $131 million in June, or $36 million higher than the year before. The worst situation existed in the United States where in June inventories were $21 million higher than the year before, much of it in the form of obsolete Massey-Harris tractors and partially completed Ferguson tractors, and sales were 3 per cent lower. Bank advances which in February had been $17 million higher than the year before, and which it was thought, would be reduced with the proceeds of the new debenture issue, stood at $50 million, or $36 million over 1955, at the end of June. Estimates of cash requirements proved to be very inaccurate, and in that sense the company could be thought of as having been financially out of control.

What was the reaction of the Board and of management to these developments? The Executive Committee of the Board gradually began to take a closer interest in the operations of the company. It will be recalled that Duncan appointed Thornbrough Executive Vice President in the fall of 1955 after the Executive Committee had insisted that someone be appointed to such a position. They followed this up with enquiries to satisfy themselves that Thornbrough had in fact been given adequate authority in that new position. Then in December 1955, while Duncan was recuperating in Jamaica, E. P. Taylor asked Thornbrough to join him, M. W. McCutcheon, and J. A. McDougald for lunch. At that meeting the four discussed the company's many problems in a general way. Thornbrough, knowing how such a meeting might be interpreted by the President, asked permission to inform Duncan of it. Permission was granted and Thornbrough wrote to him. Fairly soon after that Duncan returned from Jamaica, only to fall ill again. In early March he went again to Jamaica and he was able to return before the end of the month. Taylor, meanwhile was becoming increasingly concerned over the inventory build-up. Following an Executive Committee meeting in April, one attended also by Duncan, he took the slightly unusual step of going to Thornbrough's office and asking Thornbrough to come to see him soon and tell him about the inventory problem. That meeting never took place since Thornbrough had to leave for Europe, but the request for it, together with Taylor's December meeting with Thornbrough, seem to have convinced Duncan that the Board was by-passing him. This undoubtedly complicated the situation, in the same way as it had when Thornbrough came to the conclusion that senior executives were by-passing him and going directly to Duncan.

Nor was this all. Duncan and Thornbrough began to disagree on what should be done to meet the problem of excessive inventories. It was Duncan's firm view that sharp price reductions were not the answer. In this view he was supported by all of his long-time Massey-Harris executives. In early April, at a meeting of his senior executives, with Thornbrough sitting to his right, he went around the table asking each one for his views. All, except Thornbrough, supported him. Thornbrough favoured major price reductions, particularly on obsolete tractors, as an attempt to move excessive inventories in the early selling season. He was over-ruled and with that his position as Executive Vice President became more or less meaningless. A policy of generally minor price reductions and substantial cuts in production and in expenses was pursued. As noted earlier, this did not lead to visible improvement in the company's sales or in its cash position.

In the meantime Phillips had become quite concerned over the company's performance in manufacturing and production control. He had favoured bringing in a firm of consultants to consider the matter, and it is at least a reasonable guess that he did so partly to obtain information in an indirect way about the state of the company. In a Board meeting in January he also suggested that the financial information presented to the Board might be arranged so as to give an indication of the manufacturing performance of the various operating units. It is true that the format of the reports to the Board made it difficult to form a reliable impression. But in an environment of growing suspicion it was also true that any such suggestions could be interpreted by management as an encroachment on its prerogatives.

On February 22, 1956, Thornbrough wrote a memorandum to the President in which he made recommendations about the company's manufacturing operations. These recommendations envisaged an end to Batavia as a manufacturing centre, a new plant at Brantford for the manufacture of combines, eventual abandonment of Woodstock, and also of the Market Street factory at Brantford. Later he recommended establishing a position of Vice President Manufacturing, and filling it with an outside executive. It is indeed curious that these recommendations, which would have been as relevant in 1952 as they were in 1956, had to wait until 1956 to be made. Worse still when made in 1956 they appear not to have been considered seriously. The suggestion of bringing in an outside man would certainly not have been well received.

The President unfortunately became ill again in mid-April and underwent an operation. A month later he had recovered sufficiently to attend a meeting of the Executive Committee of the Board convened on May 16 at the suggestion of Taylor. It was at that meeting that the directors really had to decide whether they would support Duncan or Thornbrough, for Taylor had called it especially to discuss whether or not the company should attempt to reduce its inventories and improve its cash position by cutting prices.

Thornbrough had not been asked to attend, although at least one executive junior to him had been. Duncan reiterated his view that sharp reductions in prices at that time would not be desirable for various reasons, including the risk of stimulating a price war. Taylor, who of course knew that Thornbrough disagreed, asked that he be brought in to explain his views. This was done. The meeting then adjourned. In retrospect it seems that the Argus group decided at that point that they favoured the Thornbrough approach and that there would have to be a change in top management.

But they did not act immediately, and for an interesting reason. Just about that time Massey-Ferguson had learned that Deere and Company might be interested in a merger. Meetings were held, although Duncan was again ill and could not preside, but in the end nothing came of it. One may at this point justifiably ask a question: why, if the position of the company was as unsatisfactory as they believed, did the members of the Executive Committee not act sooner? After all, the financial deterioration of the United States company can be traced as far back as 1951 or 1952 and on its regular business it was running a loss in 1954. There are a number of possible reasons for this. As long as profits from manufacturing and sales outside of North America were offsetting deteriorating results in North America, it was easy to argue that such was a pattern that an international company should expect. Also, the kind of information provided to the Executive Committee and the form in which it was presented made appraisal of performance difficult and the Executive Committee did not press for more meaningful information until the situation had become serious. Perhaps most important the Committee did not choose to become intimately acquainted with the details of either the great opportunities or the potentially fatal "pitfalls" created by the 1953 merger. Nor is it a simple matter for an Executive Committee to become involved in operations unless someone in it is prepared to accept the burdens of becoming Chief Executive Officer. At sixty-three Phillips did not relish the thought. A change in top management also involves "nasty" publicity for a company, troublesome and costly separation arrangements, and a blow to team spirit.

But finally the financial difficulties of the company in North America brought an end to rationalization. There emerged the serious possibility that a failure to act might cast grave doubts over the ability of the Argus group to provide necessary management guidance to this or any other operation. So the decision was made. Phillips was delegated to tell the President. On Friday, June 22, Phillips telephoned Duncan at his home, where he was convalescing, to arrange for an appointment at 11.00 a.m. on Monday, June 25. On Monday they met, these two men, who in the dominance of their personalities had more in common than either would probably ever admit. At that meeting Phillips indicated among other things that he and his colleagues had come to the conclusion that he, Duncan, no longer enjoyed the good health necessary for him to carry on, and that they had decided that another person should be brought in. Duncan for his part replied that he had been expecting the move from the Argus group for some time, that he felt his health would enable him to carry on, and that he could not carry on as Chairman without authority. That settled it, for Phillips was not about to change his mind.

A meeting of the Board of Directors was called for July 6 at 10.00 a.m. Separation and retirement agreements were approved at that meeting. Following that, James S. Duncan resigned as Chairman, President, and Director, and left the company after forty-six and a half years of service. He had made great contributions to the company. They were made in earlier years. He succeeded in keeping the company in the United States market when the Board wished to pull out. He supported the development, the production, and the marketing of the self-propelled combine. He maintained the international fields at a time when even the Board favoured restricting its operations to North America. He stretched the company's facilities to catch the post-war demand. He succeeded in catching the fancy of Harry Ferguson in a way that paved the road to merger.

But as the company increased in size and complexity, particularly after the merger in 1953, the whole approach of another age to management, to organization, and to control no longer sufficed. Times change and organizations must change with the times. Individuals often cannot do so, and they cannot see that they are not doing so. It is a common failure. After the Board meeting of July 6, W. E. Phillips was appointed Chairman and accepted the new burdens of Chief Executive Officer; E. P. Taylor became Chairman of the Executive Committee; and A. A. Thornbrough was confirmed in his position as Executive Vice President. Within several months a number of the company's Vice Presidents had gone, and before the process was complete many other members of management had left, their places gradually taken by a new breed of men.

CHAPTER 10

New foundations: Organization and control

What does one do first after the sudden, and to the majority of the company's employees, unexpected and surprising, departure of a President? The various operating divisions had of course immediately to be notified. W. E. Phillips, the new Chairman and Chief Executive Officer, sent a cable to them on July 6, 1956, the day of the management change. The one to the Eastern Hemisphere Division read:

MEETING BOARD OF DIRECTORS TODAY REGRETFULLY ACCEPTED MR J S DUNCAN'S RESIGNATION AS CHAIRMAN, PRESIDENT AND DIRECTOR OF MASSEY-HARRIS-FERGUSON STOP NO PRESIDENT HAS YET BEEN APPOINTED STOP I HAVE BEEN ELECTED CHAIRMAN OF THE BOARD AND PENDING THE APPOINTMENT OF A PRESIDENT WILL ACT AS CHIEF EXECUTIVE OFFICER OF THE COMPANY STOP THE ORGANIZATION WILL CONTINUE TO FUNCTION THROUGH THE USUAL CHANNELS OF COMMUNICATION STOP I BESPEAK YOUR CONTINUED LOYALTY TO THE COMPANY STOP TOGETHER WITH J A MCDOUGALD WILL MEET YOU SHORTLY AFTER JULY FIFTEENTH.

On the same day A. A. Thornbrough was confirmed in his position of Executive Vice President, was made a director of the company, and was placed in charge of the company's direct operations. At first the Executive Committee had thoughts of bringing in an outside executive as President, but they soon changed their mind and on December 13 Thornbrough was appointed President. Thornbrough was forty-four years old. He was a graduate in agricultural economics of Kansas State College, had completed his course work for the Ph.D. degree at Harvard University and then, after a period during the war with the Office of Price Administration in Washington and the army (Lieutenant-Colonel), had chosen to go to Harry Ferguson Incorporated at Detroit.

The telegram quoted above correctly suggested that Phillips intended to fill the position of Chief Executive Officer for only a short period. He was sixty-three years old at that time and was looking forward to spending more time at his Nassau home. But he later changed his mind. This was because he became increasingly fascinated with, and intellectually stimulated by, the reorganization of the company, and wished to be as near as he could to that process, and because he found he was able to delegate almost all of the usual responsibilities of Chief Executive Officer to Thornbrough. His influence was exercised not through active day-to-day participation, or even usually through the formulation of detailed policy, but rather through the impact of his ideas, particularly those relating to managerial concepts, on policy formation, and on operation. He remained as Chairman of the Board and Chief Executive Officer until his death on December 26, 1964.

Lines of communication between the parent company and the operating divisions had to be clarified. Thornbrough sent a cable for this purpose on July 9:

ALL COMMUNICATIONS AND RECOMMENDATIONS PREVIOUSLY FORWARDED TO EXECUTIVES OF HEAD OFFICE OTHER THAN TO THE PRESIDENT AND EXECUTIVE VICE-PRESIDENT SHOULD CONTINUE ON THE SAME BASIS; WITH REGARD TO THE COMMUNICATIONS PREVIOUSLY FORWARDED TO THE PRESIDENT HAVING TO DO WITH THE INTERNAL OPERATION OF THE COMPANY THESE SHOULD NOW NORMALLY BE FORWARDED TO THE EXECUTIVE VICE-PRESIDENT UNTIL FURTHER NOTICE.

Steps were soon taken to make separation agreements with H. H. Bloom, First Vice President of the parent company and President of the United States company, and E. G. Burgess, Vice President of the parent company, and the company's senior official specializing in manufacturing matters. Both Burgess and Bloom had lost the confidence of Thornbrough long before July 6.

The preliminary moves were now over, but the problems were still there, only more so, because of the mind-boggling personnel difficulties created by the shattering of previous patterns of personal relations and loyalties. Three years before it had been the "red" versus the "grey." Now, even before that conflict had disappeared, a new one had arisen. It was between executives who favoured major changes and found it easy to accept new top management, and those who were not convinced that major changes were necessary and found it difficult to accept new top management, new policies, and new procedures. Both groups had supporters deep down in the organization and they both included old Massey-Harris personnel. It soon became apparent that the speed with which advanced management techniques and policies could be introduced depended not just on the speed with which they could be formulated but also on the speed with which they were accepted as sensible and desirable by employees of all ranks.

The course of actual development soon became exceedingly complex because of the many areas in which action was simultaneously taken – including inventories, organization, control, planning, manufacturing facilities, marketing, and engineering policy and programmes. It was further complicated by the many regions or countries in which action was taken and in which significant developments in manufacturing operations occurred. To avoid the danger of losing our way among these developments it seems prudent to outline briefly what the major developments were and the order in which they took place.

One of the first tasks was to improve the liquidity of the organization and this required prompt action in North America. The Eastern Hemisphere Division was making profits and seemed relatively well organized, which fortunately permitted parent company management to concentrate on operations in North America. A pattern of action that seldom varied began to emerge: the introduction of new concepts, procedures and policies in North American operations and their gradual application after that to the company's operations outside North America.

First, within North American operations there was the application of rational concepts of organization, the construction of a suitable organizational structure, and the development of an adequate system of planning, supervision, and control. These had to be "rational," "suitable," and "adequate" in the sense that they had to be appropriate for an organization that was internationally dispersed. Appropriate policies and procedures had also to be formulated and implemented for the major functional divisions – marketing, manufacturing, engineering, finance, planning and procurement, personnel and industrial relations and public relations.

A major change occurred in manufacturing in that the company greatly increased its control over the manufacture of its own products through expanding its Detroit tractor plant, acquiring a transmission and axle plant at Detroit, acquiring the tractor manufacturing facilities of the Standard Motor Company at Coventry, England, and the diesel engine manufacturing company of F. Perkins Limited, at Peterborough, England.

By 1959 conditions in North American operations had greatly improved, and these acquisitions and other developments made it essential to concentrate attention on development in European operations. Also in 1959, there appeared the Special Operations Division of the company concerned with the "frontier" of the company's international development, including its projects in Brazil, India, Spain and Turkey, and the Landini operation in Italy.

New management did not have a detailed plan ready for implementation. But some difficulties obviously required immediate attention. When Phillips and Thornbrough first met after the change in management it was agreed that the company's cash and inventory problems, its organizational problems, and the problems generated by the two-line policy had to receive high priority. This in effect meant that short-term and long-term policies and changes would be implemented simultaneously, beginning immediately. Almost immediately Phillips and Thornbrough decided to kill the contemplated MH-50 tractor project at Manchester, thereby effectively halting the move toward two permanent product lines in the U.K. and in export markets. In succeeding months the much more painful move towards one distribution system in North American operations was begun, as we shall see in the succeeding chapter.

Phillips, as his cable shows, decided immediately that he would go to London to investigate the state of the Eastern Hemisphere Division. Thornbrough would stay behind to oversee the company's operations, to improve its liquidity immediately, and to begin to accumulate information and advice about what to do about its demoralized operations in North America. While no comprehensive guide or detailed plan for fundamental reorganization existed, certain views that were later to influence significantly the direction that new management would take, came to the fore even before Phillips went to London. He told his Vice Presidents: "in our business we want less romance and more recompense. . . ." It was put in a way that seemed to set a professional tone for the organization, and its meaning soon became unmistakably clear. Above all it meant that in both the short run and the long run single-minded attention would be given to achieving a significant improvement in the profit position of the company in North America and to using profits as a guide for making decisions. The phrase "profit impact" was to become a favourite one in conversation among company employees and in the planning and control procedures.

The pointedness of this emphasis was new. Back of it lay a feeling about the nature of the company's difficulties that was in marked contrast to the view that had previously predominated. Phillips believed, instinctively at first, that internal mismanagement as well as external adversity explained the unenviable financial position of the company and he further believed that if the former were remedied, profits could be made in North American operations even at existing sales levels.

Putting first things first, the company's liquidity had to be improved, losses in North American operations had to disappear, and a break-even point had to be established. A policy of price reduction and very attractive incentive schemes was introduced, particularly relating to the 5,000 or so obsolete Massey-Harris #33, #44, and #55 tractors. By early September of 1956 Thornbrough could report to the Executive Committee that capital expenditures had been "frozen" (which Phillips defined as "cancelled"); hiring of employees had been forbidden and a policy of staff reduction was under way; overtime was banned; Batavia, Woodstock, and Market Street, Brantford, plants were to be permanently closed in 1957; and operating units were to plan for profitable operations. On September 5, Phillips outlined this objective of a break-even point to the Executive Committee:

The immediate task in the North American organization is to set the establishment in terms of personnel and facilities at such a level that a break-even point shall be reduced to a possible sales volume of 10 million a month in the United States and 2 million per month in Canada.

The operating policy for 1957 had already been defined by Thornbrough in this way:

In summary, the budgeting policy for 1957 will be to achieve maximum sales with major reductions in inventory and a substantial increase in liquidity, and to budget operations on a profitable basis.

By early November 1956 a significant increase in liquidity had been achieved. From June 30 to October 31 North American inventories were reduced by $29 million, much more than over the same period the year before, and so not explained by seasonal factors; and the company's net cash position in North America was improved by over $40 million. Its total bank loans were down by $50 million. The liquidity objective was in sight of being achieved.

But as it turned out the profit objective referred to in the statement on 1957 operating policy was not nearly achieved. Losses in North American operations amounted to $15.5 million in 1957 of which $14.4 million were experienced in the United States business. Realization of a "substantial increase in liquidity" was far more costly than had been expected. This was because the state of the company's inventory was worse than had at first been thought. At the Detroit plant some 5,000 Ferguson tractors had been stored in a field, most of them either incomplete or rejected and all deteriorated by the elements. This had resulted from a decision in early 1956 to clear the Detroit lines and begin production of the new MH-50 tractor. Several thousand combines were stored at Toronto (apparently the result of a policy of stabilizing production), which were later difficult to sell. The side-mounted Ferguson balers and foragers suffered from major conceptual deficiencies, manufacturing faults, and from lack of customer acceptance. Massey-Harris tractors still in stock were rapidly becoming completely obsolescent because of the Ferguson-system trend in the company's tractor models.

One price reduction after another was required to liquidate this inventory, accompanied by much "fire sale" advertising. When the year 1957 was out North American operations inventory had been written down by $13.6 million, and this followed a write-down of $6.5 million in 1956. Additional expenditures of $2 million were required to terminate the franchises of the independent Ferguson distributors, a move which resulted from a basic change in distribution policy. These two items explain most of the loss in North American operations in 1957, and while that year's operations were a disappointment it did appear as if non-recurring factors largely were responsible. It had not been possible to soften losses by utilizing "inventory reserves," for those reserves had all been transferred to surplus in the previous year.

The fact that the losses were non-recurring and that the company had at least experienced a substantial improvement in liquidity explains why the 1957 results appear not to have upset unduly either the company's management or its Board of Directors. Plans for reorganizing the functional divisions of the company and formulating policies for them could be proceeded with. But discussion of this we leave to the next chapter.

The second item on the list of top priorities was organization. It was here that Phillips' influence was particularly apparent. He approached his new tasks with the conviction that the farm machinery business was not fundamentally different from other businesses and that Massey-Ferguson was not fundamentally different from other industrial organizations. A quite opposite view had prevailed in the company for many years, a view probably not uncommon among business-men who have not been exposed intensively to the rapidly developing science of business management and administration.

By believing both that external factors alone did not explain the company's adverse financial position and that the company was not unique, Phillips could simply conclude that the major solution to its ills lay in applying known concepts of organization and management. He believed that the direction to be taken was decentralization of authority with adequate supervision and control. So on July 12, less than a week after he assumed his new position, he presented the Executive Committee with a new organization chart for their approval, here reproduced as Organization Chart 3. This was only a small beginning to what was subsequently undertaken. When released generally within the company on July 24, the chart served the purpose of making it clear that management regarded matters of organization, supervision, and control as being vitally important to the company; and, that in such matters, it was going to pursue more formal procedures than it had previously done. Moreover, as we shall see, that chart embodied several significant new concepts as far as the company was concerned.

Another important principle was adopted by the new management group in those first days, although it was not at that time specifically defined as a principle. The view was held that each position in the company carried with it specific responsibilities, that individuals with ability and experience adequate to meet those responsibilities had to be found to fill the positions, and that those positions should carry with them remuneration sufficient to attract individuals with the necessary qualifications. This principle was subsequently applied in three stages: an increase in salaries to levels reflecting current demand for the kinds of executives that were required; termination of employment of individuals who were thought not to be meeting minimum requirements; and appointment of other individuals. The view that the company was not unique later permitted the new group to go not only outside the company but also outside the industry for new management personnel. Most of its new senior executives were subsequently recruited from other industries – a move that would previously

ORGANIZATION CHART 3

Massey-Harris-Ferguson Limited
July 12, 1956

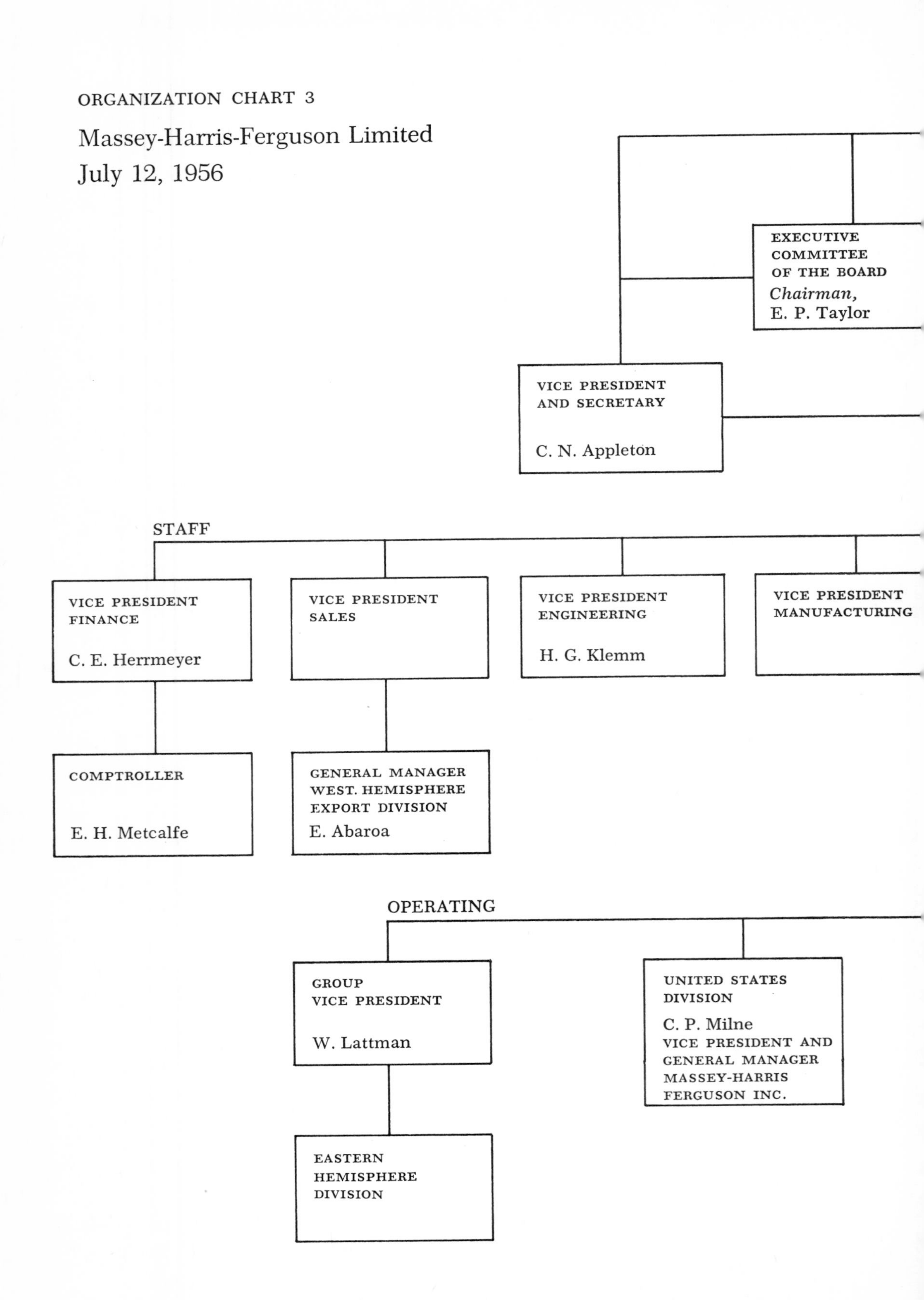

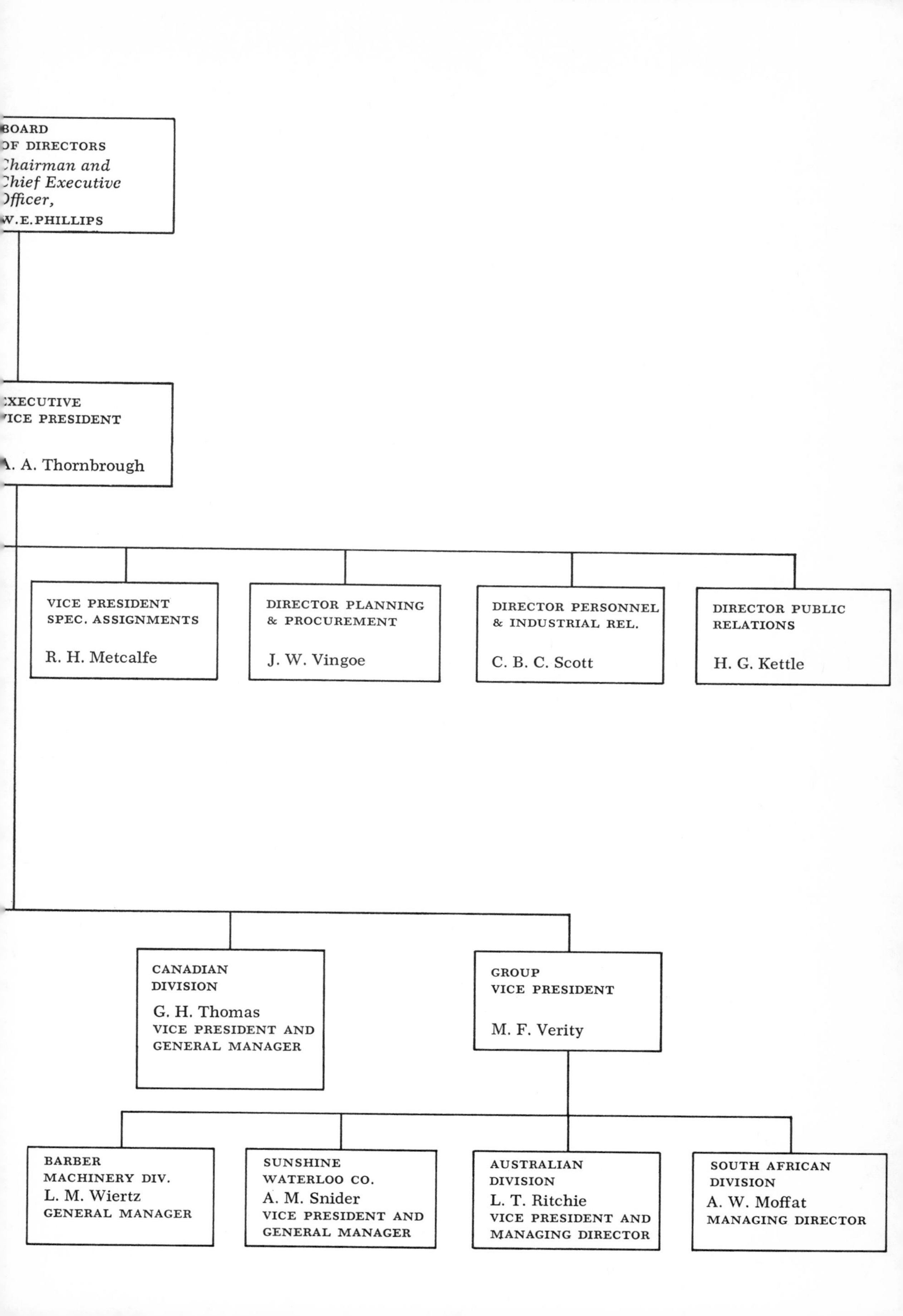

BOARD OF DIRECTORS
Chairman and Chief Executive Officer,
W.E. PHILLIPS
EXECUTIVE VICE PRESIDENT
A. A. Thornbrough
VICE PRESIDENT SPEC. ASSIGNMENTS
R. H. Metcalfe
DIRECTOR PLANNING & PROCUREMENT
J. W. Vingoe
DIRECTOR PERSONNEL & INDUSTRIAL REL.
C. B. C. Scott
DIRECTOR PUBLIC RELATIONS
H. G. Kettle
CANADIAN DIVISION
G. H. Thomas
VICE PRESIDENT AND GENERAL MANAGER
GROUP VICE PRESIDENT
M. F. Verity
BARBER MACHINERY DIV.
L. M. Wiertz
GENERAL MANAGER
SUNSHINE WATERLOO CO.
A. M. Snider
VICE PRESIDENT AND GENERAL MANAGER
AUSTRALIAN DIVISION
L. T. Ritchie
VICE PRESIDENT AND MANAGING DIRECTOR
SOUTH AFRICAN DIVISION
A. W. Moffat
MANAGING DIRECTOR

have been unthinkable. The first step was taken on July 12, for on that day the Executive Committee of the Board approved substantial salary increases for senior executives.

The second step, the termination of employment of executives who were found wanting, was in every way more difficult than the first stage. Its implementation began as one senior executive after another was required to leave, and it filtered through to lower levels until perhaps well over one hundred members of the managerial ranks, one way or another, had departed. The process, before it was over, had extended to every one of the company's operating divisions around the world, and had taken at least six years to complete.

Necessarily, of course, the third stage, that of bringing in new management personnel, moved closely with the second, although finding new executives was a most difficult task and introducing programmes to develop them internally provided no immediate relief.

There was another, most significant attitude of the members of the new group that influenced the approach they took to solving the problems of the company. This was their willingness, even eagerness, to go outside the company for advice and assistance if their own resources or abilities appeared inadequate. Not only were executives freely sought outside the company, but on a number of occasions consulting firms were commissioned to undertake major studies for the company. In the end the company was greatly affected by these "outside" influences.

The introduction of new concepts of organization and control began as already noted with the appearance of an organization chart on July 12. That chart was important first of all simply because it appeared, for none had been issued for some time and none existed that portrayed the then existing structure of the company. Indeed, when the matter of the need for a chart arose immediately after the change in management, A. A. Thornbrough, the Executive Vice President, had to sketch one in his own hand based on a general impression of the actual, as distinct from the formal, structure of the company and of the responsibilities of its senior executives.

But there were other reasons why it was important. It introduced into the company for the first time the concept that staff (or advisory) and line (or operational) responsibilities should be separated as explicitly as seemed possible. Previously the parent company had attempted to distinguish, in its own structure, between co-ordinating and policy divisions and operating divisions with the former roughly equivalent to the staff positions shown on the July 1956 chart, and the operating divisions identical to those shown on that chart. But since, for example, the operating divisions of the Sunshine Waterloo Company, Australia, South Africa, and Barber Machinery Company were shown in 1954 as being the responsibility of one of the Vice Presidents holding a "co-ordinating and policy" or "staff" position, the confusion between line and staff concepts was embodied in the organizational structure. In July 1956 this was clarified by appointing a Group Vice President who was responsible in an operational or "line" sense for that group of operations, and another one for the Eastern Hemisphere. Those operations could have been made directly responsible to the Executive Vice President but the course chosen had the advantage of easing the burden on that executive.

The July 1956 chart also introduced the idea of defining parent company "staff" positions in functional terms so as to include all the major functional divisions. This involved establishing the new positions of Vice President Sales (later called Marketing) and Vice President Manufacturing. In 1954 there had been a Vice President Manufacturing, but the position seems simply to have vanished with the reassignment of the individual holding the position – an example of the way the organizational structure had tended to revolve around individuals rather than individuals around the organizational structure. The staff positions, unlike the operating divisions, shown on the July 1956 chart, changed little after that – Marketing, Manufacturing, Engineering, Finance, Planning and Procurement, Personnel and Industrial Relations, and Public Relations becoming the "hard core" functional divisions of both the parent company and of the various operating divisions. The standardization of functional divisions between the parent company and each of the operating divisions was eventually achieved and proved to be a most useful technique for the parent company to train executives to co-ordinate and control the operations of the various divisions and to render advice to them.

Finally the chart shows that the company retained its long-standing practice of having major policy and certain operational decisions passed on to management through an Executive Committee of the Board rather than through the Board of Directors itself. The Executive Committee operated generally by approving or rejecting management proposals that were required to be submitted to it, or by referring them to the Board of Directors. Matters to be presented to the Executive Committee in 1956 and later on included advertising budgets, donations, audit and legal fees, certain price changes, budgets, labour agreements, progress with new products, use of consultants, investments, certain salary matters, and all capital expenditure proposals.

While the July 1956 chart introduced important principles of organization, it was merely one step in establishing a rational organizational structure. Its major short-coming was that it was not accompanied by any memorandum outlining the theory of organization that it represented. There had not been time for that. Charts by themselves do not ensure that a corporation will have an adequate system of organization. Positions needed specifically to be defined so that individuals could clearly understand the nature and area of their responsibilities; staff-line relationships had to be defined, as clearly as possible, to ensure that responsibility for performance would not fall between the stools of "adviser" and "operating" executives; arrangements that would ensure co-ordination between the functional divisions and operating divisions had to be evolved; criteria for performance had to be established to permit judgments to be made of results achieved by individual executives; decisions had to be made relating to the division of responsibility between the parent company and the operating divisions. Of overwhelming importance, a system of planning, supervision, and control with the necessary flow of information that that implied, had to be devised to permit the parent company to plan future operations and continuously to appraise the performance of the operating divisions against over-all corporate plans and objectives.

It will be seen from Organization Chart 3 that the Eastern Hemisphere Division was regarded as one operating division with a Managing Director in charge answering to a Vice President of the parent company, who in turn answered to the Executive Vice President (after December 1956, the President). Since the Eastern Hemisphere Division consisted of the United Kingdom, French, and German operations and of Eastern Hemisphere Exports, and was partly responsible for South African operations, it really meant that there were a number of executive layers between those divisions and the parent company – decentralization with a vengeance, in a sense. It could work only if the parent company had confidence in Eastern Hemisphere executives. After Phillips came back from London he reported that, in his view, that division would be able to operate with a minimum of parent company supervision under clear-cut policy. This greatly simplified immediate problems of organization and operation, for it meant that the parent company executives could devote the greater part of their time to North American problems.

A further major step in developing the organizational structure of the company was taken on July 30. On that day Thornbrough announced the formation of a Central Co-ordinating Committee composed of senior parent company executives with the Executive Vice President (since no President had yet been appointed) as Chairman. The purpose of this committee was to advise and assist the Executive Vice President (*a*) in establishing over-all company policy, (*b*) in co-ordinating the execution of established policies and programmes, (*c*) in facilitating communication to responsible executives and to all levels in the world-wide operations. There had been parent company operating and co-ordinating committees long before this, beginning in 1944, but the difference was that this new one placed greater emphasis on the role of co-ordination, which it was able to do since its members represented all the "functional" divisions in the organization. Nor was it a sort of club for the elite; it was intended (and so it happened) that individuals other than members of the committee would attend and contribute from time to time. It was also much more permanent than its predecessors, for it was to meet quite frequently and regularly, and minutes of proceedings were to be kept. Both its character and its procedures therefore indicated that it was intended to fulfil a definite and permanent role in the organizational structure of the company. In time all operating divisions of the company as well as the parent company adopted the device of the "Co-ordinating Committee."

It is important to note that the use of this committee, and any other committee in the company, at the insistence of Phillips and also Thornbrough, was never permitted to blurr lines of responsibility. Its purpose was to communicate information and opinions and generally to facilitate co-ordination between the functional divisions and between the various operating divisions. Individuals, not committees were regarded as vitally important for improving the quality of decisions ultimately taken.

The Chairman chose on September 5, 1956, to report to the Executive Committee on matters of organization. The report is interesting because it reflects views that were beginning to prevail, and also, because it indicates his views on failings of the company over past years:

In matters relating to organization, we have made good progress. Much remains to be done, but, at any rate, it can be reported that the general principles of organization have been stated and appear to be understood and acceptable to the organization world-wide.

It would be my basic intention to decentralize the organization as far as that is reasonably practical and as long as we have at our disposal effective and practical means of supervision and control. I repeat what I said at the meeting of August 10th. . . . that this problem in the Eastern Hemisphere organization appears to be not only attainable but desirable.

I have, however, some reservations in regard to the U.S. Company. At the moment, our view is hardening that the U.S. Company, whilst maintaining in every apparent sense its characteristics as an American Company, must in reality be operated as a division of this Company. Further recommendations in this connection will be brought to the Committee. . . .

Based on my experience of the last two months, I am unable to attribute the present difficult position of the Company solely to the decline in the market for farm machinery. This has, of course, played an important part, but, in my opinion, the following additional factors have made the major contribution:

1. The lack of any clearly stated, general policy, long and short term, stating the objectives and governing the operation of the Company.
2. The almost complete failure to apply any of the well known and generally accepted principles of business organization.
3. The absence of any effective communication between the various executive levels within the Company and its many divisions.

The reference to the United States company was related to a growing dissatisfaction with the way management there was responding to suggestions being made, and of the results they were achieving, and also to a fundamental organizational matter – whether there should be separate Canadian and United States operating divisions as shown on the July 26 and earlier charts. It was becoming clear that the problem of individuals adjusting to the policies and procedures of new management was going to be much more difficult than at the time of the merger.

Thornbrough had not been at all content with the attitude that a number of long-term employees were taking toward new policies, and on the day before this meeting of the Executive Committee he wrote a memorandum to Phillips on that subject:

For approximately one month following first stages of waiting out major problems upon the retirement of Mr. Duncan, I have been pushing things to subordinates, giving policies and decisions to the fullest extent possible.

I shall now enter into a new phase of relationship to subordinates. This would require first of all to step up the pace and to weed out those who are not going to go along. Secondly, it is required in order to get results.

Phillips agreed that "... the time has now arrived when action must take the place of inquiry or investigation.... What we urgently need are a few competent people with some talent for the executive functions."

An immediate trouble spot was the Detroit factory and Phillips, on September 10, made a suggestion to Thornbrough that had far-reaching organizational implications, however innocent it may have sounded. He wrote in a memorandum:

Am I right in my conclusion that the state of affairs at Detroit is really little better now? If so, surely the case demands substitution of a first-class new man with management talent, including, of course, the adequate technical competence, and if that you should give consideration to placing the new manager under either Head Office or some jurisdiction other than that of Racine....

If you see fit... make a recommendation that Detroit, for at least a time, should be under completely new management and that management itself responsible to you.

This statement reflected growing lack of confidence in Detroit factory management and also in management of the whole United States company at Racine; but also it, in effect, recommended what came to be the first step toward unification of United States and Canadian management through parent company executives assuming "line" as well as "staff" positions. Some factory management changes were made at Detroit and the factory was made directly responsible to the parent company at Toronto, as was the Batavia factory. For several months the process did not go further, and little progress was made in improving the operations at the Detroit and Racine factories. This was partly because the parent company had not yet been able to hire a senior manufacturing executive.

During this period consideration was being given to writing and distributing within the company a memorandum that would comprehensively outline the principles of organization that the company intended to pursue. One area that caused considerable debate and much personal anguish for those directly involved was the staff-line relationship in engineering. The *need* for close co-ordination of the company's activity in this area had long been recognized, and after the 1953 merger an Eastern and Western Hemisphere Engineering Co-ordinating Committee had been formed. In September 1956 the name was changed to Engineering Co-ordinating Committee since all engineering departments, including those of Australia and South Africa, were now to attend its meetings. The minutes of the September 1956 meeting of the Committee outline its general purpose and objectives:

1. Provide information on the Company's world-wide Engineering and Research activities;
2. Review the status and progress of Engineering projects;
3. Review future Engineering or Research programmes and decide what Engineering Division was best suited to undertake them;
4. Exchange experience on the performance of new equipment;
5. Discuss Engineering problems that might have arisen;
6. Acquaint members with new prototypes and demonstrate new equipment for them for appraisal and consideration in their respective markets;
7. Co-ordinate changes of design on tractors and implements in order to maintain interchangeability of the same products produced by the various operating divisions;
8. Discuss service problems which may affect other divisions;
9. Discuss and appraise bold and novel engineering concepts submitted by the Chief Engineer of the various divisions;
10. Exchange of information on new competitive developments.

This made good sense and caused little concern, but differences of opinion began to emerge when the crucial matter of procedures for the approval of engineering projects were discussed. Parent company management had committed itself to the concept of a decentralized organization and had made it clear that the relationship between the Vice President Engineering of the parent company and the engineering departments of the operating divisions was of an advisory staff, not line or decision-making character. Phillips wrote to Thornbrough from London on October 18:

I am anxious that you should make it abundantly clear to him [i.e., Vice President Engineering, parent company] that if he thinks it necessary to issue anything approximating an order, it must be done through the Executive Head of the Eastern Hemisphere. . . . I think the key to the easy working of this engineering programme lies in rigid adherence to that responsibility. It is, in reality, emphasis on the staff responsibility.

Actually behind that statement lay a serious and long-standing problem of personalities. The animosity between Ferguson engineers at Coventry and those at Detroit that had existed prior to the merger, continued after it, and was exacerbated when for a period Harry Ferguson denied Klemm permission to produce the new #35 tractor, and when Klemm recommended that the Coventry-designed #60 tractor should not be produced. Also Klemm was never too tactful and he did not have confidence in the non-engineer heading the Ferguson engineers in the United Kingdom. For this reason it was not feasible to place Klemm over the Ferguson engineers in the United Kingdom, particularly not until a strong engineering head had been placed there.

At the same time Phillips did make it clear that expenditures involved in engineering would come under parent company control. He wrote:

> All engineering effort throughout the Company world-wide requires money. It therefore follows simply that all engineering appropriations of every nature are subject to budget control.
>
> The final decision in approval of budgets lies with top management and nowhere else.

Phillips had become instinctively suspicious that there had been no positive sense of control of such expenditures in relation to return on capital. While Phillips favoured parent company control of engineering through the budget, he did consider the possibility of distinguishing between "major" and "minor" projects, perhaps on the basis of amounts of money involved. All this seemed more or less to settle the issue but, as we shall see, this it did not do, and a major change in the approach to lines of authority in engineering activity came later.

The execution of the plan to define clearly the individual responsibilities of the management group began formally on October 26, when the first of a series of memoranda was circulated outlining such responsibilities.

By the end of October, all the major pieces of a more comprehensive management reorganization plan than the one represented by the July organization chart had appeared. And in early November Phillips had begun to circulate for comment a draft copy of such a plan among senior executives. Some success, too, was finally achieved in hiring senior executives.

On November 14, the Executive Committee ratified the appointment of J. H. Shiner as Vice President Marketing, and H. A. Wallace as Vice President Manufacturing, effective December 1, 1956, thereby giving substance to the concept introduced in July of parent company staff positions being based on functional divisions. Shiner came from the Ford Motor Company and was a University of Colorado graduate, and Wallace, a graduate of the University of Illinois, had many years of manufacturing experience at Allis-Chalmers and Ethicon Incorporated. For the next two months Shiner and Wallace concentrated on becoming acquainted with the company's problems.

At last on December 13 there was released to management a "Memorandum on Organization" which was to become the most important organizational document ever produced by the company. It included an introduction by Phillips, an outline of the principles of organization that were to be pursued by the company, an organization chart, and a schedule of the responsibilities of each of the major executive positions.

The memorandum was brief, but its importance for the history of the company is such that relevant parts of it are here reproduced. It defined "staff" and "line" concepts and discussed the role of both groups of executives. It emphasized that responsibility being delegated must explicitly be defined; that it must be matched with commensurate authority and accompanied by an adequate system of supervision and control; and it was also emphasized that delegation of responsibility in no sense reduced the accountability of the executive who delegates. Both an organization chart and a schedule of responsibilities for executive positions were regarded as essential for conveying the required information. The role of committees was also outlined.

First, the introduction by the Chief Executive Officer:

MEMORANDUM ON ORGANIZATION

Whenever a group of more than two persons join together in an effort to achieve a common objective, some degree of organization must be involved. When the group is small, undoubtedly there is merely some subconscious agreement as to the division of labour, but, nevertheless, the concept of organization is involved, even if it be in a primitive sense. As the group increases in numbers, it usually follows that consideration is given to a more formal record of agreement between the persons involved as to the division of labour and authority. Considering the state of our industrial society of some 25 years ago, I am sure that what were then regarded as substantial industrial enterprises conducted their business satisfactorily with the minimum of formal organization.

The complicated conditions affecting all business today render it not only impossible to expect satisfactory operation, lacking a clear plan of management organization, but indeed render it foolish to expect success without it.

I am satisfied that the question of organization must be brought to the attention of all those who are concerned with the successful operation of Massey-Harris-Ferguson Limited.

We are gradually developing new and positive Policies, both long- and short-term. These Policies determine the methods to be used in achieving our objectives. These Policies require, if they are to be successfully applied, constant and effective supervision and control.

In the following pages, I have attempted to set out in simple language the fundamental principles which will provide the basis of our Management Organization Plan. I urge you to study them, to ask any questions which may come to mind, and to use them as a basis for the development of your own detailed Management Organization Plan.

As we become more practiced in the use of the new Management Organization Plan, there will be a need for a greater degree of decentralization within our world-wide organization. Any such decentralization will depend upon the proof that the individual executives concerned are able to discharge successfully their respective responsibilities.

Spread as we are across the world, it becomes more than ever important that the procedures which establish the methods which are to be used in the conduct of our business shall be, as far as possible, common throughout the whole organization.

We are now committed to present our annual statement in the form of a Consolidated Balance Sheet covering the world-wide operations of the Company, and this consideration alone must over-ride certain local conditions, which hitherto may have prevented the application of identical procedures. Indeed, under the conditions which now prevail, we must direct that Common Procedures are to be the rule.

Progress towards the ideal goal of Common Procedures will, to a large extent, govern the degree of possible decentralization. It should do much to eliminate certain problems which now exist and tend towards the simplification of day-to-day routine.

On two occasions in the past few months, when discussing the question of decentralization, I have used the word "autonomy." Apparently some misinterpret what I mean by "autonomy." I do not mean self-sovereignty in any sense. I intend to imply that decentralization, subject to effective control, should bring with it the freedom to make the correct decisions. It certainly is not a licence to make any haphazard decisions. All decisions so made must fall within the limits of the budget, which is authorized for all Divisions by Head Office.

Do not forget that the best executives are those who make "paper" their servant, not their master.

W. E. Phillips
Chairman and Chief Executive Officer

Toronto, Dec. 13th, 1956

MANAGEMENT ORGANIZATION PLAN

The successful operation of any Industrial Enterprise depends primarily upon the quality of the men who constitute the company.

It is clearly not possible to operate Massey-Harris-Ferguson as a Monarchy, subject to the executive decisions of any individual at Head Office.

It is intended that the Company shall be operated by a TEAM and that executive authority shall be decentralized to the extent that it may be practical.

The rules of the game within which this TEAM will operate are, in effect, laid down by the MANAGEMENT ORGANIZATION PLAN.

The satisfactory operation of the MANAGEMENT ORGANIZATION PLAN depends entirely upon the successful delegation of RESPONSIBILITY and AUTHORITY and, above all, its CONTINUOUS SUPERVISION AND CONTROL.

The delegation of responsibility within the MANAGEMENT ORGANIZATION PLAN must follow simple but fundamental principles:

1. The delegation of RESPONSIBILITY must be set out in explicit terms.
2. The AUTHORITY delegated must in every respect match the responsibility.
3. The operation of the RESPONSIBILITY and AUTHORITY so delegated must be subject to continuous supervision and control.
4. The fact that RESPONSIBILITY has been delegated does not in any way lessen the accountability of the Executive who delegates.

Within the structure of the MANAGEMENT ORGANIZATION PLAN two types of authority are subject to delegation:

(a) LINE AUTHORITY

The chain of responsibility, perhaps best indicated by the organization chart, determines the flow of executive directives from the Chief Executive Officer downwards to the Head of the smallest Department.

EXECUTIVE ACTION MAY BE INITIATED ONLY
THROUGH THIS LINE AUTHORITY STRUCTURE

(b) As the complexity of management function increases, the need for STAFF EXECUTIVES to handle certain of them must be recognized.

In general, the functions of the staff executive relate to investigation, interpretation, information, advice and co-ordination. In his capacity as a staff executive he is NOT responsible for the execution of any line authority functions.

The leadership to be exercised by the staff executive is essentially one of ideas based on his specialized ability, knowledge and experience. In reality, the staff executive supplements and extends the competence of the line executive.

Broadly speaking, every line executive is entitled to expect, and if necessary to request, the services of the appropriate staff executive, but the important principle involved is that the staff executive in that capacity HAS NO AUTHORITY TO ISSUE EXECUTIVE ORDERS TO ANY LINE EXECUTIVE.

The LINE EXECUTIVE is not compelled to accept even good advice from the staff executive, but it would be most unfortunate if he did otherwise, because the entire responsibility for error would fall on him. On the other hand, the staff executive is bound to give good advice and it would be most unfortunate if the advice were otherwise.

It is, of course, expected that many executives whose appointment in certain areas is of a staff nature, will, within their own departments, or in specified areas, function as a LINE EXECUTIVE.

In a world-wide organization, such as ours, engineering and administrative functions are constantly intermingled and the vital flow of ADVICE and GUIDANCE, particularly in the field of the engineering function, is to be exercised within the structure of the staff authority.

I would place as the first essential to the smooth working of any MANAGEMENT ORGANIZATION PLAN that priceless ingredient: GOOD WILL.

Next and equally important is a clear understanding of our MANAGEMENT ORGANIZATION PLAN and its avowed objectives.

To ensure that the details concerning the delegation of responsibility are known to all concerned, use is made of:

(a) A GENERAL ORGANIZATION CHART in graphic form;

(b) A SCHEDULE OF RESPONSIBILITY set forth in explicit form.

The graphic organization chart corresponds to what might be described as a "general assembly drawing"; the schedule of responsibilities corresponds to a set of "engineering specifications." They must be read together.

The "committee" provides a most useful instrument for management purposes, but it is important that the inherent limitations which attach to any committee be clearly recognized.

The committee even at its best is a poor substitute for an effective MANAGEMENT ORGANIZATION PLAN. It is an ideal vehicle when used as a means of consultation and review but it cannot of itself function as a vehicle of EXECUTIVE RESPONSIBILITY.

The responsibility for executive action arising out of the proceedings of any committee must be vested in the Chairman of the Committee, or the appropriate Executive to whom the Committee reports.

Once the MANAGEMENT ORGANIZATION PLAN has been established, it cannot be changed in any respect other than by the recommendation of the COMMITTEE ON ORGANIZATION AND PROCEDURES.

This is not to suggest that the established plan will not need change. I am certain that many changes will be desirable but such changes CANNOT be made at the will of ANY ONE INDIVIDUAL.

THE COMMITTEE ON ORGANIZATION AND PROCEDURES is charged with the issue of the details of the MANAGEMENT ORGANIZATION PLAN and the PROCEDURES which set out the methods and practices which are to be followed in the day-to-day administration of the Company and all subsidiaries and divisions. Such directives relating to organization and procedure and all changes and alterations will be issued within the authority of the President and in such manner as he may decide.

The organization chart accompanying the Memorandum was little changed from the July chart, and perpetuated the separation of the corporate staff positions from all line responsibilities in operating divisions, including the Vice President Engineering. Actually the memorandum did not define sufficiently clearly the responsibilities of those corporate staff positions, and did not really recognize that important line responsibilities in some of the functional divisions – such as Engineering and Finance – would have to be retained by corporate executives. Later this was more explicitly realized, as we shall see, and it was subsequently recognized that a policy of decentralization was really one of degree, based on identification of responsibilities and the delegation of certain of those specifically identified responsibilities to the operations units.

Since the Detroit and Batavia factories were in effect under parent company control, the chart in that sense did not even reflect reality of the day, but this was because the arrangement was at first regarded as being a temporary expedient.

It was an arrangement that turned out to be much less temporary than had originally been expected. Parent company management became increasingly dissatisfied with results being achieved in both Canadian and United States divisions, and increasingly convinced that because of the close and complex relations between those divisions, their operations should be combined. Yet it did not seem possible to replace existing members of management quickly with personnel in whom parent company management would have confidence. At the same time by early February Wallace and Shiner felt they knew enough about the company's problems in manufacturing and marketing to assume line responsibilities. Therefore the solution adopted was to place temporarily in abeyance a major principle included in the Memorandum on Organization – the separation of parent company functional executives from most line responsibilities in operating divisions.

On February 5, the company released internally its plan for placing the United States and Canadian operations under direct parent company control. As a result, parent company executives assumed line positions with respect to North American operations while retaining their staff positions in relations with the Eastern Hemisphere Division. This move also had the effect of combining the Canadian Division and the United States Division into one North American Division. As of that date the parent company Vice Presidents of Marketing, Manufacturing, Finance, Engineering, Planning and Procurement, Personnel and Industrial Relations, and Public Relations became directly responsible for operations in North America. Most important of all the President became in effect the General Manager for operations in North America. Each executive therefore wore two hats. This was regarded as a temporary move, but it in itself introduced an approach to problems that was used on future occasions. The approach was that of parent company or "corporate" executives assuming line positions when an operating division encountered serious problems and relinquishing those positions when the problems were under control. It was an approach that was to be used in United Kingdom operations in 1959, after the acquisition of F. Perkins Limited and of the Standard Motors tractor facilities, in France when serious management problems arose in that division, and also in Australia to a limited extent.

So for a while, far from decentralizing operations in North America, these became highly centralized, except in one important sense: each senior operating executive was given a remarkably free hand in establishing new policies and procedures, that is, as long as his performance enabled him to retain the confidence of superiors. When, for whatever specific reason, that confidence was lost the individual involved was required to leave and another executive was appointed who once again enjoyed relatively great freedom of action. This was the logical outcome of the adopted policy of defining responsibilities for managerial positions and of seeking executives who were capable of carrying those defined responsibilities, rather than one of shrinking or expanding responsibilities of a position or creating positions to suit particular individuals.

The February 1957 reorganization in North American operations caused the greatest exodus of senior personnel the company was to experience – some because they did not enjoy the confidence of parent company management, others because reorganization had made them redundant. It was an unsettling experience for many. Some entirely well-intentioned people, from within and from outside the company, came to the President to tell him that his policy of shrinking the executive ranks and replacing some with outside executives would destroy the organization. The General Manager of one European operating company even wondered what his position would be as an "old Massey-Harris man." "More Shake-ups at Massey-Harris," reported the *Toronto Star* of February 9. The human tragedies that these upheavals involved were the consequence usually not of active opposition to new policies and procedures, but of passive opposition to them, often arising from a failure to understand why changes were being made.

The North American reorganization did not, of course, modify the *principles* of staff and line, and it did not change the way the company was functionally divided. Nor did it change the fundamental principle of *delegation of defined responsibility with commensurate authority, under continuous supervision and control without dilution of accountability through delegation.* In other words it did not halt the application of the recently adopted concepts of organization.

If common plans, policies and procedures were to be adopted, as the December 1956 Memorandum on Organization instructed, it was essential to develop a system of manuals for communicating such information. Preliminary methods for this were first outlined in a policy and procedure memorandum dated June 10, 1957, on the subject, "Manuals on Organization, Policy and Procedure, and Personnel Appointments." Eventually internal memoranda were divided essentially into three groups – organization, procedure, policy, each group collectively constituting a manual.

In May 1957 J. G. Staiger, who was to play a major role in introducing new administrative and planning procedures, came to the company as assistant to the President in matters of administration. He was a graduate of the University of Dubuque, had taken post-graduate training in Accounting and Business Administration at the universities of Marquette, Wisconsin, Michigan, and Chicago, and prior to joining Massey-Harris-Ferguson had been assistant to the Executive Vice President of American Motors.

A major move forward in defining explicitly areas of responsibility of division heads, department heads, and other managerial personnel was made when surveys were undertaken in North American operations to complete a salary determination plan. This plan was essential for developing a scale of salaries that would be competitive and would reflect the responsibilities inherent in the various positions. It also, however, provided much useful data for ensuring effective utilization of manpower, in other words for controlling overhead costs. A memorandum of August 19, 1958, by J. G. Staiger, gives an impression of the company's approach to this whole area:

> It is not always feasible to get a proper perspective upon the many functions in a large enterprise. In fact, companies often find it necessary to secure the services of management consultants to establish, by means of surveys, interviews, etc., the inter-relationships and status of the several types of organizational units comprising their enterprise.
>
> Massey-Ferguson is fortunate to have available, at this time, a management tool that provides relatively accurate measurements for ascertaining the current status of the Company's overhead organization. This management tool is available as the result of the completion of the Salary Determination Plan surveys. The purpose of these surveys was to establish consistent, equitable and properly co-ordinated salary levels, by analysis and evaluation of the job descriptions for existing divisional, departmental and branch activities. It was required that these analyses and evaluations be based upon careful, detailed charting of the divisional, departmental and Branch organization structures, and upon personal interviews with a high percentage of employees.
>
> These surveys point up many things that will be of great value in a successful overhead cost control programme, by providing the bases for determining improper organizational concepts, and for assuring effective manpower utilization. . . . To assure consistent results, all staff division vice-presidents and directors, and the operations head for other North American operating units, are being asked to implement the following programme:

(a) Define division, department and other objectives, duties and functions.
(b) Examine division, department and other organization charts and evaluate with respect to objectives, duties and functions.
(c) Evaluate and reconcile inconsistencies or duplications and determine non-essential activities.
(d) Establish the organization, policy, methods, procedures or personnel changes required to secure the most effective manpower utilization. . . .

Some specific considerations in organization planning that will contribute to the lasting success of a manpower utilization programme are these:
(a) Elimination of functions that are not required for the successful operation of the Company. At branches, in particular, this means simplifying and clarifying the "line of command."
(b) Reduction of functions to levels consistent with good operating practices.
(c) Substitution of mechanical for manual applications or manual for mechanical applications, depending upon the results of methods analyses.
(d) Consolidation of similar or continuous functions into one complete function either within a single division, or by transfer of responsibility from one division to another division.

To assist division and department heads in making determinations of these types, the services of Head Office Procedures and Systems Department are available.

Execution of this programme resulted in pushing the chosen concepts of organization deep into the company. It also confirmed once again that implementation of new concepts was a quite different task than defining them. The character of the company on both the Massey-Harris and Ferguson side had been such that principles of organization had largely been unknown. Adoption of them by all levels of management required not merely a certain willingness but also a certain intellectual effort. The complexities of old loyalties sometimes stood in the way of the former, while in some cases the necessary intellectual effort was not forthcoming – traceable in part to the overwhelming emphasis by previous management on the "self-made man."

As reorganization in North American operations proceeded management began to believe that a new corporation was being created, and wished to tell the public about it, particularly its customers, shareholders, and creditors. Public relations therefore assumed a prominent role. H. G. Kettle, a graduate of Oxford University, who had become Director of Public Relations just prior to the change in management, was responsible for this area. The approach taken was to highlight change in the corporation. In March 1958 the name of the company was changed to Massey-Ferguson Limited, "Harris" being dropped, and a new corporation symbol, the triple triangle, was introduced. Greatly increased financial and other information was made available to the public, and the company's annual reports soon began winning prizes.

In late 1958 management began to consider applying its new organizational concepts in detail in its operations outside of North America, and it was concluded that to accomplish this a special effort would be required.

The parent company organized a conference for the general managers of the Eastern Hemisphere operating divisions – United Kingdom, France, Germany – in October 1958. Its specific purpose was to make them more acquainted with the new management and organizational concepts, procedures and practices that were being developed in the company's North American operations to achieve greater control over operations and ensure the company's competitive position. Suddenly there was increased urgency for this to be done and the foreword to the conference programme explains why:

> Substantial profitable operations in the Eastern Hemisphere enabled Management to give priority to establishing control over less successful Western Hemisphere operations. Recent changes in Europe to a much more competitive market situation make it most desirable that our new management concepts, which have now been sufficiently tested, be introduced at the earliest date in Eastern Hemisphere as well as in all our other divisions.
>
> The very character of our operations makes it, indeed, a matter of urgency to have our entire world-wide organization operate under a unified concept and with unified practices.

Thornbrough made a point of attending the first meeting of the newly formed Eastern Hemisphere Co-ordinating Committee in London on October 21, and he chose again to present the fundamentals of the organizational concepts that the company was adopting. He stressed the concepts of delegation of responsibility with matching authority, formulation of plans through expert and informed staff functions, communication through co-ordinating committees to ensure that staff and line functions would be aware of all phases of the business and would participate in the formulation of plans.

Both the Toronto management conference and the aforementioned committee meeting illustrated that applying new concepts and procedures was not merely a matter of issuing plans, and issuing orders for the implementation of plans, but that it also involved a process of education. Indeed, the principle that "the more time available for patient instruction the smaller the change in management as a result of reorganization" could be regarded as one of the major principles of reorganization; and it certainly implies correctly that constant attention to organizational matters with resulting evolutionary changes is likely to be less disturbing to everyone concerned than infrequent attention to them with consequent infrequent but abrupt reorganization.

Before much had been done in actually applying the new organizational approach to the Eastern Hemisphere Division, it began to be evident that there were deficiencies in the very concept of the Eastern Hemisphere Division. Difficulties in maintaining this concept had in early 1959 been multiplied by the acquisition of F. Perkins Limited, manufacturers of diesel engines at Peterborough, England, and by the imminent acquisition of the tractor manufacturing facilities of the Standard Motor Company Limited. At the same time, both the French and German companies had increased the extent of their manufacturing operations. In addition, a fundamental problem was the increasing need to co-ordinate manufacturing, marketing, engineering, and inter-company movement of finished and component parts between *centres of operations* – particularly since the policy of North America using United Kingdom tractor components was by that time well advanced. The Eastern Hemisphere Division was not one centre of operation but three – United Kingdom, France, Germany. It also supervised the South African companies and Eastern Hemisphere export territories. Even within the existing Eastern Hemisphere Division conflicts between operating centres had begun to arise – as when one centre modified its small combine to gain an edge over the other two in exports. This suggested the need for much closer co-ordination and control. Was this to be gained by building up staff at the London head office of the Eastern Hemisphere Division?

The more the problem was considered the more anomalous it became to build up a "corporate group" in London in addition to another corporate group in Toronto, particularly since this would not enhance urgently needed co-ordination between *centres of operations*. Thornbrough discussed the problem in a memorandum to Phillips dated March 9, 1959. It is of particular interest to the organizational development of the company:

> The Eastern Hemisphere Division was established following the amalgamation of Massey-Harris and Ferguson Companies. Its purpose was on one hand to be an operating division reporting to Toronto and in other respects it was considered as an adjunct to the Toronto management. I believe it is a fair statement to say that the inter-relationships created and implicit in such a concept have not as yet been resolved.
>
> There has been the desire to follow the principle of decentralization in order to fix greater local responsibility for results. The geographical nature and distances of these operations from Toronto made such action all the more appealing and apparently necessary. Yet the basic policy decisions since amalgamation to interrelate production insofar as possible and thus to a great degree engineering design and specification, have set in force necessities for much more specific and closer integration of operations between Toronto and the Eastern Hemisphere.
>
> In general principle, unless we are completely to reverse our course toward component manufacturing in England for North America, it appears we are currently in the position of decentralizing operations which cannot successfully be decentralized in the true sense of a more or less self-contained and independent unit. Faster plane service and improved telephone service are tending to reduce geographical obstacles. In addition to Engineering and Manufacturing, integration of Marketing concepts and efforts also tends toward lesser decentralization. . . .
>
> It is proposed to rearrange and reorganize operations of the Eastern Hemisphere as follows:
>
> A. To establish three operating units (i.e. United Kingdom, France, Germany) to replace the single Eastern Hemisphere Organization.
>
> B. To institute operationally for each of these three new units direct channels

of authority and responsibility with the Parent Company in Toronto. . . .

C. To organize each of the three new operating units basically on the Parent Company functional form and pattern; each Managing Director will be responsible directly to the President in such manner as the President establishes. Staff of the President will have staff authority as regards each Managing Director.

D. . . . to establish an export trading company as soon as feasible for trading operations between and within Eastern and Western Hemispheres.

Once again there was the matter of the impact of reorganization on personnel. Thornbrough said in the same memorandum:

It is inevitable that these major rearrangements and reorganization will create problems of impact on many individuals. Great tact in handling will be required and the greatest of understanding and fairness will be necessary. There may, however, be certain instances in which individuals will ask alterations or revisions to the reorganization to the extent that the plan itself may be determined or prejudiced. In these instances, reasonable and fair termination will have to be considered as a necessary alternative. . . .

The recommendations in this memorandum obviously required most serious consideration. But other organizational issues also presented themselves at about this time. The company seemed to be beginning to spend increasingly more time on proposals for establishing additional manufacturing facilities in various countries, or proposals for becoming associated with local manufacturers as a minority shareholder or through licensing agreements. In this respect – as we shall see again – interest in Italy, Brazil, Spain, India, Turkey and elsewhere suddenly increased. Yet the company had no formal machinery for dealing with these proposals.

Yet another development in 1959 was the greatly improved position in North America which suggested that parent company executives might be able to relinquish their line responsibilities in operations there, provided that a suitable operational structure adequately staffed could be introduced. This would have the advantage that parent company executives could begin to turn their attention to world-wide problems, including the development of procedures for controlling the organization with its internationally dispersed operations units.

All these developments suggested that an over-all review and summary of the company's world-wide organizational structure was again desirable. A firm of consultants, McKinsey and Company Incorporated, was engaged to assist in this project. The result was the completion and internal distribution in November 1959 of a manual entitled "Realignment of Corporate Management and Disengagement from North American Operations." In an introduction to it Phillips wrote:

Since July 1956, there has been a vast improvement in the position of the Company. In North America, the unfavourable conditions which had characterized our operations there have been, to a large extent, reversed. Furthermore, we have established a foundation for continued improvement of our competitive position in that market.

Portrait by Sir James Gunn, R.A.

W. Eric Phillips, CBE, DSO, MC, LL.D., Chairman of the Board and Chief Executive Officer from July 6, 1956, until his death on December 26, 1964.

In the Eastern Hemisphere, the purchase of Perkins and the acquisition of the Standard tractor assets have at long last given us full control of our own engine and tractor production.

We are now, for the first time in our long history, a well-integrated, well-equipped manufacturing enterprise in our own right. This changed situation, together with our well-established world-wide distribution system, brings us to the very threshold of great, indeed, unique opportunities, which challenge us in every market.

If we are to exploit effectively these opportunities, we must equip ourselves in every respect with the best available management tools and above all, in matters of organization, embrace a concept of world-wide management.

The broad principles of organization relating to defining responsibility, delegating authority, staff-line relationships, accountability, control, use of committees were those of the December 1956 memorandum. Other aspects of organization were outlined in greater detail than they had been before. The memorandum outlined the objectives of the enterprise:

1. To produce and market agricultural and light industrial equipment.
2. To be world-wide in scope.
3. To have a full product line.
4. To be an integrated producer.
5. To become the dominant factor in the industry.
6. To improve profits significantly.

Actually it would have been more accurate to regard (6) as the objective, (1) as the chosen area of operations, and the remainder as means for achieving (6) for they presumably were not ends in themselves.

The crucial issue of the degree of decentralization of management, an issue that had not been covered adequately in the 1956 memorandum, was treated this way, in summary:

The organizational structure that best serves total corporate interests will be a blend of decentralized and centralized management. Marketing and manufacturing activities together with some supporting service functions, should be organized in a way that would bring them as close as possible to the local market situation. On the other hand, the activities that determine the long-range character of the Company – such as control of product line, facilities, and money, and planning the strategy of reacting to changes in the patterns of international trade – should be handled on a centralized basis.

This statement clearly recognized that decentralization was a relative, not absolute, concept and that in the area of engineering and finance especially, it could not go very far, while in marketing and manufacturing it could.

It was recognized that this approach to organization would require the establishment of a small group of people with technical skills located at the corporate or parent company level who would be available to assume defined line responsibilities, who would render advice to the operations units, and who could also counsel the President on world-wide problems and assist him in appraising results achieved by operations units. It was also recognized that suitable means for communicating between corporate and the operations units would have to be established – which soon was to become the next major organizational development.

To encourage initiative and enterprise on the local level and to maximize local responsibility for operations, functional divisions (with the exception of Export Activity, Engineering and, to a great extent, Finance) in each operation were made responsible to the local Managing Director and he reported directly to the President of the parent company. Also, it may be noted in passing that by about this time the policy had begun to emerge of staffing management positions, and of course other positions as well, with executives who were nationals of the countries in which the operations units were located. This was a decisive break from a policy that had existed since the early days of the Masseys, of frequently placing a Canadian executive in charge of operations abroad. This new policy was based on the belief that local executives would, in time, operate more effectively than would executives who had not grown up in and become part of the local environment. The President, in the 1959 reorganization, was regarded in the fullest sense of the word as being a participating President maintaining personal contact with heads of the various operations at home and abroad.

These considerations led to several far-reaching organizational changes. First, a North American operations unit (the term "operations unit" replaced the term "division") was created encompassing all operations in Canada and the United States. Second, corporate executives who in February 1957 assumed line responsibility for operations in North America were disengaged from those activities, and after that their relations with management of the North American operations unit were fundamentally no different than their relations with overseas operations units. This change permitted them to devote their time to staff (and some line) activities relating to all the company's operations units. Third, the Eastern Hemisphere Division was abolished and its place taken by separate French, German, and United Kingdom operations units responsible directly to the President at Toronto. All export activity was centred in "International Export Operations" located at Coventry but responsible to the Vice President of the Marketing Division at the corporate level.

The principle that each country should have its own operations unit, and should generally be run by nationals, was violated in the case of North America. This was curious, reflecting perhaps the view that Canadians and Americans were essentially alike and that it was possible to minimize overhead costs by having one North American operation.

As time went on those responsible for marketing performance in the United States began increasingly to believe that a more specific identification of the company with the United States market was necessary. While the concept of the North American operations unit largely disregarded the international boundary, that boundary could not be disregarded entirely.

Earlier, when the company had permanently closed the Batavia, New York, factory and had stopped manufacturing at Racine, moving much of its production to Canada (see chapter 11), it had reduced substantially the force of its "corporate presence" in the United States. The idea of a North American operations unit, in other words, was administratively very convenient, but management later on felt more and more that it was not an effective vehicle for projecting the company's image in the United States market. Yet its sales in that market were three times as large as in Canada.

In the early 1960s executives began to outline an advertising programme to sharpen that image. When the amount of the advertising and public relations expenditures was estimated the President felt that a different approach should be considered. In August 1965 it was announced that the executive headquarters of the United States subsidiary, Massey-Ferguson Incorporated, would be moved from Detroit to Des Moines, Iowa, that most of the senior executives of North American operations, including marketing executives, would move from Toronto to Des Moines, and that an assembly operation would be established at Des Moines. These moves reflected some uneasiness over the concept of "North American operations" and further evolution of it is not at all unlikely.

But returning to the 1959 reorganization, a new department, Special Operations Division, was created. Its Director was made directly responsible to the President, and it was centred in London, England. This division was now available to deal with proposals for establishing new manufacturing centres in countries where the company did not already have an operations unit, or to negotiate manufacturing licensing agreements, and to supervise new manufacturing operations until such operations were ready to become full-fledged "operations units."

It will be remembered that the company earlier had argued back and forth over the way engineering activity should be organized. In this "Realignment" Manual it was revealed that the company's engineering activity – like its export activity and certain of its financial operations – would be centrally controlled. Such control really meant control over the company's product line. Difficult personality problems in engineering that had earlier inhibited change had diminished by 1959. The fundamental reasons for adopting the new approach are significant for they reflect the problems of a world-wide organization in engineering matters:

Centralized control offers real economies of time, money and engineering talent. Few of the operations units have sufficient volume to support, individually, an engineering staff with the wide range of skills and facilities necessary to handle requirements of their full product line. Centralized control of engineering means that these skills are available to the Company as a whole, avoiding unnecessary duplication at the local level. It also means that the possibility of duplicating engineering projects in more than one of the operations units is eliminated. In addition, the best engineering talent in the Company can be

ORGANIZATION CHART 4

Massey-Ferguson Limited/Corporate Organizational Plan
June 8, 1964

CHAIRMAN OF
THE EXECUTIVE
COMMITTEE

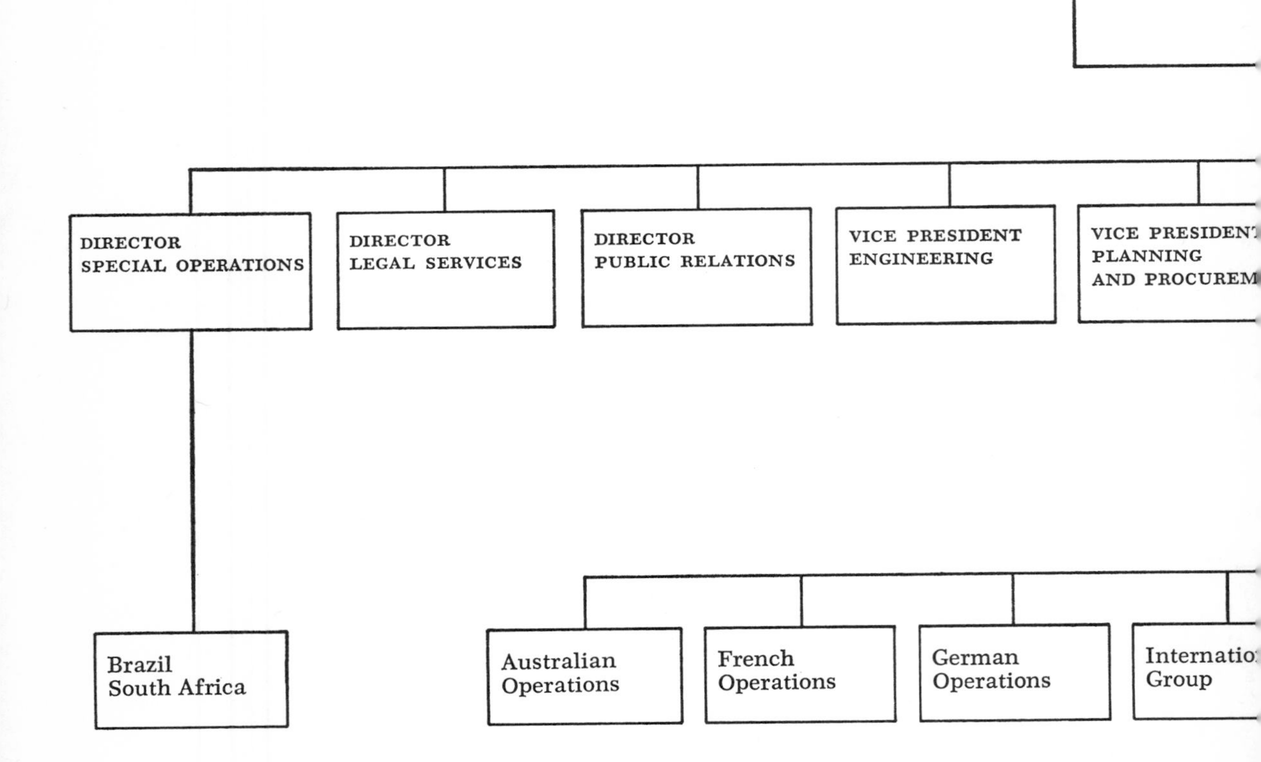

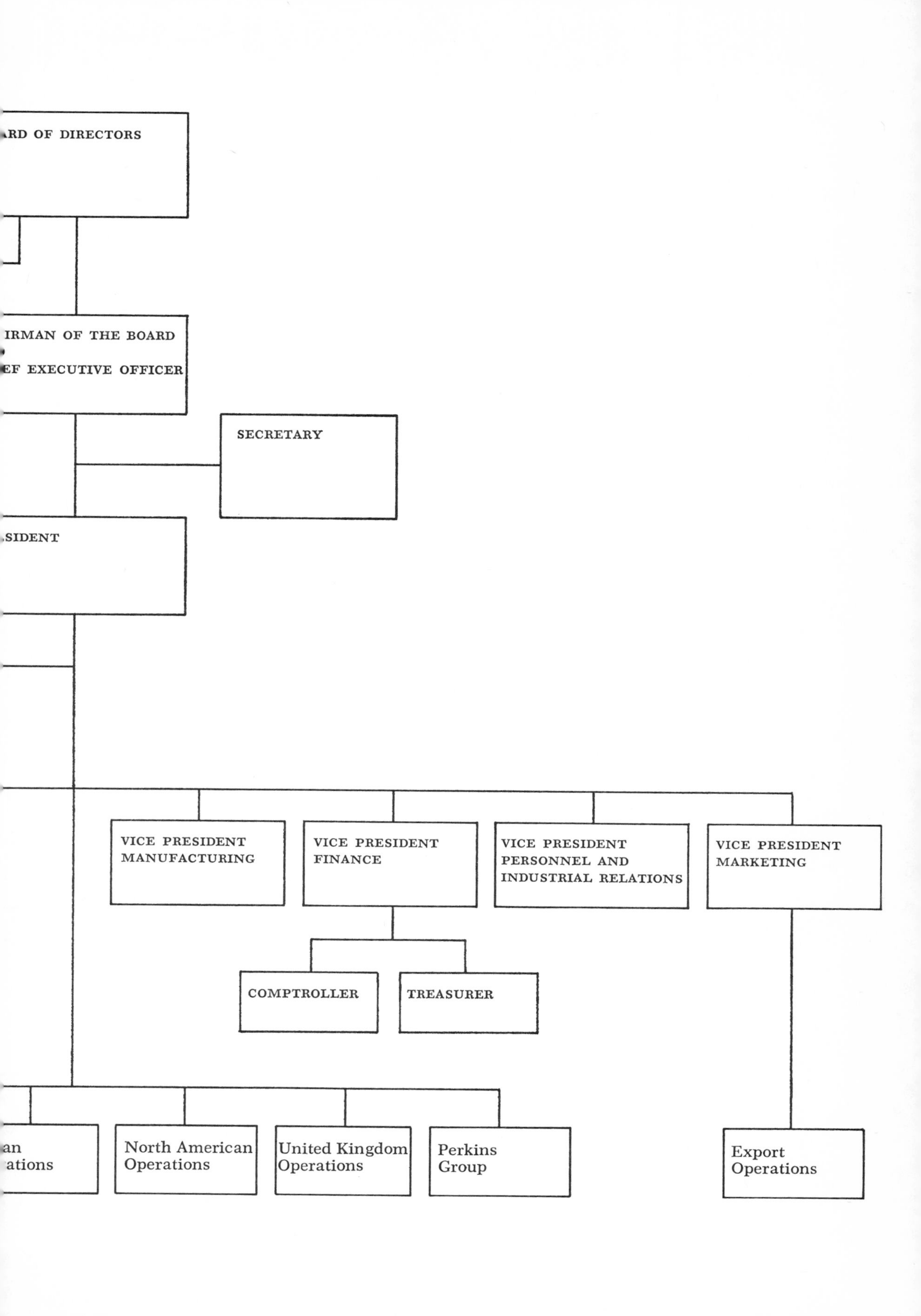

ARD OF DIRECTORS
IRMAN OF THE BOARD
EF EXECUTIVE OFFICER
SECRETARY
SIDENT
VICE PRESIDENT
MANUFACTURING
VICE PRESIDENT
FINANCE
VICE PRESIDENT
PERSONNEL AND
INDUSTRIAL RELATIONS
VICE PRESIDENT
MARKETING
COMPTROLLER
TREASURER
an
ations
North American
Operations
United Kingdom
Operations
Perkins
Group
Export
Operations

brought to bear on the most difficult problems, total engineering effort can be directed at those projects with the greatest potential value and the requirements for engineering development time should be reduced.

In addition, centralized control of engineering is a logical move in respect of world-wide farm equipment needs. While different national markets are characterized by a diversity of crops, soils, climates and farm practices, a wide range of market needs can be satisfied with one basic line of farm equipment. Naturally, certain local developments and modifications are required to meet local market requirements or to take advantage of local manufacturing or purchasing economies. These should not be a major problem, however, since in many cases they can be anticipated, planned for, and accommodated by the basic design. . . .

Central control should not be confused with physical centralization or consolidation of engineering. Supervision and direction should come from one source, but engineering personnel will be located where they are readily accessible to each of the operations units and intimately aware of local requirements . . . in all likelihood . . . located in the several operations units.

In practice all engineering facilities as such were located in the operations units.

The Finance Division, divided into a comptrollership and treasury, was headed by a Vice President who was given both line and staff responsibilities in operations and he was to exercise these through a Comptroller and Treasurer. The Comptroller was given line responsibility for matters such as accounting practices, the format for annual plans, and inter-operations unit prices. The Treasurer was given line responsibility principally for company borrowing activities; for the movement of funds between the various operations units, and between the parent company and the operations units; for investment of company funds; and for insurance matters. Both were, of course, also to render advice to operations units on all matters relating to their functions.

The organizational structure of the company that resulted from the November 1959 reorganization is represented on Organization Chart 4. But one important part of the organization is not shown – this is the new "International Group." International corporations, for tax and other reasons, find it desirable to form "offshore" subsidiary companies to perform certain functions. Massey-Ferguson has found it beneficial to incorporate three such companies. On September 21, 1959, it incorporated Massey-Ferguson Services N.V. in Curaçao, Netherlands Antilles. The principal function of that corporation is to engage in the purchase and sale of patents, and to enter into licensing and technical service agreements with other companies and with the various Massey-Ferguson companies. It is subject to a very low rate of taxation.

Massey-Ferguson International A.G., under another name, was incorporated on November 9, 1960, in Zug, Switzerland, and it has come to be responsible essentially for purchasing whole machines from outside suppliers and re-selling these to the operations units wanting them. In this way the various operations units need not negotiate separately with the same supplier.

Then there is Agrotrac S.A., a subsidiary company incorporated under another name on August 25, 1959, in the City of Panama. Its activities are centred largely on borrowing short-term funds in the various international money markets and re-lending these funds to operations units which would otherwise have to obtain them at higher cost in local markets. It is in practice the company's international banker. Most of its funds are obtained in the Euro-currency markets of London, Frankfurt, Zurich, Amsterdam, and other cities. The United States dollar is the principal currency it borrows, but it also borrows German marks, Swiss francs, and other currencies. Through it Massey-Ferguson can lower its average cost of short-term funds and with relative exchange rate risks in mind can allocate its short-term liabilities among the various currencies. Profits made by Agrotrac are also, of course, subject to a low rate of corporation tax.

With the development of these international companies the organizational structure of the company was largely complete and for a number of years after the 1959 reorganization, development in organization related mainly to co-ordination between the activities of the various operations units. It became crystal clear that to preserve maximum decentralization of the operating divisions of this global corporation demanded close co-ordination of its geographically dispersed activities. Failure in co-ordination would inevitably result in a drift of line responsibilities to corporate executives.

The company, however, was undergoing change in its internal character and was encountering a changed external environment, both of which soon illustrated again that the corporation's organizational structure, no less than its policies and operations, had to react to change. Furthermore, in order to avert the risk of "growing like topsy," it seemed once more that organizational changes should be explicit and controlled, not haphazard. After an extended period of internal discussion the corporation, in November 1966, released to its executives a new confidential manual, a successor to its 1959 manual. It carried the title "World-Wide Management Organization."

Since 1959 the company had increased the geographical dispersion of its operations, and the balance of its line of major products (agricultural, industrial and construction, diesel engines) had changed. The increase in the number of operations units had greatly increased the number of individuals that reported directly to the President on operating matters. As a consequence the President, particularly in his role as Chief Executive Officer, was left with inadequate time to develop long-range strategies and policies and to co-ordinate world-wide activities. This was happening at a time when long-range planning was becoming even more necessary, for the industry as a whole was acquiring a new generation of formally

ORGANIZATION CHART 5

Massey-Ferguson Limited
World-Wide Management Organization

November 1966

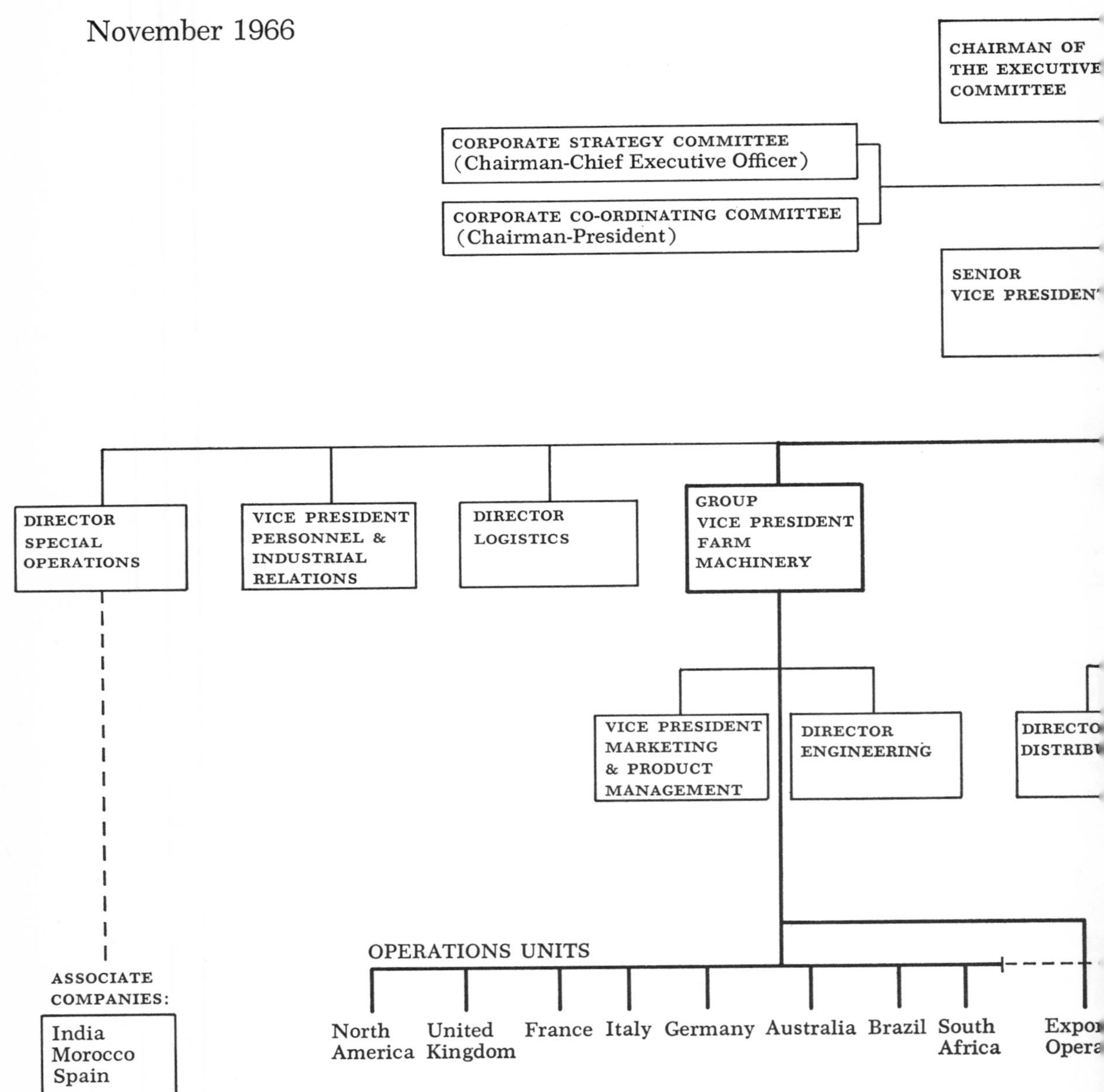

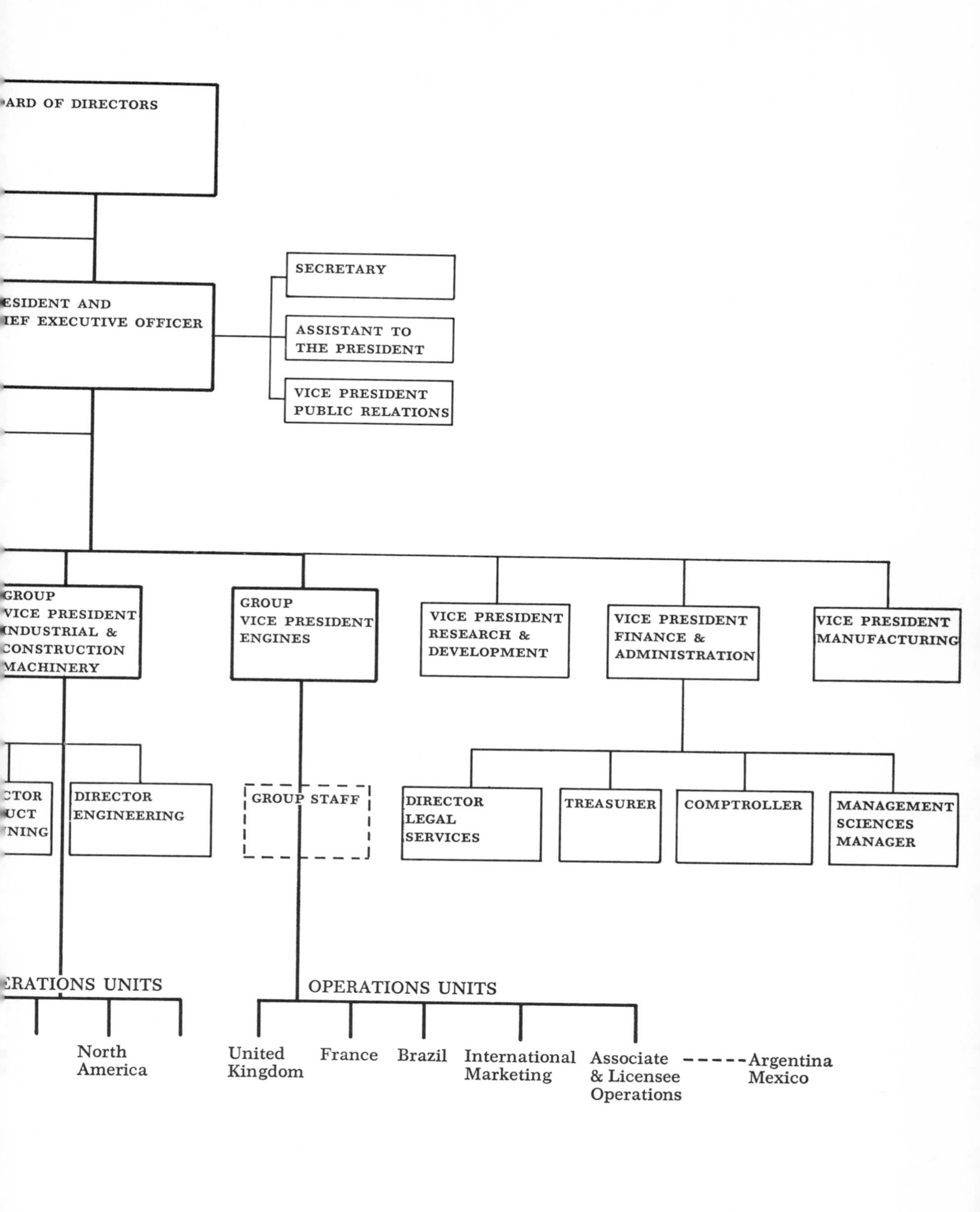

ARD OF DIRECTORS
ESIDENT AND
IEF EXECUTIVE OFFICER
SECRETARY
ASSISTANT TO
THE PRESIDENT
VICE PRESIDENT
PUBLIC RELATIONS
GROUP
VICE PRESIDENT
INDUSTRIAL &
CONSTRUCTION
MACHINERY
GROUP
VICE PRESIDENT
ENGINES
VICE PRESIDENT
RESEARCH &
DEVELOPMENT
VICE PRESIDENT
FINANCE &
ADMINISTRATION
VICE PRESIDENT
MANUFACTURING
CTOR
UCT
NING
DIRECTOR
ENGINEERING
GROUP STAFF
DIRECTOR
LEGAL
SERVICES
TREASURER
COMPTROLLER
MANAGEMENT
SCIENCES
MANAGER
ERATIONS UNITS
North
America
OPERATIONS UNITS
United
Kingdom
France
Brazil
International
Marketing
Associate
& Licensee
Operations
Argentina
Mexico

trained executives, and its customers, including farmers, were becoming increasingly sophisticated in making equipment decisions on a business-like basis.

The growing importance to the company of its industrial and construction machinery business, as well as its diesel engine business, also seemed to demand organizational changes to ensure that growth in this area would not be inhibited by the still preponderant agricultural machinery orientation of the corporation's executives.

To accommodate these developments the corporation introduced the following major changes into its organizational structure (see Organization Chart 5). First it divided its operations into three world-wide product groups – farm machinery, industrial and construction machinery, engines – each headed by a corporate Group Vice President, with his own Group Staff, to whom the President delegated over-all line responsibility for the product group concerned. The actual move towards separate operations units for each product group in the various countries in which the company has manufacturing facilities will occur gradually as conditions justify it, apart from North America where it was introduced immediately. Second, in addition to the Corporate Group Staff that was to concern itself with the product and market needs of each product group, there was created a corporate General Staff dealing generally with strategic planning, policy formulation, and operational co-operation activities encompassing all product groups. The details of organization of the corporate General and Group Staffs were designed to avoid duplication of activities and to minimize pyramiding of staff – principally through their members operating as one team at the corporate level. Third, a new Corporate Strategy Committee was established. It is composed of the Chief Executive Officer (Chairman), Senior Vice President (when appointed), Group Vice Presidents and Vice President Finance and Administration, thereby representing all segments of the corporation and able to take an over-all view of strategy formulation and review. The Corporate Co-ordinating Committee, with the President as its chairman, was retained to concern itself specifically with co-ordination and communication in the implementation of strategy. Several other changes may be mentioned: corporate Vice President Planning and Procurement now became Vice President Logistics; a corporate Vice President Research and Development was introduced, reflecting increased emphasis on that area; the previous Vice President Finance became Vice President Finance and Administration, and his office now included legal services, and the increasingly important area of management services, as well as the offices of the Treasurer and Comptroller; the corporate Vice President Marketing became Vice President Marketing and Product Management, and instead of reporting to the President he now reported to the relevant Group Vice President; the staff functions of Secretary, Public Relations and Assistant to the President were attached specifically to the office of the President; and, finally, there was emphasis on the need for completed staff work to accompany recommendations being made to superiors to hasten decision-making.

Some corner stones of the preceding organizational structure and approach did not change at all. There continued to be emphasis on maximum decentralization of marketing and manufacturing decision-making; and centralized control of product line strategy and design, location of facilities, allocation of funds, and decisions relating to the flow of trade. The operations units continued to be the largely self-contained operational entities in the geographical areas where the company has manufacturing facilities. Committees continued to be regarded as a management mechanism for communication, co-ordination and review, not for making decisions. And again there was the emphasis on defining executive responsibility, delegating authority commensurate with responsibility, accountability for results achieved with authority enjoyed, and the importance of subjecting delegated responsibility and authority to supervision and control.

The 1966 organization manual emphasized the importance of the human element, perhaps even more so than did the earlier organizational memoranda. In the letter that accompanied the new manual Thornbrough endorsed W. E. Phillips' 1956 statement to the effect that "I would place as the first essential to the smooth working of any management organization plan that priceless ingredient: Goodwill". In the manual itself there was included a section on "Courtesy" which pointed out that " . . . experience has shown that good relationships – and effective communications – depend as much on courtesy as on the will to make them work. . . ." Elsewhere in the manual it was emphasized that " . . . One of the most effective means of preserving vitality is to encourage direct spontaneous communications that supplement the more formal structural relationships. It is for this reason that the need for personal communications has been stressed again and again in this manual."

The development of an adequate system of supervision and control proved to be a time-consuming and painstaking process. It was after completion of the 1959 plan for reorganization that serious attention could be turned to devising a system of control procedures that would permit parent company executives – or "corporate," in the terminology of the company – to plan and control the system of far-flung individual operating centres or "operations units" that had been created. One thing seemed certain at that time: the type of information which parent company executives had been receiving when they occupied line positions in North American operations was not the type of information needed for controlling the company's world-wide operations. Yet no one really knew what detailed information was available, could be available, or should be available to enable the new corporate group to fulfil its functions. Not all corporate executives were even certain of what, in detail, their new role should consist.

Consequently the company commissioned McKinsey and Company to undertake a study for "Identifying Information Needs of the Corporate Group." It was completed and presented by April 1960. The study helped to clarify the various roles of the members of the new corporate group and it concluded that "corporate" would require a great deal of information – going well beyond the type of data found in the traditional budget approach to control. Several years later all this was revised beyond recognition, and in retrospect it seems to have been a considerable waste of money and effort. But at the time it helped confirm that the company intended to base its planning and control on comprehensive data. It also helped to identify data that were *not* needed and should *not* be collected.

The study gave corporate executives some impression of the kind of information required for control purposes. But there remained the crucial and most difficult task of devising an actual system involving annual plans reports, and more frequent control reports, that would enable corporate executives to appraise and approve plans for future operations and to appraise results achieved. Again McKinsey and Company were engaged. The only mandate they were given was to devise a system involving reports and procedures that would permit a *relatively small* corporate group to exercise adequate control over the world-wide operations of the company.

It was decided that a start should be made by constructing a set of plans for the 1961 operations of one operations unit. It was not in any sense to be an "abstract" system. The United Kingdom operations unit was chosen as the "guinea pig" principally because Thornbrough and Staiger were already over there and perhaps because North American operations had become somewhat weary of seeing so many consultants in such a short period of time.

The study was begun at Coventry in March 1960 and McKinsey and Company were permitted to alter whatever procedures they felt should be altered to create a system they regarded as desirable. The approach taken was to require all of the departments, down to the very smallest, to construct a plan of operations for the following year. A central part of the outline of these individual plans was (*a*) an explicit statement of objectives to be achieved, (*b*) an outline of the action programmes designed to achieve the stated objectives and (*c*) an estimate of what the profit impact of the action programme was likely to be, as a means for justifying action planned. When all the individual plans were co-ordinated and summarized, the result was an over-all plan for the operations unit. Conceptually the most important development was first the format of the planning procedure – objective, action programme, profit impact. – Also important was the break with the traditional and narrowly financial approach to planning and its replacement with an approach emphasizing all-important operational factors.

Actually one important procedure for making this approach work was recognized and adopted after the first United Kingdom plans were presented for approval. The consolidated plans produced a profit result of

£2.3 million. Thornbrough examined the plans, regarded the expected profit results as unsatisfactory, with firm insistence requested executives to resubmit plans, and instructed the planners to revise their action programme to conform with the original profit objective of £5 million. This was done, and it must be recorded that the profit objective was achieved in 1961. Ever since it has been the practice to begin with annual over-all goals and objectives for each operations unit and then begin the planning procedure.

This approach meant that when the President came to review the plans of the operations unit he would see not merely an estimate of final results that were anticipated but also an outline of the action programmes that were being introduced to achieve those results. It was a major step also in permitting top management to appraise the performance of its managers for it emphasized that planning included recognizing and defining desirable objectives, and that "managing" consisted not of *letting* things happen but of *making* them happen, that is, of introducing action programmes that would lead to the achievement of stated objectives.

This whole process represented a complete break from tradition. The traditional budget control process, quite sophisticated in many respects, suffered from the defect that it expressed plans solely in terms of results hoped for and not in terms of action necessary to achieve the results. Financial results are difficult to express in terms of what a manager actually has to do, and so he tended to look on budget control as being separate from day-to-day operations. The new approach sees results arising out of operations, and more particularly out of the execution of planned programmes. In the budget control process there seems to have been a tendency to focus attention on *allocating blame for undesirable events;* while the new approach made it natural to focus attention on whose responsibility it was to *devise and execute off-setting action.*

Because it was a new approach, world-wide in scope and including all functional divisions, it required a new name. It was called Integrated Planning and Control (I.P.C.). Why the term "integrated" was chosen is not clear. Comprehensive planning and control might even have been a more accurate description. But it is a small point and the name has now become firmly established in company language. It was Staiger who more than any other executive, saw to it that I.P.C. was developed and adopted by the company.

As a by-product of preparing the United Kingdom plans for 1961 a "planning guide" was produced which included an outline of the many tables and charts that an operations unit should submit to corporate executives in Toronto as its annual plan. The objective was to create identical systems of reports and procedures in all the operations units. It was immediately apparent that the kind of information that would flow from the operations unit to the corporate group would differ greatly from that contained in the normal balance sheet and income statement format, both in terms of what was and what was not included. This, and related matters, caused a serious division among senior executives, to the point that the I.P.C. approach was in danger of being jeopardized. The Vice President of Finance, C. F. Herrmeyer, a long-time Ferguson executive, opposed the approach and in the end retired from the company. His place was eventually taken by K. C. Tiffany.

Annual plans for the United Kingdom operations unit in their new format took eight months to complete. It was then realized that little attention had yet been paid to constructing a system of monthly control reports that would provide the corporate group with information about the operations units over the planning period. A system of control reports was hurriedly devised. The most important concept adopted here was to make all comparisons of current operations against planned operations, and not against realized operations of the preceding year. The view was that once the corporate group had approved plans for the following year, judgments relating to performance should be based on planned and not past performance. From the periodic control reports the corporate group could tell at a glance the extent to which plans were or were not being achieved by the various functional divisions. Variances from plan, whether over or under, attracted attention, invited explanation, and induced consideration of remedial action and such material came to be the substance of the control reports sent to the corporate group. Attention could be limited to important variances as far as these reports were concerned – an important consideration when it is remembered that the corporate group was to be kept small.

The planning system as developed in the United Kingdom in the first year was incomplete and too detailed. It was too detailed principally because it was not recognized at the time that the planning process had to be confined to those individuals who occupied managerial positions – that is, to those who were in a position to initiate change within defined areas. There was the added difficulty that no one was sure which positions were and which positions were not essentially managerial in character. Consequently too many individuals were asked to make plans and too much detail was included. The overwhelming significance of the conceptional distinction between managerial and other positions was not fully realized until the same mistakes had been made in introducing I.P.C. into the Australian and North American operations units.

As far as the incompleteness of the approach was concerned, this was realized when I.P.C. was introduced into the Australian operations unit in 1961 and it led to another important conceptual break-through. Previously, the concept of "profit impact" was used essentially to judge whether specific action programmes were justified. It was now decided that the approach should be extended so that it could be used to exercise control over the operation as a whole. Plans should be outlined so that they would show the details of each division's expected impact on profits, with the profit impact divided into (1) the portion arising from various factors beyond the control of management, that is the portion which would have arisen even if management had taken no *new* action and (2) the portion arising from management actions (action programmes) planned. The first was in turn subdivided into that arising from external factors and that arising from decisions taken in the past – that is, the carry-over effect of past action programmes.

Each division in the operations unit would show the details of its plans in terms of their expected impact on profits, divided into factors over which it had no control, and those directly related to new action programmes it was introducing. For example, increased efficiency in a department as indicated by a reduction in expenses without a decline in its "output" would be shown as having a positive impact on profits. All the individual departmental or divisional plans would then be added together to provide an over-all statement of profit impact for the operations unit. The total of the profit impact could then be added to the realized profits of the preceding year to obtain an estimate of expected profits for the new year. The spotlight, as it were, was in each functional division and in each department turned on "profit impact." This emphasis on "profit impact" of managerial action taken could, in a sense, be regarded as a technique for encouraging the growth of the entrepreneurial spirit in the managerial ranks of a massive organization.

The approach also had the great advantage that the performance of executives was separated from fortuitous developments for which executives could take no credit. It completely replaced the traditional budgeting approach. Instead of arriving at profits by showing revenues and expenditures in detail, the approach was to show the profit impact of action to be taken, the profit impact of factors over which management had no control, and total profit impact. No longer was a healthy total profit position a sufficient criterion for appraising the performance of executives. The control reports at regular intervals could, of course, show realized profits resulting from management action against planned results from that source.

In early 1964 it suddenly became obvious that one characteristic aspect of the chosen planning and control reporting approach would have to be changed. This was the practice of requiring plans and reports of the operations unit to be submitted on a standardized system of tables and charts grouped on a functional division basis. Plans were becoming stereotyped because they consisted essentially of filling in "blank spaces." Besides this, the plans of one operations unit following that approach suddenly totalled 336 pages, and it became almost certain that no one would have time to examine reports of that size. A further disadvantage was that by dividing the plans information on a functional basis it was difficult to appraise the total effect of a major programme – say the introduction of a new machine. A new approach was taken and 336 pages were reduced to 77, and of these only 50 needed to be read, the remainder being for reference only.

Several things made this possible, particularly the use of the "impact on profit" concept. The functional division of material in plans and control reports was discarded. In its place came (1) a summary of the details of planned profit impact, (2) a discussion of the assumptions underlying that portion of expected profits which was seen as arising from influences other than management action, (3) a discussion of the major action programmes – including their individual impacts on profits, and (4) a discussion of the areas where the company might be vulnerable during the year. The massive amount of details that the "information needs" study had recommended to corporate were therefore reduced to a few crucial and sensitive details presented in the form of "profit impact."

From this intensive review of the form that plans presentation should take, there also arose a new concept for the control reports. Instead of basing control reports essentially on a comparison of realized results against planned results, it was decided to base them on an estimate of realized results for the year as a whole against planned results for the year as a whole. "Latest look at the future" instead of "latest look at the past" was the criterion adopted. One interesting result of this approach was to show the amount by which estimated profits arising from management actions for the year as a whole varied from planned profits arising from such actions; and this in turn would tell the General Manager at a glance, whenever the latter was larger than the former, the extent to which he would have to try to revise his action programme if he was to meet his plan.

The simple concept that the function of a manager is to make things happen pervades this approach, and the reporting procedure is designed to show when he is or is not successful. It is in this way that the somewhat esoteric qualities of imagination, initiative, and ambition are translated into identifiable quantities related directly to profit maximization.

Also the approach described above gives increased meaning to the company's policy of decentralization. To say that the company's policy is one of decentralization is, in an operational sense, quite meaningless since full decentralization is impossible in any large organization. What is required is to define the division of responsibility between parent company and operations unit – as has been done – and then to apply criteria of performance to those areas where the operations unit enjoys responsibility. The planning and control procedures adopted emphasize the areas where the operations unit does enjoy responsibility, and it shows the impact on profits from actions for which it is responsible. Decentralization can be successful only when an operations unit knows precisely what its area of responsibility is and when it and the parent company are able to see what contribution to profits its own actions have made.

The planning procedure generally speaking involves (*a*) the President of the company setting profit goals for the world-wide organization as a whole for the succeeding year, (*b*) the General Managers of the various operations units submitting to the President a tentative plan of what the contribution of their individual operations unit will be to that over-all objective, (*c*) the General Managers and President agreeing on the tentative profit objective of the individual operations unit, (*d*) completion in detail

8 / Opening of the North American combine plant, Brantford, Ontario, on June 9, 1964. Left to right, Tom Carroll, Australian-born pioneer combine engineer, Honourable Mitchell Sharp, Canadian cabinet minister, and John G. Staiger, Group Vice President Farm Machinery.

9 / Listing of Massey-Ferguson Limited shares on the New York Stock Exchange on March 7, 1966. Ticker symbol is MSE. Left to right, R. W. Main, Secretary; K. C. Tiffany, Senior Vice President and Vice President Finance and Administration; Keith Funston, President NYSE; A. A. Thornbrough, MF President; Robert G. Stott, of Wagner, Stott and Company, specialist in MF stock; and H. G. Kettle, Vice President Public Relations.

10 / Assembled in a conference room at the new ICM plant in Aprilia, Italy, are senior members of MF management from around the world.
A. A. Thornbrough, President, sits in the centre of the far table. Going round the room from his left, and then down and back the aisle in front of him, are: K. C. Tiffany, Senior Vice President and Vice President Finance and Administration; J. E. Mitchell, Group Vice President Industrial and Construction Machinery; G. K. Blair, Special Assistant to the President; Dr. Ursula Brinkmann, Special Assistant to the Vice President Finance and Administration; H. G. Kettle, Vice President Public Relations; J. W. Beith, Vice President and Managing Director United Kingdom; Dr. B. F. Willetts, former Deputy General Manager United Kingdom; R. W. Main, Secretary; J. J. Jaeger, Vice President Research and Development; C. L. Baker, Director Engineering Farm Machinery Group; J. E. Williams, General Manager Brazil; W. Reed-Lewis, General Manager Mexico; S. R. Wilson, Director Management Structure and Processes and Director Logistics; J. Winstanley, Director Marketing Engines Group; T. H. R. Perkins, General Manager Northern Europe, Engines Group; G. E. Smith, Director Engineering, Engines Group; H. Lymath, General Manager International Operations, Engines Group; W. K. Mounfield, Assistant Secretary; J. A. Evans, Director Legal Services; P. Poniatowski, General Manager Southern Europe, Engines Group; S. Wallach, Special Assistant to the Group Vice President Farm Machinery; V. O. Griffin, General Manager Latin America, Engines Group; Dr. L. B. Knoll, General Manager South Africa; H. P. Weber, General Manager Australia; H. A. R. Powell, Managing Director, Massey-Ferguson Holdings Limited; Dr. F. Fadda, General Manager Italy; M. I. Prichard, Group Vice President Engines; J. G. Staiger, Group Vice President Farm Machinery; B. C. Bell, Co-ordinator Spanish Operations; H. A. Wallace, Vice President Manufacturing; H. Vajk, General Manager France; P. J. Wright, Vice President and General Manager Export; Dr. R. Durrer, Assistant General Manager Germany; R. A. Diez, General Manager Germany; J. P. Wleugel, Treasurer; L. J. Boon, Director Special Operations; and J. D. Goodson, Director Marketing ICM Group.

11 / Dr. Flavio Fadda, General Manager Italy (left), conducts Italy's Minister of Industry, Giulio Andreotti, on a tour of the Aprilia ICM plant.

12 / Members of MF's Corporate Strategy Committee pause in Rome's Piazza Navona for a rare picture showing all five of the company's top policymakers together. From left to right are: M. I. Prichard, Group Vice President Engines; J. G. Staiger, Group Vice President Farm Machinery; Albert A. Thornbrough, President of Massey-Ferguson Limited; J. E. Mitchell, Group Vice President Industrial and Construction Machinery; and K. C. Tiffany, Senior Vice President and Vice President Finance and Administration.

13 / Signing agreements with Motor Iberica, December 6, 1966 in the conference room of the Barcelona Chamber of Commerce. Left to right: K. C. Tiffany; A. A. Thornbrough; Don Gerardo Salvador, Chairman of Motor Iberica; and J. F. Sonnett, a Managing Director of Massey-Ferguson Services NV.

by each operations unit of its plan, (*e*) submission of the plans of all the operations units to the President, their review at a "plans review" conference at Toronto, and approval of them by the Executive Committee of the Board. Once plans are approved subsequent performance is judged against those plans.

Similarly, an annual "product planning conference" is held which is attended by the General Managers and, of course, by engineering and marketing representatives of the various operations units. The purpose of this conference is to review plans for developing new or improved products, to approve such plans, and to make all operations units familiar with the over-all current and planned product engineering activity of the company. Centralization of control over engineering activity makes such a conference essential.

Throughout the year, that is between annual conferences, co-ordination of the company's various operations is provided through the executives of the parent company. The President talks frequently to the various General Managers and visits them personally during the year. The Vice Presidents, each being concerned with specific functions, ensure co-ordination within their respective function in each operations unit (that is, for example, the Vice President Manufacturing of the parent company would at all times be able to render advice to and receive reports on manufacturing divisions in each of the operations units). This necessarily involves frequent visits to operations units and to special operations scattered around the globe; and it is not unusual for a corporate Vice President to look back at the end of the year and find that he has spent half the year away from Toronto. Inevitably this means that the corporate group emerges as a collection of men unusually well oriented in the international business environment.

Finally there was the matter of appraising the various capital projects presented to the parent company for approval. A capital expenditure request procedure was formulated. On the relevant request forms an estimate appears of the annual rate of return to be expected from the project – whether a new product, a cost-saving expenditure, replacement expenditure, or capacity expansion expenditure – and also of the pay-back period. The emphasis is on the addition to profits, that is on marginal profits, that capital expenditure will produce, even though accurate estimates of it are not available. In appraising the cost of new machines or of expansion of output of existing machines account is taken of the reduction in unit overhead costs that would be involved in increasing the volume of total output. That is, account is taken of the concept of marginal or variable costs. A capital expenditure request must be approved first by the General Manager of the operations unit, then by corporate executives, then by the President of the parent company and finally by the Executive Committee of the Board of Directors of the parent company. The Executive Committee had always been required to approve capital expenditures, but now, procedures prior to the committee receiving the request are rationalized and formalized, and information is generated that enables projects to be appraised with the criterion of relative profitability. Further refinement in appraising the attractiveness of alternative opportunities for investing the company's capital may be expected.

There was yet another aspect to developing a comprehensive system of planning, and of supervision and control; it related to appraising the over-all financial performance of the company and of its individual operations units and to planning for the provision and allocation of capital resources between operations units. On the operational level, as we have seen, Integrated Planning and Control provided control through application of the variance principle (i.e. showing the variance of actual from planned results) in both environmental and managerial areas, and it included information on action taken with respect to important variances. But it did not provide data on the relative efficiency with which an operations unit was utilizing its assets, or capital, as indicated, for example, by its over-all rate of return on assets. Obviously this was the kind of information ultimately required if the corporate group was to allocate its capital in a way that would lead to global profit maximization.

But the problem here was not merely one of devising an appropriate flow of data. It was necessary first to create a capital structure for each operations unit that would reflect the resources it required for its own operations. This task fell to K. C. Tiffany, who in 1961 became Vice President Finance of the parent company. In the past the practice of a particular subsidiary making loans or granting credits to others had obscured the cost of capital that should rightly be attributed to a particular subsidiary; in this way it compromised the company's avowed policy of granting each subsidiary or operations unit as much operational independence as possible. Only if the assets and debt of an operations unit properly reflected its own volume of business could that unit be held operationally responsible for efficient utilization of its assets and for servicing its debt.

These problems were particularly urgent for the United States company, which was part of the North American operations units, and for the United Kingdom operations unit. Projection of future capital requirements of the United States company as well as the nature of its existing capital structure showed it to be heavily under-capitalized. Its short-term obligations to banks, to the parent company, and to other operations units, particularly the United Kingdom, were much too large. Short-term funds were being used for long-term purposes and were being diverted from the United Kingdom operation where they were needed for capital programmes and expansion. Nor was there any possibility that future operations would reverse the position, since the United States company was not expected to be a major exporter to the United Kingdom; and the internal funds of the parent organization were insufficient to meet all capital requirements after paying dividends, servicing its long-term debt and convertible preferred stock, and financing its operations in Canada – there being no separate Canadian operating company.

A plan was devised that involved relations between operations units being placed on an orthodox commercial basis, and providing each operations unit with a capital structure adequate for its operations; giving managers of the operations units increased responsibility for the utilization of and return on capital of their companies; and transforming Massey-Ferguson Limited into a holding company, but with its management – the "corporate group" – continuing to be responsible for over-all control and management of debt and assets.

Implementation of the plan began with a reorganization of the capital structure of the United States company. It was estimated that that company would require $60 million of long-term funds and it was realized only too well that the net asset position of the United States company ($80 million) and its earnings history did not constitute a strong base for borrowing such an amount of money. But in early 1963 the United States company (Massey-Ferguson Incorporated) was able to sell through direct placement $35 million 5¼ per cent serial notes with repayment extending from 1966 to 1982, and $25 million 5⅞ per cent subordinated notes due in 1984. To make this possible the parent company guaranteed to maintain net assets of its United States subsidiary at $80 million, to hold its own debt together with that of its subsidiaries at or below 40 per cent of net consolidated tangible assets, and to continue to hold all the capital stock of its United States subsidiary company. The borrowing ability of the Massey-Ferguson companies was augmented by a rights issue of common stock at the rate of one share for each ten outstanding.

Next management turned its attention to the capital structure of its United Kingdom operations. Repayment of loans by the United States company and the new practice of immediate settlement by operations units for goods received from operations units gave the United Kingdom operation new funds. By transferring to the parent company many subsidiaries that had been owned through the United Kingdom operation, the total future financial burden of that operation was reduced, and in any case it was a move consistent with the developing character of the parent company.

However, since an existing three-year bank loan was within eighteen months of its maturity date, further financing seemed desirable. Investigations into placing a long-term loan in the London capital market indicated that such an approach would not only be costly but would also, in terms of necessary warranties and conditions, be undesirably restrictive. Instead, the company negotiated a £9 million loan with its London bankers and arranged with them for a £9 million line of credit for seasonal purposes, with permission to obtain an additional £4 million of either type from any source. Conditions agreed to were the same as those relating to borrowing of the United States company.

Then there was the interest-free loan of the parent company to the United Kingdom operation. With the agreement of their London bankers and of the Bank of England, the United Kingdom operation repaid about half of it, partly in cash and partly through the transfer of subsidiaries, to the parent company. An interest charge was imposed on the remainder of the loan, and repayment of it could be effected whenever the company wished.

In this way the United States company and the United Kingdom operations unit were given capital structures that reflected their individual requirements. But there remained the operation in Canada, which had always been reflected in the accounts of the parent company. In early November 1964, a new Canadian company, Massey-Ferguson Industries Limited, was formed. It is responsible for all operations in Canada, and to it have been transferred all the Canadian operating assets formerly on the books of the parent company. At the same time plans were made, and were soon executed, to place an appropriate amount of outstanding obligations with the Canadian company and, since that amount was smaller than the debt of the parent company, to retire remaining parent company long-term debt. The new Canadian company and the United States company constitute the company's North American Operations Unit, that is, they remain one unit from an operational point of view.

Underlying these changes are important concepts and principles which, together with several not explicitly described, summarize the company's approach to its financial activities. The parent company, Massey-Ferguson Limited, is an international holding company which, through its executives (the corporate group) and within the organizational structure previously outlined, provides management guidance to its subsidiary companies around the world. It owns no fixed assets and has no long-term public debt outstanding, both of which appear on the balance sheets of the subsidiary companies, although it does make long-term advances to its subsidiaries. Its assets include primarily the stock of its subsidiary and associate companies. Its subsidiary, Agrotrac S.A., a Panamanian company, borrows short-term funds in various currencies and makes these available to operations units. However, the largest portion of an operations unit's short-term fund requirements are in practice satisfied by local banks.

The amount of invested capital and retained earnings of each subsidiary is generally maintained at minimum levels consistent with reasonable credit standing in the local financial markets. Where political and exchange risks are greatest, invested capital is especially minimized. While an operations unit obtains long-term funds through capital investments of the parent company (and in some cases also of minority shareholders) and retained earnings, as well as from long-term loans from the parent company, the objective is that locally contracted long-term debt should be about equal to local net fixed assets. The local operations unit is responsible for the profitability of its operations and so is responsible for obtaining a satisfactory return on its invested capital and for servicing its debt.

Since the parent company is concerned with international profit maximization, it is expected that dividend payments to the parent company by subsidiaries may have to be larger than those required for meeting the dividend payments and expenses of the parent company. The difference would then be available for allocation to new projects or to the expansion of existing ones. Operations units for the most part are not to invest in or make loans to other operations units or associated companies.

All borrowing and investment activities are the responsibility of the Finance and Administration Division of the parent company and that division must therefore ensure that the funds required for executing the approved annual plans of the operations units will be available at minimum cost commensurate with acceptable exchange risks.

As a result of the capital reorganization previously described each operations unit has the type of balanced capital structure that is appropriate for its operations and that reflects the size of its business. Consequently it is possible to appraise the relative efficiency of the way each operations unit uses its capital by examining rates of return on it. Detailed monthly reports using statistics of return on assets employed, asset turnover, return on sales, and average assets of *each operations unit* are regularly presented to the Board of Directors of the parent company, and to make these statistics even more meaningful they are expressed in terms of a moving average and presented in chart form. Such is also the case for sales data, gross profits, expenses, profits before exchange adjustments and taxes, net income, and earnings per share. One page of the regular report shows at a glance the extent to which the sales and profits of each operations unit vary from plan, and a few additional pages tell why. Thus if the company, or any of its operations units, encounters serious difficulty in future it will be the most comprehensively documented difficulty that it has ever experienced in its long history. This places great responsibility on the Executive Committee of the Board of Directors, and on the Board itself, for with very little effort they can now be intimately informed on current and expected developments.

In addition to receiving comprehensive reports on operations, the Executive Committee of the Board continues to retain responsibility over specific areas. The new planning procedures just discussed made it desirable to redefine these in 1964. The Executive Committee is now required to approve the annual plans of the company and its subsidiaries, including capital expenditure plans. Through the year it authorizes and releases all individual capital expenditure projects in excess of a certain minimum amount. Its specific approval is also required for matters such as salaries to and incentive payments for senior executives, and for donations, labour agreement policies, hiring of consultants, legal fee accounts, investments in subsidiary or associate companies, product rectification expenses in excess of a stated minimum, appointment of directors and officers of subsidiaries, fees of directors of subsidiaries, medium- and long-term borrowing of subsidiaries, and major contracts or agreements. The Executive Committee enables the controlling shareholders to be more intimately involved in policy and certain operational decisions than would be possible through the Board of Directors alone, for all its members are expected to take a deeper interest in the company's affairs than is normally expected of Board members.

It will be observed from this discussion, that the organizational structure and system of controls that the company has adopted for its world-wide operations were the product of evolution and of trial and error; they did not arise through the application of a detailed preconceived plan. But, significantly, evolution in the application of the principles of organization and control was made possible by the strong emphasis placed on the importance of such matters by Phillips and Thornbrough. It is difficult to see what could initially have been accomplished in this area without the strong support of Phillips. No company is likely to make rapid progress in organization and control without such support because so many other areas are affected. This is why a company with severe organizational and control problems is not likely to overcome them without a change in management.

It is certain that the company's organizational structure and its control procedures will change in future. The principle that managers achieve improved results by making things happen will ensure this, and there will be new problems created by further growth and increasing complexity of the organization. The attempt to achieve increased efficiency in co-ordinating the various operations of the company will undoubtedly continue to be worthwhile. Even so the system has already attained a remarkable degree of refinement.

CHAPTER 11

New foundations: North American operations

The application of rational concepts of organization, the development of a suitable organizational structure, and the creation of an efficient system of planning and control were prerequisites for the successful operation of a global corporation. But they were not the only prerequisites for achieving a basic improvement in the financial position of Massey-Harris-Ferguson. It was essential that the organizational principles of increased management control and improved procedures be applied to the major functional divisions of the company. Earlier it was seen that policies pursued in distribution had contributed heavily to creating severe financial distress in the organization in the middle 1950s, with control over major areas of the company's marketing activity resting in the hands of independent Ferguson distributors. Manufacturing costs seemed to be too high, the company did not own or control its own source of engines anywhere in the world, nor its source of assembled Ferguson tractors outside of North America. In engineering, developmental work seems in important respects, particularly on the Ferguson side, to have become insufficiently customer-oriented, and on the Massey-Harris side it was deficient in quality control. The need to control inventories had amply been illustrated, both in the manufacturing and the marketing divisions.

Many of the major changes in the functional divisions were to be introduced into the company's North American operations first. Of major long-term significance to the company's North American operations was the decision in late 1956 to operate through one integrated North American operations unit instead of through separate Canadian and United States divisions. How this was done we have already seen in the preceding chapter, where we also saw that for the first few years after 1956 corporate management also assumed line responsibilities for operations in North

America. These decisions to consolidate North American operations gave the new senior executives of the parent company the opportunity to deal directly with the problem of reorganizing the company's most troublesome operation.

After the excessive inventories of 1956 and 1957 were liquidated, there was a long period during which the company worked out an effective system for continued inventory control. Chart 7 shows the ratio of its inventories to sales in North American operations from 1955 to 1966. That chart accurately reflects the fact that even by 1959 adequate control had not been achieved, for excessive inventories again appeared in that year, but that remarkably sustained progress was made thereafter.

CHART 7

Massey-Ferguson Limited/North American operations

Ratio of inventory to net sales 1955-1966

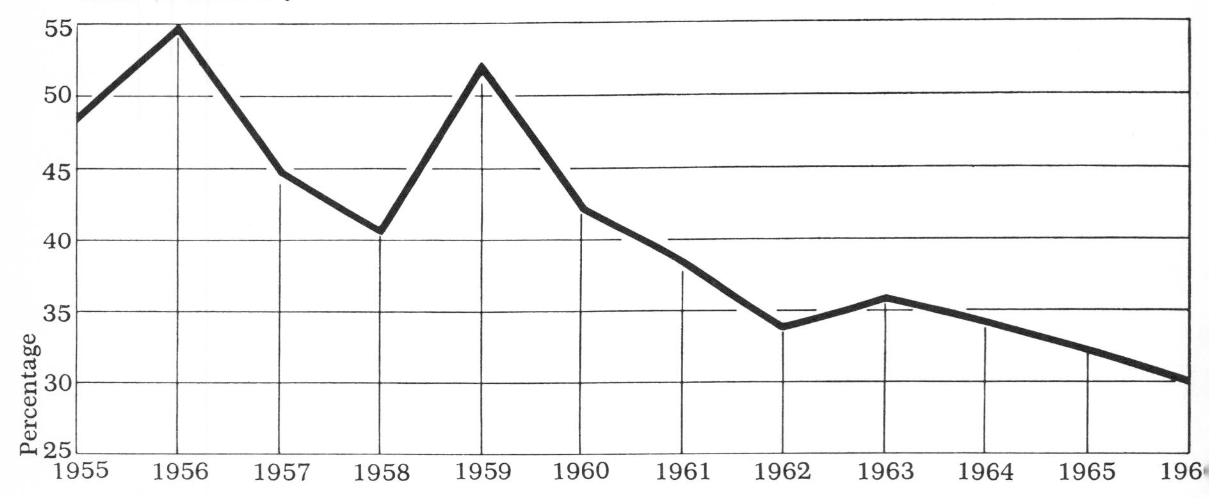

Nowhere was the need for new policy and reorganization more necessary than in the company's distribution system, where the past policy of having two competitive lines and two competitive distribution systems had created chaos. The first decision that moved the company toward a policy of having one line and only one sales outlet in each market area was taken almost immediately after the change in management, although the Executive Committee did not have an opportunity of formally approving it until August 10. The minutes of that Committee recorded the decision in this way:

Operations of the Company are to be unified so that the distinction now existing between the Massey-Harris and the Ferguson divisions of the business will be eliminated eventually. Sales-wise, the ultimate goal will be a single line of goods, or, alternately, two complementary non-competitive lines handled through a single outlet in each sales area. It may be necessary for a few years to maintain the present arrangement of dual, competitive lines in some areas such as the United States, but it will be the policy of the Company to eliminate these areas as soon as practicable.

As we noted earlier, to stop the spread of the two line concept it was also decided, immediately, that the Eastern Hemisphere would not manufacture or sell a red or "Massey-Harris" version of the new FE-35 tractor, as had been done in North America in the form of the MH-50 tractor. The ultimate objective of one line and one outlet was insisted upon in any decisions that the Eastern Hemisphere might take.

But the really difficult decisions relating to developing the "one line, one outlet" approach related to the problem of combining the Massey-Harris branch system with the Ferguson distributor system so as to form one system. No obviously easy solution was in sight and the new group did not immediately know what path to take. In commenting on the situation to the Executive Committee on September 5, 1956, Phillips said:

> The philosophy upon which our distribution system is based in North America gives rise to much concern. It is difficult to trace any common pattern, and in many cases, the justification for the present situation appears to be lacking. We are all satisfied that there is urgent necessity to clarify our thinking in regard to distribution and to give effect to it with the least possible delay. . . . I should add that our concern for the moment, in connection with distribution, is confined to the North American organization.

The problem was, should the company buy out the Ferguson distributors or should it retain both the Company's branches and the independent distributors, giving the latter the whole line of the company's machines? On October 3, Thornbrough recommended the latter policy for the present so that the new models of tractors could be announced in 1957. The Executive Committee approved the recommendation.

But uncertainty over this as a permanent solution remained and Thornbrough soon recommended that a consulting firm be engaged to study the company's distribution system and to make recommendations. Actually this move was, in part, prompted by a legal complication which had first appeared in August when Southland Tractors Incorporated, a former Ferguson distributor, took legal action against the company for damages arising from the two-line policy. It seemed that Massey-Ferguson might have to clear its intended actions in matters of distribution with the Federal Trade Commission before it could proceed. McKinsey and Company were retained in December to study the distribution system and to make recommendations which, if necessary, could be submitted to the Federal Trade Commission as outside objective opinion.

During this period (1957-58) Thornbrough, and then J. H. Shiner after he joined the company in December as Vice President Marketing, spent many hours trying to reconcile the basic and often very personal conflicts between the company's branches and dealers and the Ferguson distributors and their dealers. All schemes of dividing territories seemed operationally unworkable and impossible from an anti-trust law point of view. The really decisive turning point came when Thornbrough and some of his executives held discussions with their Utah Ferguson distributor. The distributor was losing money and it would have been impossible for the company to extend to the other distributors the assistance he wished to receive. Yet effective retail distribution was urgently needed. Thornbrough became convinced that to control its own wholesale distribution system the company would have to buy out the Ferguson distributors, and that the McKinsey officials, as third parties, might be able to develop a common approach to this problem.

By April McKinsey consultants were able to report their own view. The company, they recommended, should distribute its products only through its own branches and dealers. This was in harmony with the President's predilections and also those of Shiner. The company therefore began negotiations for terminating the independent Ferguson distributors. Thornbrough and Shiner, over a period of seven months, went from one Ferguson distributor to another using the company's airplane. It was not a pleasant task; the two men faced the tears, recriminations, and even deceptions of independent businessmen who were, after all, being asked to sell their businesses.

By fall all but one distributor had been terminated at a total cost of about $2 million. If the termination programme had been unsuccessful through, for example, intervention of the Federal Trade Commission at the request of a group of distributors, the company would have been in a most difficult position. As it had stopped production of the Massey-Harris tractors while termination was under way, the company would have been obliged to permit its branches to market the Ferguson-system tractors, a move which would have caused even greater consternation among the distributors than the earlier introduction of the MH-50 had done. A major calculated risk had successfully been taken and it for once paved the way for the company to develop its North American marketing system based solely on its own branches.

From this clarification of matters relating to distribution in North America, management subsequently evolved the policy of operating through its own branch system in every country in which it controlled manufacturing operations. In countries where the company did not have manufacturing facilities it turned, more through circumstances than design, to distributing its product almost entirely through independent distributors. From the Massey-Harris side of the organization it had inherited branch operations in Denmark, Belgium, New Zealand, Brazil, Argentine, Uruguay as far as purely export territories were concerned. New management decided to dispose of its branches in Denmark, Belgium, New Zealand, and Uruguay to local independent distributors.

The matter of disposing of these branches arose at a time when the parent company faced a shortage of cash for modernizing and expanding its manufacturing facilities and such sales generated cash. It was also one way of moving toward a single distribution system, and in some cases the local Ferguson distributor seemed to be more effective than the local company branch. Besides this, some members of management believed that the company would do better by operating through a local national organization than through its own organization. The results do not show uniformly that the company pursued the most appropriate policy. In the case of Uruguay the local Massey-Harris branch system may have been more effective than the distributor subsequently was, and in Italy sales rose sharply after a local Massey-Ferguson company was formed.

Undoubtedly one of the most important changes in the company's approach to distribution was the replacement of the somewhat narrow sales organization with a comprehensive marketing function. J. H. Shiner introduced that function into the company and also marketing techniques derived from the automobile business that had not previously been employed in the farm machinery industry. To facilitate the successful introduction of new techniques, new and younger branch managers were appointed in most of the branches. Professionally trained personnel were taken onto the staff at Head Office.

The approach Shiner took to developing the marketing function was one that concentrated detailed attention on the customer. Market analysis replaced the salesman's hunch in forecasting sales and in planning product changes. What are the various groups of customers? What product line and mix are favoured by the various groups? What kind of services – parts availability, service, credit, farm management counsel, dealer inventory – must be provided with the product? What kind of dealer is required to provide the necessary services? What kind of policies in advertising, terms, facilities, field organization, should the company have to support its dealers? What priorities should the company establish in each of the aforementioned areas so as to maximize profits? These were the considerations which were emphasized in establishing the new consumer-oriented marketing organization and which directed the research activities of the new group and of the consulting services used.

The relationships between market research, product planning, and engineering were so defined that appraisal of the character of existing and expected demand would strongly influence the company's plans for changes in its product line, and such planned changes would in turn largely decide the character of its engineering activity. It is probable that the weaknesses of the company's product line in preceding years resulted, in large part, from a failure of management to understand these relationships or, at least, from a failure to be strongly guided by them. Such a failure invited lack of co-ordination between marketing, product planning, and engineering.

A comprehensive staff-line organizational structure was outlined for the marketing division. Within that structure staff services in advertising and sales promotion, product planning, general marketing and economic research, dealer development, distribution planning and others were specifically provided for and were available to the general sales managers of Canada and of the United States. This structure emphasized that marketing involved a group of functions each with specific areas of responsibilities. The two general sales managers were directly responsible to the Vice President Marketing.

Relationships between the branches and Head Office were changed materially. Formerly the branches had, in some respects, been islands of independence in the organization. Since formal definition of their areas of responsibility, including formal guidelines to operations such as rules governing credit operations, had largely been absent, the individual branches had enjoyed considerable freedom of action; at the same time Head Office, not having defined branch functions or operational guidelines, was required to become involved in many of the details and decisions of branch activities. This approach generally was reversed. The responsibilities of branch managers were defined, and gradually guides to operations were developed. Branch managers enjoyed independence within those defined responsibilities, and Head Office supervision shifted somewhat towards examination of data reflecting the performance of branches.

Experimentation with new types of retail outlets was also undertaken. In some areas of the United States the company was simply without adequate dealers and new company-owned retail outlets were introduced. They took several forms: individual stores, a "mother" store with a group of smaller subsidiary outlets in the surrounding territory, and a store selling a large range of supplies – not merely machinery – required by farmers. These approaches to retailing were not conspicuously successful, in part because of the failure or inability of the company, for a period, to use "dealer" type management in the stores and to devise a system of control appropriate for those operations. In fact, there is no certainty of the speed and direction with which changes in retail distribution will take place in North America. Experiment serves primarily the purpose of ensuring that the company will remain at the forefront of change and, to a lesser extent, of giving the company retail outlets in agricultural areas where, for historical reasons, it has not been well represented.

The company began to make much more intensive use of national selling and advertising campaigns than it previously had done. It was Shiner's view that the way to improve the quality of dealers was to make the company franchise valuable. To do this the company needed to show the dealer that it was supporting and assisting him through advertising campaigns, product development, and advice on accounting and business planning. In return for this the company could reasonably expect performance from dealers. Volume of sales of individual dealers was soon enhanced by ambitious advertising campaigns and also by a reduction in the total number of dealers in North America from around 4,000 to less than 3,000.

In 1957 the company introduced the new MHF-65 tractor to its dealers with an "On the Move," extravaganza-type travelling show, which gave dealers the impression that there was new blood in the company. To introduce the MF-85 tractor in February 1959 the company invited all its dealers and wives to Detroit for a "Show of Progress." It involved, it is said, the biggest one-day civilian airlift ever undertaken. Ten hotels and three banquet halls were used. At that show the company introduced, with closed circuit television, a western music television programme – the Red Foley group – which it then sponsored for a period on national television in the United States. On that programme farmers were told that they would receive a $100 cheque from Red Foley for every MF-35 tractor they purchased. No implement company had previously used national television or had attempted such a promotion. It was an amazing success. Some will still argue that Red Foley and not new management put Massey-Ferguson back on its feet in North America. All this was from time to time accompanied by "Bonus Banquets," "Parade of Profits," and "Bonus Fairs." There was an unusual momentum to it all.

The momentum helped to achieve gratifying improvements in the company's position. Charts 6 and 8 show that up to and including the year 1959 the company's sales and profits in its North American operations increased greatly, as did its share of the market. However, after 1959 and up to 1961 measures of performance deteriorated again. This was partly because 1959 performance had been artificially enhanced through a

CHART 8

Massey-Ferguson Limited/North American operations

Net sales 1955-1966

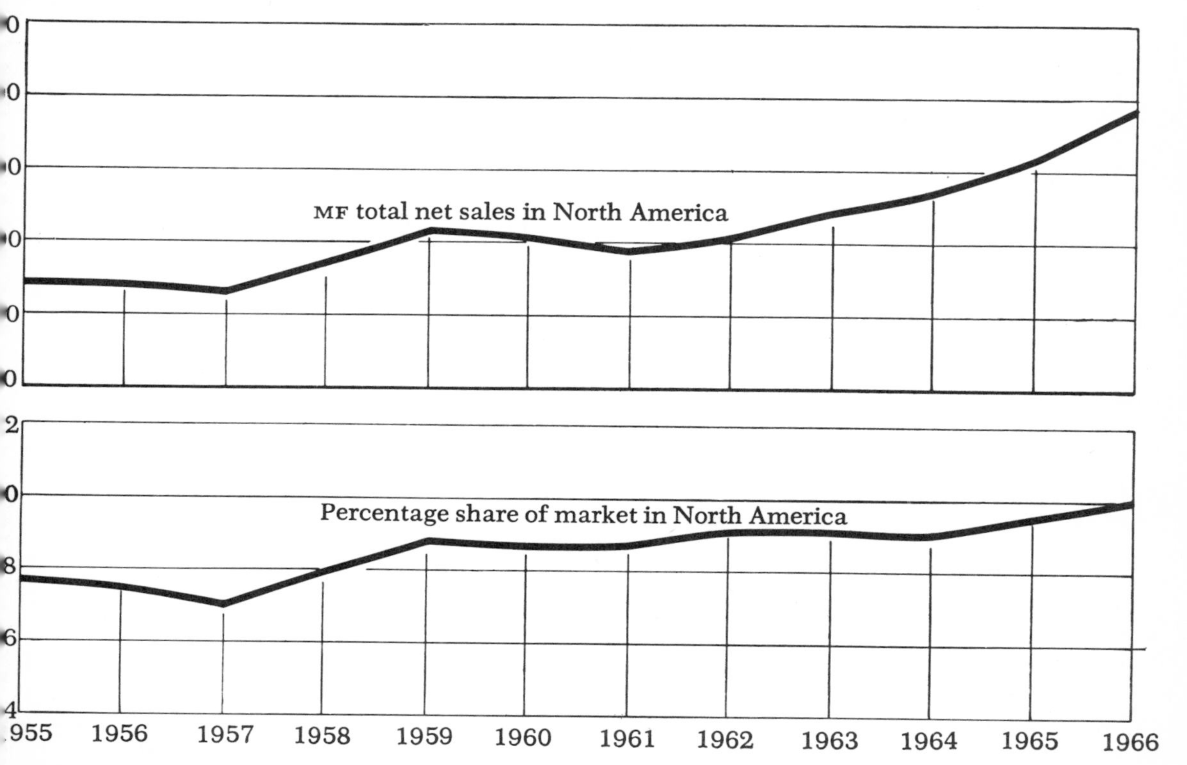

build-up of dealer inventories. But even by 1964 it could not be said that the company's sales and profit performance in North America was adequate. The over-riding reason for this was that it found it exceedingly difficult to penetrate deeply into the market for farm machinery in the mid-west United States, a region referred to as the corn belt.

For a period it was generally believed that with improvements in the company's product line, its name would become more familiar among corn-belt farmers and its products would be more readily accepted. The introduction of new combines, new tractors, and numerous other implements greatly enhanced the quality and completeness of the company's product line. Following their introduction surveys were taken to determine whether corn-belt farmers were becoming generally more acquainted with Massey-Ferguson.

The results indicated that they were not doing so in any significant way. New strategy seemed to be called for and it seemed to require moves that would cause mid-west United States farmers to regard Massey-Ferguson as a company permanently committed to the United States market – in a sense to recognize the reality of Massey-Ferguson's position, for it had in every way already become committed to that market. The strategy that was developed was essentially marketing and public relations strategy, even though it involved establishing an implement manufacturing plant in the United States.

Implementation of new strategy in marketing did not begin until 1965. Dissatisfaction with the results of the previous organizational arrangement in marketing led to a major change in that year. Four United States divisional general sales managers, in place of one United States general sales manager, were appointed. The divisions reflected major regional differences in agricultural activity and machinery demand. Each division was given its own headquarters and staff, and was made a profit centre so that its performance could be measured by profits and not just by sales volume. In this way staff and decision-making relating to profit maximization were moved closer to the market. Within North American marketing, strategy emphasis was shifted from general marketing and sales plans to individual marketing and sales plans for each major product line, i.e., tractors, combines, hay and forage machines, and general machines.

But the most dramatic changes occurred when the company decided that it would establish a new base for operation at Des Moines, Iowa. The nature of the move and the marketing consideration that led to it were explained by Staiger to his employees in this way on August 31, 1965:

The corn belt is one of our most important markets. It accounts for more than one-third of all farm equipment sales in the United States. It is the largest single market in the U.S. for combines, balers and implements and the second largest market for tractors. It is the most competitive of the U.S. regional markets and one in which our competitors are extremely strong. Five of them have their national executive offices in this region. There are 32 manufacturing facilities of our competitors located there. Massey-Ferguson does not have any manufacturing or executive facilities in this key market.

This concentration of executive offices and manufacturing facilities in the key corn belt market gives our competitors an identity with the market and with the corn belt farmer that we do not have. It provides them with a very strong market advantage.

To help counteract this and establish a similar identity in this key market for Massey-Ferguson, the American company, Massey-Ferguson Inc. has purchased a 590,000-square-foot manufacturing plant, complete with office facilities, at Des Moines, Iowa.

We are transferring to this plant, the welding and assembly of farm implements that are made primarily for sale in the corn belt. These implements are currently made in the Woodstock Plant and in Verity Plant at Brantford.

In addition, Massey-Ferguson Inc. is relocating its executive headquarters from Detroit to Des Moines. When office facilities in Des Moines are ready, the president of Massey-Ferguson Inc. and the officers of the company will move there. To bring yet another group of senior marketing staff management closer to this key corn belt market, management and professional personnel of the North American marketing staff groups will also move to the new U.S. company's executive offices. Because they are closely associated with the marketing effort, a substantial portion of the management and professional staff of the public relations department will also be relocated there. These three groups will be supported by small service staff elements.

In effect, the General Manager of North American Operations would spend most of his time at Des Moines after the move was completed. Feasibility of the shift of manufacturing facilities to Des Moines from Brantford and Woodstock rested heavily on the expected profits that would be generated by increased market penetration and on the reasonable price paid by the company for the plant at Des Moines built by the Solar Aircraft Company. Lower labour costs in Canada, as well as certain other lower costs, meant that the move involved a penalty in operating costs, but these, it was estimated, would be offset by the effect on profits of increased sales volume. The success of the marketing strategy underlying all these changes lies in the future. Its importance to the company is considerable. A further reason for moving the senior executives of North American operations to Des Moines arose from the experiences the company had had in a lawsuit in the United States. It seemed that its United States company was at a distinct disadvantage in those proceedings because its senior executives were located in Canada.

Selling the products of the company with efficiency through the development of a comprehensive marketing function in no way solved the difficulties facing the manufacturing activities. For years, management had been painfully aware that the company's cost of goods sold as a proportion of net sales was higher than that of its major competitors in North America. Profits in relation to sales, shareholders' equity, or assets suffered as a result of those high costs. After the merger with Harry Ferguson Limited the problem became dangerous.

What was wrong? While the problem had been known to management for many years, it had never been analysed with a view to developing a programme of corrective action. We have already noted that the approach essentially had been one of almost continuous exhortation by top management to "cut costs" without accompanying recommendations for management reorganization and for action relating to structural changes that would make a permanent reduction in costs possible.

There were a number of reasons why the company's costs were too high. The manufacturing division like the others, suffered from a rather loose and ill-defined organizational structure. There was an excess of inefficient capacity and a deficiency of efficient capacity. Systems of production, cost and quality control, seemed to be loose and inefficient. Violent seasonal swings in production reduced average utilization of equipment and disrupted the labour force. The lines dividing areas of responsibility of management and unions appear to have become blurred. An efficient incentive programme had not been adopted. The company appeared not to be manufacturing a sufficient portion of the goods that it sold – depending heavily on outside manufacturers as suppliers of tractors and of major components for its machinery, including engines. The advantages of inter-company shipment of component parts had not yet been realized.

Many, although not all, of these difficulties were outlined in a memorandum Thornbrough wrote to the President in March of 1956. There seems to have been a tendency over the years to assume that shortage of capital precluded basic improvements, a view that obscured the fact that a number of improvements did not require capital and some – such as closing redundant and inefficient plants – would actually generate funds. In any case the concept of "capital scarcity," unless it is related in some way to lack of capital market confidence in management and in the projects recommended by it, is not a very meaningful one.

The preliminary steps for a few important improvements were taken a number of months before the change in management – the planned shipment of tractor components from the United Kingdom operation to the Detroit factory, investigation into adoption of a standard hour incentive plan in North American factories, and commissioning a firm of consultants to devise a production control system for the company. It was to be a number of months before these initial steps resulted in improved operations.

As in all divisions, it was necessary to introduce a more efficient organization and control system into manufacturing operations in North America. The former senior official specializing in manufacturing had been required to leave immediately after the change in management and plants were shut down. Not much more was accomplished until the new Vice President Manufacturing, H. A. Wallace, assumed his responsibilities on December 1, 1956, although several less senior appointments had already been made. Wallace was given a free hand to introduce changes. Among the first changes he made related to organization, for he was surprised that the works manager previously had occupied a relatively junior position in the hierarchy. In his words:

We changed the names of these Works Managers to General Factory Managers and the General Factory Manager reported directly to me. So we just took away all this interference on decision-making or lack of it. We built up the General Managers of the factories to an executive level where they were responsible for operating their factories. These fellows grew like desert flowers that had a sprinkle of rain because of the opportunity to stand up and be counted and really do a job.

This decision to elevate the works manager to a position where he could be held responsible for the total operation of his factory led to other organizational changes. Since costs of goods produced by the factory depended in part on the success of the company in purchasing raw and semi-finished materials, factory purchasing was placed under the general factory manager. Similarly so with other functions, and liaison between manufacturing and engineering was achieved by locating the small production engineering group in the engineering division, although responsible to the general factory manager. The principal divisions in the factory, the head of each answering directly to the general factory manager, came to be the following: Purchasing, Facilities, Factory Accounting, Personnel and Industrial Relations, Quality Control, Production Control, and Industrial Safety. With this organizational structure the general factory manager had both the necessary responsibility and authority for the successful operation of his factory.

Within each one of these major subdivisions of the manufacturing division an explicit organizational structure was created, responsibilities were defined, and commensurate authority was granted. One of the important immediate problems that had to be faced was found in personnel and industrial relations. As early as July 1956 the following was reported to Thornbrough:

An immediate policy decision is required with respect to our impending labour negotiations. During and since the war, due to compromise, labour union pressure, etc., progressively our piece-work standards have become loose. In an attempt to correct this situation and reduce costs, we propose to convert our present piece-work price wage payment plan to a standard hour plan.

Not only had the piece-work standards become loose but, at the Racine factory at least, the system had really been nullified. The practice apparently had developed, with the tacit approval of management, of "policing" average hourly earnings so that all received about the same income. The incentive element in piece-work pay was thereby effectively neutralized.

The problem for the company was really one of maintaining stable labour conditions while introducing a new incentive system and the many new procedures of the production control system. All this was made somewhat easier by the company's adoption of a definite policy of paying wages that were competitive within the relevant market areas of the various operations. Previously the company's wages had not always been fully competitive. But the new approach was not accepted quietly because it raised some rather fundamental issues of the role to be occupied by management and by labour. Their respective roles had become blurred in compromise settlements made over the post-war years. All this did not really come to a head until the dispute late in 1958 between Massey-Ferguson and union locals in the Toronto, Brantford, and Woodstock plants.

J. A. Belford, a McGill University graduate and previously with Canadian National Railways, had been appointed Director of Personnel and Industrial Relations in April 1957 and had begun to define principles that the company should pursue in its negotiations. The dispute went before a conciliation board of the Ontario government. In its submission the company raised the crucial issue of the role of management by quoting a letter it had sent to all manufacturing division employees in Canada on September 2, 1958. It read in part:

> A great number of the differences between the Company and the Union hinge on two questions:
>
> What is the proper job of management in a unionized plant?
>
> What is the proper job of the Union in the plant?
>
> The Company holds that it's the job of management to manage – to direct operations and to make decisions.
>
> The Company holds that it's the job of the Union to protect the rights of employees under the agreement – to represent employees in processing grievances – to run its own affairs.
>
> We think the Union agrees with the Company that these are the Union's jobs. But the Union goes one step further. It also wants to take part in the management decision on what should be done under the agreement.
>
> For example, the present agreement requires that there should be "mutual consent" between the Company and the Union on retaining employees, because of their particular skills, during lay-off. This is the kind of mutual consent arrangement the Company thinks is wrong.
>
> Here's why:
>
> If the Union takes part in deciding what employees will be retained, how can the Union represent an employee who makes a grievance against that decision? Where the collective agreement has a mutual consent provision, *if the Union processes the grievance of an employee it is grieving against itself.*

This was the crux of the matter. Management decision may well affect particular employees adversely. If the union is party to such a decision, it cannot represent the aggrieved employees. This was the view the company emphasized in working towards a sharper delineation between the responsibilities of management and those of the union. As for resolving conflict, the company regarded settlement through economic confrontation of the strike as the ultimate procedure to be adopted when all else fails. It also followed the practice of permitting its professionally trained personnel and industrial relations executives to bargain with the labour unions.

It appears that the aforementioned principle relating to division of responsibility became fairly well understood and that it was partly responsible for the success of the subsequent introduction of the new production control system and the standard hour incentive system. The latter is a system that involves measuring the output of a qualified worker operating at normal pace using the methods approved by the company, and of paying the worker an assured minimum hourly rate but in addition an incentive payment for output in excess of normal. Each job is evaluated to determine how long it would take the worker referred to above to produce 100 pieces. If, for example, it took 8 hours the worker would receive the wage classification rate for the job, as indicated in the wage agreement, multiplied by eight. If in those eight hours, he produced 150 pieces he would be said to have worked 8 × 150/100 or 12 standard hours and would be paid for that number of hours. No limit is placed on performance. The incentive payment is a payment for performance. Management decides procedures and receives suggestions for improving procedures and if these are adopted special payment is made for them.

The standard hour incentive system was finally adopted. The company's interest in it and its importance to the company is simply explained. Consultants advised them that experience had shown output per man to be almost twice as large in factories employing the system than in factories using the traditional piece-work system. This was largely because in the latter there was no urgent reason to examine work methods thoroughly through time and motion studies, while it is a prerequisite for using the standard hour incentive system.

Introducing a comprehensive production control system proved much more difficult and took much longer than had originally been thought when consultants went to work on it in early 1956. Production control is the use of the most efficient methods available to match the output of the factories with the best forecast of demand so that products are available for delivery at the right time, in appropriate quantities, at least cost. The system therefore involves both quantitative and financial factors. The quantity and cost of labour, material, and overhead must each be so controlled as to achieve the aforementioned objective. The flow of raw materials, of inside and outside component parts, and of sub-assemblies must be controlled to ensure the maintenance of assembly schedules.

It did not help that the system initially recommended by consultants did not work. Also, at first, the view seems to have predominated that a single control system could be devised, then applied to North American plants, and finally transplanted into each one of the company's factories. This proved not to be possible and while the broad principles were universally applicable the unique problems of each factory made it necessary to design a system for each factory.

One of the first difficulties encountered in introducing a new production control system was that of passive resistance. The consulting firm that had been brought in prior to the change in management needed close internal co-operation. This was not forthcoming. Thornbrough asked that all the executives involved should be brought into a committee room. When this had been done he came in, explained the importance of the production control system, and said that opposition to it would be regarded as sufficient cause for immediate termination. This helped. But it did not completely solve the problem, and as late as 1963 major executive changes were occurring among North American factory management personnel. There is no doubt that the task of closing the gap between known technology and the ability of company personnel to understand and apply it took longer in manufacturing than in the other functional divisions.

All these developments in organization, personnel and industrial relations, and production control formed the basis for improvement in the operating efficiency of the manufacturing division. The other area where improvement was required was in the character of the manufacturing facilities.

It had been decided soon after the change in management that some of the company's North American plants – in particular, the Batavia, Market Street, Brantford, and Woodstock plants – would permanently be closed as soon as possible. This was accomplished within eighteen months, with the exception of Woodstock which remained open until 1966. Consultants were retained to plan the closing of the Batavia plant. There was considerable concern that the community reaction to the company's closing down of an operation that had been in the same location for almost fifty years and, more particularly, to transferring production out of the country to Canadian plants, would be unfavourable. No major difficulties were encountered.

It was also decided that tractor manufacturing would best be concentrated at the Detroit factory, where the Ferguson tractors had been assembled, and that the Racine plant manufacturing Massey-Harris tractors should be closed. This, too, was accomplished; some of the Racine property was sold, and the remainder was used as a central parts warehouse, or North American Parts Operation as it came to be called. With this done, the process of reducing overhead through disposing of inefficient and redundant plants had largely been completed.

The need to increase the capacity of efficient plants was soon felt in both tractor and combine manufacturing operations. In the case of tractors it was closely related to another development – that of using major tractor components manufactured in the United Kingdom in tractor assembly operations at Detroit to reduce the cost of the finished product. The Detroit factory had since 1955 been manufacturing the TO-35 tractor using components manufactured in the United States. It was realized that if the FE-35 tractor planned for 1957 production in the Coventry plant of the Standard Motor Company and the TO-35 were designed so as to make their major components interchangeable, cost reduction through volume production of components at one plant, and at the least expensive plant, might be considerable. The result was the implementation of the plan which, as we have already seen, had first appeared in the Ferguson company, that shipping tractor components to Detroit from Coventry. In August this policy of interchangeability of major tractor components was beginning to be broadened to include other tractor plans. One executive reported on August 30, 1956:

> I also should like to record the suggestion made by Herman Klemm and concurred in by the Central Coordinating Committee that the Eastern and Western Hemisphere styling be made identical, if technically possible and economically justified, in order to assure both flexibility regarding interchange of components between producing units and maximum similarity of products where selling units will receive identical tractors and, perhaps, in different producing units.

Having committed itself to using Coventry components at Detroit, it became of the greatest importance to the company that the "mixing" of United States and United Kingdom parts should go smoothly. An engineer wrote a memorandum on September 3 on the company's progress with tests at Coventry, and Thornbrough wrote "Very Important" beside the following paragraph of that memorandum:

> The complete exchange of English parts into American housings and vice versa has been accomplished with no trouble being experienced. The two resulting transmission assemblies and two axle assemblies were then motored on existing test brakes at Standard and were free of interference and excess noise and functioned satisfactorily. This of course does not give assurance that completely random assembly of mixed American and English details would give satisfaction. . . .

Indeed it did not, as later experience was to show, but in the meantime the rewards in the form of lower costs of such a policy, and its increasing practicability, made management decide tentatively that its proposed larger tractors – eventually the MHF-65 and MHF-85 – should also be assembled at Detroit with United Kingdom components, and even that whole MHF-35 tractors with diesel engines should be imported into the United States market. A more detailed study of the relevant facts, however, greatly modified these tentative plans.

To ensure successful introduction of the new assembly policy at Detroit, a new factory manager responsible to the parent company was appointed. Thornbrough outlined for him the objectives he should seek. These were:

1. Reduction in manpower to standards in effect prior to amalgamation, adjusted for difficulties of handling a much larger number of models.
2. Improvement in quality and appearance of the tractors.
3. Elimination of programming which will result in a large number of finished tractors in the field.
4. Installation of procedures and equipment which will permit production of tractors to order.

5. Correction of assembly errors and other planning which will result in continuing to run the assembly line even though a large portion or all of the tractors are not acceptable by inspection or are incomplete.

The memorandum was heavily influenced by mistakes which had in the past left the Detroit factory with a huge inventory of partially finished and defective tractors. By October 15, 1956, the President could write to the new factory manager: "I should like to compliment you and your staff in bringing the total of tractors accepted at final inspection from zero percentage on and prior to September 17th to 100% on October 5th. . . . This perfect record for October 5th was the first in over two years."

With the policy of interchangeability of components beginning to emerge, it also became important to define rules for deciding when components would be purchased from outside the company and when not. A memorandum written by Thornbrough on November 1, 1956, emphasized that "maximum utilization of company facilities is of extreme importance. Equally important is the supply of all components at the lowest possible cost." He went on to elaborate rules relating to interdivisional quotations on components that would lead to that result.

Planning for the new intermediate-sized tractor, the MHF-65 (later the MF-65) had begun in early 1956, and they took full account of the principle of interchangeability. That principle was pushed a big distance forward in December 1956 when the Eastern Hemisphere Sales Division accepted the MHF-65 with planetary rear axle reduction unit and thereby made world-wide standardization of that sized tractor possible. Standard Motors tentatively agreed to begin producing North American components for the MHF-65 in November 1957.

The first Detroit MHF-35 tractors using Coventry components were assembled in early February 1957. It became apparent soon after that that the company's difficulties were not over, for quality problems began to appear. Noisy ground speed gears, knocking hydraulic pumps, and faulty roller bearings were some of the problem areas. It was discovered, for example, that quality differed between roller bearings manufactured in the United Kingdom and those manufactured by the same company in the United States. The company had to begin imposing quality standards demanded by the North American market on its United Kingdom suppliers of components – just one consequence of its policy of interchangeability. The programme of interchangeability of components had also to overcome difficulties arising from the great distance between primary manufacturing operations and assembly operations. Scheduling had to allow for longer periods in transit, inventory policy had also to be adjusted, and if parts servicing became deficient – as it did for a number of reasons for a period – the company could expect that its customers would regard this as a consequence of its using imported items.

In spite of these difficulties the basic soundness of the approach was fully confirmed by cost analyses, and this meant that the Detroit plant would indeed develop into a major assembly operation. Its capacity for this was inadequate, and in early 1957 plans were completed for a major extension. The cost of the expansion was estimated at $3.4 million which, it was estimated, was about half of what it would cost to obtain adequate facilities at Racine. In addition, the "pulling power" of Detroit from the point of view of the quality of labour and costs generally, especially those relating to products of the sheet metal industry, were important considerations in the decision to expand in Detroit.

The Board considered the project on May 2, 1957, and approved it. But during the course of the discussion the Chairman reiterated a view he had held for some time, that the company did not manufacture a large enough proportion of its products and that concentrating tractor production at Detroit was a step in the right direction. While numerous specific incidents encouraged and helped precipitate future factory acquisitions of the company, it was this view from the top that ultimately ensured that such a policy would be implemented.

The company had planned to terminate its contract for transmissions with the Detroit Gear Division of Borg-Warner Corporation – one going back to the days of Harry Ferguson Incorporated in 1947 – as soon as its supplies from Coventry were assured. It had been assumed that components for the proposed large MHF-85 tractor, as well as the MHF-65 and 35 tractors, would also be supplied by Coventry. In May 1957 it appeared that Standards would not be able to supply components for the #85 tractor in time for the 1959 sales season. Attention had again to be turned to using North American components for that tractor – apart from those that were interchangeable with the MHF-65 tractor. This led to an interesting discovery: since demand for the tractor would largely be limited to North America, which would limit its total volume, the cost advantages of producing it at Coventry would be much less than had earlier been thought. It was also thought that if North American components were used the tractor would be available for the 1959 season. So it was decided that it should be produced largely from North American components.

It was just about this time that the President received an offer from Borg-Warner to sell its gear and transmission plant in Detroit, known as the "Kercheval plant." The price was to be about $1 million for a plant of 198,000 square feet, with a liquidation value of $450,000 and a replacement value of approximately $4,500,000. It was a rare opportunity for the company to make some progress with its intention of moving further into the manufacture of its own products. It could also be exceedingly attractive financially if the price paid could be made to cover the company's outstanding liabilities under existing contracts for components with Borg-Wagner. To the President's pleasant surprise, Borg-Wagner officials agreed to such a settlement. In recommending the purchase to the Board on November 14, 1957, Thornbrough outlined these arguments in favour of it:

1. Our Company's liability under the existing contract with Borg-Warner would be covered in the purchase price;
2. Borg-Warner would release key men and factory personnel for employment by our Company;
3. There would be a substantial saving in tractor costs as well as greater manufacturing flexibility through ownership of gear and transmission facilities;
4. The Kercheval plant has a hydraulic pump department which could be utilized in our manufacturing operations;
5. Capital appropriation for a new tractor was approved by the Executive Committee on October 9th of which $1,545,000 was intended to cover production of Borg-Warner parts. By the diversion of this sum and by the expenditure of an additional $385,000 it will be possible to purchase the Kercheval plant and make some necessary related expenditures on the Verity plant in Brantford. This would give our Company facilities to manufacture not only gears and transmissions, but also hydraulic pumps.

The Board was convinced, it approved the purchase, and the company acquired a facility that improved its ability to manufacture a larger share of its own products. Financially, it soon proved to be a most judicious acquisition. Within several months the plant was able to relieve the pressure on Standard Motors at Coventry by producing some components for the MHF-35 tractor – a possibility, by the way, that would not have existed if a policy of interchangeable components had not earlier been adopted.

While consideration was being given to acquiring the Kercheval plant, the company proceeded to acquire a company that manufactured light industrial equipment, particularly loaders and back-hoes. Mid-Western Industries Incorporated of Wichita, Kansas, had been manufacturing such equipment for the company. In 1948 Massey-Harris had had an opportunity to acquire that company but decided against it. The opportunity appeared again, and in early 1957 the Central Co-ordinating Committee discussed taking up the option that the company had received in May 1955 to purchase the facilities.

John Shiner, speaking for marketing, and Harold Wallace, for manufacturing, favoured doing so. Herman Klemm of engineering pointed out that since the company had committed itself to the light industrial field, it was logical for it to acquire a source for industrial equipment. He referred to the saving in engineering costs that would be involved in acquiring the "Davis" line of equipment, named after Charles J. Davis, founder, owner and manager of Mid-Western Industries. Charles Herrmeyer, the Vice President Finance, feared that such a "one-man" organization might not fit into the Massey-Ferguson operation. Thornbrough noted the relationship between the tractor business and industrial equipment, and also that the major competitors were becoming increasingly active in the light industrial equipment market. It was agreed that McKinsey and Company should be retained to appraise the potential of

the light industrial market in general and the Davis line in particular.
McKinsey and Company reported as follows at the end of April:

> In summary, our conclusion is that Mid-Western Industries occupies a prominent position in the light industrial market. It is selling a well-regarded product, has maintained an aggressive and successful expansion program, has developed excellent working relations between the Company and the distributing organizations, and is widely regarded as one of the top three companies in the field.

McKinsey, however, went on to point out that Massey-Ferguson " . . . is confronted with an enormous consolidation and rebuilding program . . . and in a widespread reorganization of its agricultural operations. In our opinion . . . [it] . . . is not in a position to give Mid-Western the attention which the latter company would require operated as an internal, closely regulated division. Only if Mid-Western is operated as an independent subsidiary with broad authority to establish its own progress, can the right environment for meeting competitive conditions and expanding industry positions be established." Nor did McKinsey think that industrial equipment should be sold through the Massey-Ferguson branches since the latter concerned themselves essentially with farm machinery. Given these conditions, and a few others, McKinsey recommended acquisition of Mid-Western Industries. Massey-Ferguson proceeded to do so in May 1957, for a total cost of $3.4 million. Davis agreed to work for the company. On July 1, 1957, Mid-Western Industries became the Industrial Division of the United States Massey-Ferguson company.

Massey-Ferguson began by running the Industrial Division as a quite separate division and it marketed industrial products through both the former Davis distributors and Massey-Ferguson branches. However, the company did not analyse carefully the problems of operating the Division on a relatively independent basis. Its "workability" as a relatively independent division rested more on the presence of Davis than on defined principles of organization and control. So when Davis decided to re-enter business for himself the Division was not really organized to continue operating the way it had done. The issue at that stage was whether the company should work out the difficulties of developing a relatively independent operation or should incorporate the organization into its farm machinery business.

There was a difference of opinion in the company as to what should be done. The President generally favoured operating it as an independent division. Others favoured integrating it with the agricultural business. It may have been that the influence of the second group was enhanced when in 1959 the President was heavily occupied with matters in the United Kingdom. Also McKinsey made a second study in which they favoured breaking up the Division and maintaining separate marketing strategies. In any case, the decision was taken to move the whole marketing function and the engineering function at Wichita to Detroit, leaving behind only a factory operation that was subordinate to the manufacturing management at Detroit.

In 1961 it was decided to move even the manufacturing operation from Wichita to Detroit partly because it was thought that this would save transportation costs involved in equipping tractors assembled at Detroit with industrial machinery manufactured at Wichita. However, since wages were lower at Wichita than at Detroit it is not clear that a net saving was involved. The management of the North American operations unit felt at the time that the company should operate from fewer centres and this may have been the deciding influence. But there is little doubt that the company had failed to establish a comprehensive industrial equipment programme. Sales of the Industrial Division over this period were not impressive.

In 1964 new purpose and direction began to appear. It was decided to return to the original concept of a relatively independent division with its own retail distribution system. An Industrial Products Operation was established. To obtain facilities for manufacturing a growing line of industrial equipment the company purchased the 331,400 square foot Vickers Incorporated plant near Detroit, on Oakman Boulevard, in May 1964. The move of the industrial equipment manufacturing facilities out of the Detroit tractor plant also gave the tractor assembly operation additional space; this it needed in spite of the expansion in 1964 of that plant by 34,600 square feet.

From 1964 to 1966 the company's total sales of industrial and construction machinery increased from $53 million to $75 million, and the nature of the industrial tractor changed so that it no longer was simply a modified agricultural tractor. By that time, too, the company's Landini operation in Italy was producing two models of crawler tractors. Estimates indicated that in future the growth of industrial and construction machinery would be high and that profit margins would be better than on agricultural machinery. For all these reasons the company, in 1966, decided to establish a corporate industrial and construction machinery (ICM) group, quite separate (in due time) from the agricultural machinery and Perkins engines operations. In North America and Italy the separation would come quickly. The decision was taken in 1967 to move the company's ICM assembly operation and management from Detroit to Cuyahoga Falls, Ohio, where existing facilities were purchased for that purpose. It may also be noted that in 1966 the company decided to build a new industrial and construction machinery plant in Italy just south of Rome. The ICM operation seemed, finally, to be on the move.

The next major improvement in the company's control over the manufacture of its products after the acquisition of Mid-Western Industries in 1957 occurred in the United Kingdom, a move that was highly significant for North American operations. In early 1959 the company acquired F. Perkins Limited of Peterborough, England, manufacturers of diesel engines. Not since the late 1930s, when the facilities at Racine of the old J. I. Case Plow Works ceased manufacturing engines, had the company owned its own source of engines. The acquisition was timely for it took place just when a shift towards diesel engines was occurring in the North American tractor market. With it the company acquired an organization with its own history and its own international activities and one that soon assisted Massey-Ferguson decisively in establishing manufacturing facilities in lesser developed countries. The history of the Perkins organization and the story of its acquisition are traced in chapter 14.

Long before negotiations had begun for acquiring Perkins, Massey-Ferguson had decided that it would have to obtain increased control over the tractor manufacturing operations of the Standard Motor Company. Its policy of using some Coventry-made components for tractors assembled at Detroit meant that control over the manufacture of the tractor was passing increasingly into the hands of Standard. While disagreement, misunderstanding, and suspicion were seldom absent in the relations between Massey-Ferguson and Standard, it was this absence of control that worried Massey-Ferguson management above all else. The way Massey-Ferguson acquired the Standard tractor facilities, beginning with a secret purchase of Standard shares in 1956, is itself an intriguing story and it is told in chapter 13.

Having acquired these engine and tractor manufacturing facilities, a transmission and axle plant at Detroit, and industrial equipment manufacturing facilities, the company before the end of 1959 had taken the major steps to acquire control over the manufacture of its product in North America and United Kingdom. Only the Brantford foundry had still to be expanded to permit it to make large castings for tractors. In France, as a succeeding chapter will show, much remained to be done in acquiring control of manufacturing operations there and this was not completed until 1963.

While a combination of enlarging the Detroit tractor factory, importing Coventry-made components, and acquiring the Kercheval transmission and axle plant had given the company the efficient capacity it required for manufacturing and assembling tractors in North America, there remained the important problem of acquiring efficient combine manufacturing facilities in North America. The old combine manufacturing facilities on King Street in Toronto were becoming increasingly inadequate. By 1958 it was obvious that if the company was to be in a position to supply the expected market for a new line of combines it was designing, it would need a new combine plant. Land for such a plant had been purchased by the company at Brantford in 1955, but it was decided that a new plant location study should be undertaken.

Consultants were retained in March 1960 to determine the most economically appropriate location for the plant. No limitations were imposed on locations to be investigated in Canada and the United States. The study was based essentially on an appraisal of relative freight costs (inbound and outbound), utility costs, labour costs, and taxes at Denver, Wichita, Des Moines, Memphis, Minneapolis, Chicago, Cincinnati, and the Toronto area. Results of the study showed clearly that the relatively lower wage costs of the Toronto area in each case more than offset any disadvantages of freight and other costs. A study of historical trends in relative wage rates suggested that the differential would not soon disappear. Detailed study of the Toronto-London area pointed to net advantages of locating at Brantford – where, of course, the company already had a large foundry and factory. It was therefore decided to build a plant costing $13.5 million and covering 477,500 square feet of the land purchased at Brantford in 1955.

The project took eighteen months to be completed and by October 1963 some new MF #510, #410 and #300 self-propelled combines were coming off its assembly line. The way the project was managed is of particular interest because it was the first time that the company had utilized highly sophisticated managerial control techniques suitable for planning and executing large individual projects.

The system, termed "critical path scheduling," was designed to minimize cost and time necessary for completion of the project and to ensure that it would be completed by a pre-determined date. Failure to complete the project on schedule would be a serious matter, for it could impair the company's position in the market for combines. The highly formalized approach to project management involved in critical path scheduling had first appeared in the United States government Polaris atomic submarine project and, under various names, had been adapted to fit the requirements of a variety of other large projects, including those of individual industrial companies.

Planning the combine project involved scheduling the design and construction of the building and the ordering and installation of equipment, and also the scheduling of the design and manufacture of components for the new combine, of the transfer of personnel from Toronto to Brantford, and of the marketing activities surrounding the introduction of the new machines. Essentially, critical path scheduling involved devising a plan that would show the sequence in which individual tasks would be performed; outlining a schedule showing an estimate of the time necessary and the deadline date for completing each task; and supervising operations to ensure that individual tasks would be completed on schedule so that they would not delay commencement of succeeding tasks.

The number of calculations necessary for determining the dates by which individual tasks would have to be completed, for introducing changes into the plan if particular tasks went unavoidably off schedule, and for conveying necessary and current information to each manager responsible for individual tasks, is exceedingly large. Use of a computer was essential. With it, each individual responsible for the execution of the various tasks constantly received current information regarding the calendar date by which each task would have to be completed.

The approach seems to have been very successful, for the plant began production one day ahead of schedule. Later it was to be used for another project of great importance to the company – the world-wide introduction of a new line of tractors. It was an approach that typified the general approach to problems that the new management services group was developing, a group concerned with operations research, general systems, organization analysis, project analysis, and administrative services.

The most recent major change in the company's North American manufacturing facilities, as we have already noted, was announced in August 1965. It involved closing down the old plant at Woodstock, which had been contemplated ever since 1956, reducing the assembly of implements at the old Verity plant at Brantford and transferring that activity to a new operation to be established at Des Moines, in the heart of the United States corn belt. For that purpose a plant of 590,000 square feet, complete with office facilities, was acquired. In it the company planned to weld and assemble farm implements to be sold primarily in the corn-belt area. While the move of assembly operations to Des Moines was prompted primarily by marketing considerations, the company would have had to replace the Woodstock and Brantford facilities in any case – they were over half a century old. Consequently, the move to Des Moines effected a substantial improvement in the company's implement manufacturing facilities.

The improvements in the company's manufacturing facilities and in its control over manufacturing should finally be reflected in the cost of the goods it sells. Chart 9 shows that the ratio of cost of goods sold in North America to net sales has declined substantially from 1955 (the

CHART 9

Massey-Ferguson Limited/North American operations

Ratio cost of sales to sales 1955-1966

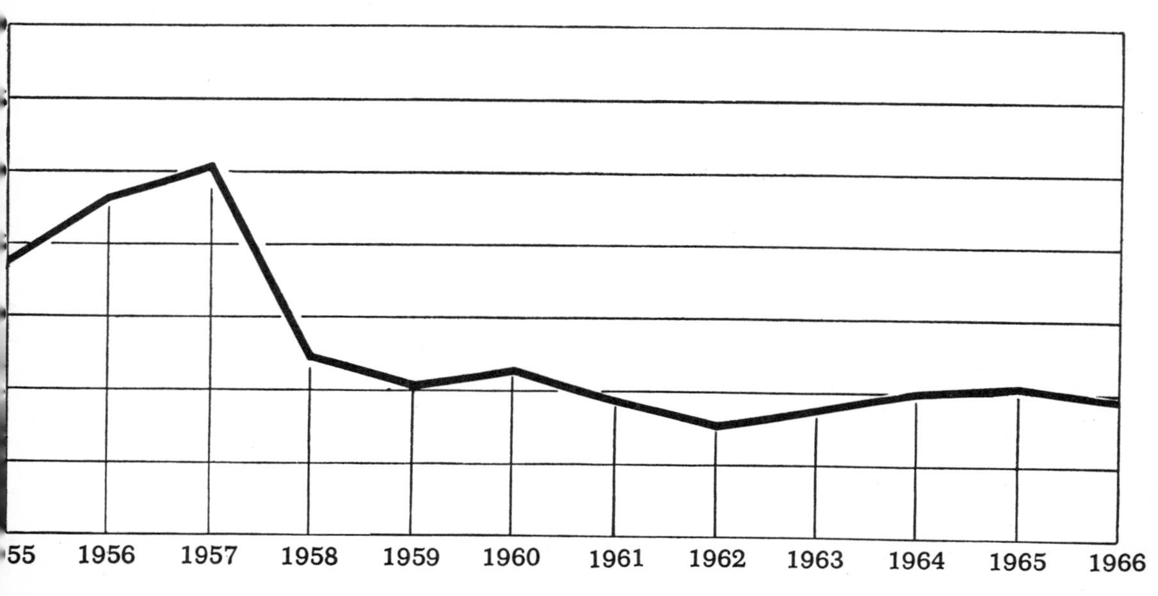

1956 and 1957 figures were heavily influenced by inventory write-offs). In 1963 and 1964 it was thought by management that the figures were influenced by the extraordinary expenditures involved in introducing a new line of combines and starting up the new Brantford combine plant. This was true, but subsequent experience also shows that future progress in reducing costs is likely to be gradual. The major improvements had been achieved by about 1962.

Since 1948 the company has owned a small implement factory at Fowler, California, which, as explained earlier, had grown from the locally important Goble disc. For a number of years this small operation was profitable because Goble discs continued to be leaders in the market. Unfortunately, that technological lead was not maintained and the company's efforts at Fowler were partially directed towards fields where it enjoyed no particular advantage – such as the development of the hillside combine. Losses inevitably began to appear.

In 1960 management at Toronto considered more seriously than previously future plans for the Fowler Division. It was realized that Fowler was located in a rich agricultural region of the United States, but one with a large variety of farm products from fruits to wheat. It was a market that demanded a large number of specialized machines, and one in which new machines appeared and disappeared rather quickly. Flexibility in manufacturing was essential, and this was a characteristic that local "short-line" companies enjoyed. Massey-Ferguson began to believe that it might be possible to operate the Fowler Division as a sort of short-line company with greater flexibility in the character of its internal organizational structure, and in its engineering and marketing activity, than existed in the rest of the company.

It is a concept that has not yet been fully developed, but still an interesting one because of the prevalence of competition from "short-line" companies in many markets in which the company is interested. New engineering talent and new products have been introduced into the Fowler Division, as well as new management, and a start has been made for it regaining a more prominent position in certain types of locally important farm machinery.

Sales of spare parts are vitally important to all farm machinery companies. They are important because they are large and relatively profitable. But more than that, failure to supply efficient spare parts service to farmers can have a devastating effect on sales of new machines. When machinery requires repair parts, particularly harvesting machinery, farmers want to be able to obtain them without delay. Delay could mean loss of a crop.

With the Ferguson merger, the company's North American parts organization became very complicated. In fact, two such organizations were established. One was centred at Racine, controlling warehouses at Batavia, Racine, and Fowler, and this organization shipped parts to branches in the United States and Canada. The other was at Toronto.

It controlled warehouses at Toronto, Brantford, and Woodstock, and also shipped parts to all the branches. Duplication of effort was inevitable and standardization of procedure was difficult.

When J. G. Staiger first came to the company he began to examine the problem of reorganizing the parts operation. His attention was soon required elsewhere, and the reorganization was continued by J. J. Chluski. The object of having one North American operations unit was achieved in the parts operation through the establishment of a North American Parts Organization (NAPO). It was located at Racine where the company, after transferring all tractor production to Detroit, had surplus facilities. Stocks of parts were transferred from warehouses at Batavia, Toronto, and Woodstock to Racine, but depots were maintained at Brantford and Fowler.

The system of inventory control of spare parts was changed. Previously the branches had maintained perpetual inventory control records and ordered parts from Racine and Toronto. The new system was one of centralized inventory control and it was probably the first of its kind in the industry. Under it each branch reports sales of parts by part number every day to NAPO, and the latter maintains control of the size of inventories at each of the branches. In this way duplication of effort is avoided, the relative level of inventories is reduced, and service to customers is eventually improved. By maintaining records of the trend of sales in the various regions, automatic stocking of the branches with spare parts can be based on projection of sales, and replenishment orders can be sent automatically to Massey-Ferguson factories and outside suppliers. The problem of maintaining records of over 100,000 different parts located in about two dozen different places, and of collecting relevant sales data, is enormous. At first a 650 IBM computer was utilized, but later this was changed to a 1401, and in 1966 to an even larger computer, the IBM360.

Introduction of the system was by no means smooth. Besides the usual difficulties of installing automated procedures, the company misjudged the relative size of the stock of parts that would have to be carried. From past experience management had become sensitive to excess inventories and ended by carrying an insufficient volume of spare parts. For a period service to customers suffered and complaints were numerous. By 1964, however, the advantages of the new system were becoming obvious, even though switching to improved procedures affected parts service for a period after that. It had been possible through centralized control to reduce the number of parts warehouses and to increase the proportion of dealer orders that branches could fill immediately. The best order fill attainable prior to 1958 had been 85 per cent, while in 1966 it was 94 per cent. Similar centralized systems were also introduced into the United Kingdom and French operations units. There was no longer any thought of reverting to the system of simpler days.

The character of the company's engineering activity after 1956 was heavily influenced by five major decisions: to centralize responsibility for such activity under the Vice President Engineering of the parent company; to move to one line of farm machinery; to relate engineering activity closely to revealed market preferences; to maximize the interchangeability of machine components between operations units; and to locate engineering activity as such in each of the operations units.

We have already seen that centralization of responsibility for engineering activity did not come immediately after 1956. However, the 1959 "Realignment" manual outlined the role of engineering in this way:

The Vice President Engineering is a line executive with direct responsibility for all the engineering activities of the Company. This includes research and advance engineering, design, the development of prototypes and field testing.

His responsibility can be divided into three broad categories according to the magnitude of basic changes in the product.

First, he is responsible for improving current Company products. This will usually involve re-designing individual parts or components. The objective may be to gain manufacturing cost advantages or to eliminate mechanical defects that are causing high warranty charges or sales resistance. On the other hand, the objective may be an increase in performance that would gain a competitive advantage in the market place. These changes can be characterized by their relative simplicity, their low cost, and by the short time that elapses from their inception to their practical application in the field.

The second category of engineering responsibility involves the design and development of new and improved models of current Company's products or the development of new types of product that are directly related to the Company's present activities. The changes are significantly more complex, are more costly to effect and have a greater impact on the Company's competitive position. They are projects that will take several years to complete, will require a wide range of engineering skills and will probably require major expenditure for new tools and for changes in manufacturing facilities.

The third category of engineering responsibility necessitates a critical analysis of the basic job the Company's products are required to perform and a challenge of the engineering concepts that have been followed. For example, a project in this area might involve developing a new method of tractor locomotion, rather than the narrower project of developing an improved tractor tire. Another project might be answering the even more fundamental question of whether present tractor-type vehicles and certain implements really provide the best solution to farm machine needs. Efforts in this area might result in the creation of totally new and different equipment. These changes can be characterized by the length of time required for their development, the uncertainty of their successful completion and their potential impact in over-all corporate operations.

Within these general responsibilities, the Vice President Engineering works closely with marketing executives in defining new product characteristics. This is an area of great importance to the Company and one in which potential misunderstanding can easily develop. The product planning groups in the individual operations units are responsible for developing recommendations for basic product characteristics for their own markets. On a baler, for example, the characteristics might be such things as the method of tying (wire or twine) and the size of the bale. On a tractor it might be the belt horsepower and the over-all width. Compromises may have to be made in order to allow one product to serve more than one market. The corporate product planning group will co-ordinate the recommendations of the product planning groups in the various operations units where marketing characteristics for products to serve multiple markets are concerned.

The decision to relate engineering activity closely to the demand revealed by market research seems, on the face of it, a rather obvious one to make. But it represented a greater change in the company's approach to engineering than might be thought. Engineering effort may be allocated in different ways between projects involving "revolutionary" design changes and those involving "evolutionary" changes. The former always hold out the prospect of substantial rewards from technological monopoly, at least for a period, but the risk of failure is great; the latter seldom lead to such rewards but they succeed in keeping the company competitive in existing machines.

Ferguson engineers had been influenced by Harry Ferguson who was obsessed with the first approach. It was an approach that had given him the revolutionary Ferguson System tractor and he was convinced it would provide further rewards of that magnitude. Consequently much engineering time and effort was directed by Ferguson engineers, both before and after the merger, towards developing machines such as mounted and side-mounted combines, balers, and foragers. Among Massey-Harris engineers the emphasis was less on "pure" research.

New management stressed the need to remain competitive in existing machines, even though by permitting operations units to engage in some advancd design work and scrutinizing closely developments in the industry in general, it also planned to participate fully in the development of machines embodying major technological advances.

The decision to locate engineering activity in each of the operations units did not represent a change from the past. The arrangement had numerous advantages. Machines can more easily be adapted to local markets, new ideas for changes in and additions to the company's product line are more likely to come forward, and an operations unit is able to enjoy the sense of pride and accomplishment that can come only from the successful design of a new machine.

Engineering more than any other functional division is affected by the company's policy of maximizing the interchangeability of components for machines manufactured in various of its operations units. The concept is a much more difficult one than might be thought. Because of the requirements of local markets, differences in materials available in different countries, and changes in production techniques necessary between operations units with substantially different production volumes, complete interchangeability is impossible. In some cases it would serve no useful purpose as in components that, because of relative costs, would never be cross-sourced between operations units. Engineering and Planning and Procurement have had to co-operate closely in working, component by component, towards a policy of minimizing cost through achieving an economically optimum, rather than technically absolute, degree of interchangeability.

The one-line policy greatly eased the burden on the Engineering Division. Development of "duplicate" tractors was halted and those already in production were phased out, as were all the old Massey-Harris lugging-type tractors. By the end of 1958 the Massey-Harris #555, #444, and #333 were no longer being produced. The Ferguson type F-40 tractor sold by Ferguson dealers, and its equivalent for Massey-Harris dealers, the MH-50, went out of production in 1958, their places being taken by the MF-50 tractor.

The company then concentrated on the MF-35 (called the TO-35 until 1960, and those built at Coventry, the FE-35), the MF-50, the MF-85, and beginning in 1958 the MF-65 tractor, while a new small tractor, the MF-25, also began to be planned for production solely in France (see illustration p. 401). The story of the development of the MF-35 has already been told, for that development occurred prior to 1956. As for the MF-65 tractor, which proved to be a very successful machine, the story of its development is interesting primarily because of the unusual speed with which it was designed.

In the mid-1950s the Ford Motor Company had introduced a new series of tractors, one of which had horsepower above the MF-35 tractor. Ford was achieving great success with that particular model. During this period, Herman Klemm had been proceeding with the design of the MF-85 tractor which left a very large horsepower gap between it and the MF-35. The Ford tractor fell into that gap. Klemm was apprehensive about it, but hoped that from a marketing viewpoint it would not be necessary to do anything. It became increasingly clear, however, that this would be a mistake.

In early 1956, after considerable discussions, Thornbrough flew to Detroit and met with the engineers, Klemm, Lee E. Elfes, and R. W. King. For the first time they addressed themselves totally to what the company had to do to develop a tractor larger than the MF-35 and of lesser horsepower than the MF-85. Out of that one meeting came engineering suggestions from which evolved the MF-65 tractor.

Basically, the MF-65 depended upon the excellent engineering concept of the MF-35 tractor. The structure of the MF-35 transmission and axle housing would permit building a larger, more powerful tractor. The peripheral speeds of the small tractor's transmission were very low, and the transmission could be speeded up at the input side which would enable carrying greater horsepower and torque through the transmission without redesign. Fortunately the design of the transmission on its input side had great flexibility in changing the ratio of speeds of the transmission. The greatest speed through a new differential and pinion would be carried to the rear wheel outboard where a final reduction would be made. Brakes would be internal in the housing between the differential and the final reduction. Engineering in Detroit, following this meeting, proceeded forthwith to develop a mock-up which was ready for full consideration at the meetings and field demonstrations that took place in late May 1956.

At those meetings there was full concurrence that the company should proceed full speed on developing the MF-65. But there was no engine. It was recognized that a diesel engine would undoubtedly be more in demand on this size of tractor than a gasoline engine. At the same time it was known that Continental Motors had an engine that would meet the gasoline requirements. It was left to Thornbrough, as Executive Vice President, to get in touch with F. Perkins Limited to see whether or not they had a suitable engine. Upon returning to Toronto, Thornbrough found waiting for him a routine inquiry from Monty Prichard, then Co-Managing Director of Perkins, who was passing through. Prichard cancelled some of his itinerary and went to see Thornbrough, the first time the two had met. They discussed the problem of an engine for the MF-65, and Prichard then went to Detroit to examine in greater detail the possibility of Perkins furnishing it. Agreements were made, the design of the tractor developed rapidly, Perkins became a part of the project, and the MF-65 was on its way.

Even with the MF-65 the company was still weak in the higher power range of tractors, a weakness from which both the Massey-Harris and Ferguson organizations had suffered for years. After the San Antonio conference it had been decided to proceed with the development of a larger tractor, the FE-80, for Ferguson dealers, and the equivalent for Massey-Harris dealers, the MH-90. With the change in management the two were combined into one project, the MF-85, a model we have already mentioned. A prototype was tested in 1957 and while its appearance and performance initially seemed satisfactory, there was some feeling that it was not heavy enough to be sufficiently stable when pulling wide-level implements. Not only did modification seem necessary, but even with modification it was evident that MF-85 would still leave a large gap in the company's line of tractors.

Thought was given to modifying the old MH-555 but sales prospects did not appear attractive. The company then decided to develop a western version of the MF-85, calling it the MF-88, which would have extra weight. It also decided that to speed the introduction of a larger tractor, a model would be purchased from another company. In 1958 Massey-Ferguson obtained a large tractor from the Minneapolis-Moline company (later Motec). This was sold as the MF-95, and after 1961 as the MF-97. Five hundred tractors were also purchased from the Oliver Corporation in 1959 and 1960 and sold in those years as the MF-98. Industrial tractors, principally the MF-202, 203, 204, 205, and 356 were assembled at the company's plant at Detroit. (See statistics of tractor production in Table 11, pp. 283-5.) The MF-35 and the MF-65 tractors were the chief source of tractor strength, while the MF-85 and MF-88 proved disappointing from both design and market points of view. Their successor, the MF-90, however, was quite successful. The MF-25, developed for the French company, encountered painful problems after being introduced in 1961 (see chapter 12).

The rapid pace of technological development and the gaps in the company's tractor line made it clear that Massey-Ferguson would have to plan for a whole new range of tractors. It was decided in a Corporate Coordinating Committee meeting in late 1962 that the company would not continue to acquire large tractors from outside suppliers and that it would not "warm over" the MF-90; it would move itself into the manufacture of large tractors and incorporate an advanced Ferguson System into them. This proved to be a very important decision because of the shift in the North American market to large tractors. The DX tractor programme was therefore launched.

From the start, emphasis was placed on developing tractors that could be assembled in various ways from components manufactured in different countries, so as to make them suitable for the company's different operations unit, should the need arise. The line itself was composed essentially of the MF-135, the MF-150, the MF-165, the MF-175, and the large MF-1100 and MF-1130 tractors. Much of the development work on these tractors was done by L. E. Elfes and B. F. Willetts under J. J. Jaeger, Vice President Engineering. Jaeger, a graduate of the Drexel Institute of Technology and of the Massachusetts Institute of Technology, came to the company in 1961, prior to which he had been President and director of Pratt and Whitney Co. Inc. He replaced Herman Klemm who had plans for early retirement. An impression of the changes introduced by the new tractors is given by the number of new drawings involved. The MF-1100 and MF-1130 tractors were completely new, and so all drawings were new. The MF-135 had 598 changes unique to it, and another 454 changes that were introduced in common with the bigger tractors; while the MF-175 had 601 new drawings peculiar to it, and 632 changes that were in common with the smaller tractors. Over four years, the Engineering Division allocated one million man hours to the designing, prototype building and testing of the new line, at a cost of about $7½ million. Capital expenditure necessary to manufacture the new line amounted to about $22 million, and over $10 million was required to develop appropriate engines.

One of the important characteristics of the new line was that it extended the Ferguson System to very large tractors by modification that made it applicable to trailed as well as mounted implements. It is interesting that this modification of the Ferguson System, in some elementary respects, seems to have had a parallel development in the "Swedish hitch," an arrangement developed by the company's highly efficient and inventive Swedish distributor A. B. Nykopings Automobilfabrik, at Nykoping. That hitch provided an external means to apply weight transference principles to trailed implements. It may also be mentioned in passing that the same distributor has from time to time introduced ingenious modifications to Massey-Ferguson tractors so as to make them more suitable for forestry operations – an adaptation made necessary because many farmers in Sweden receive about half of their income from forestry work.

Jaeger described the technical changes in this way to an internal group of non-technical personnel.

I should like now to review with you some of the design changes so that you are aware of the many plus values that are being offered. Starting with the engines, all of these will now be direct injection engines with high fuel economy, powerful back-up torques, and smooth running with many revisions to eliminate old service complaints. The AD.152 engine in the MF 135 Tractor replaces the old indirect combustion engine. This family of engines, from which this new engine was derived, is in use in more tractors in the world than any other engine ever built. The reliable AD.203 in the MF 165 Tractor is an outstanding performer and has the best fuel economy of any engine in this power range. The completely new AD.236 engine in the MF 175 Tractor includes a built-in balancer to ensure smooth operation at all speeds and is another outstanding performer. In the 1100 series, the AD.354 and its turbo-charged companion, provide power plants matched to the performance requirements of these big machines. The AD.107 engine has been redesigned for use in the MF 1130 so that this, too, will be a frameless tractor matching all of the other tractors in the line. Particular attention has been given to the accessory equipment of the engine in terms of reliability and serviceability.

In the following remarks, I will be treating the MF 135, 165, and 175 tractors as a group to outline some of the changes and features. The front axles have been strengthened by increasing the stub axle diameters and their bearings. Strengthened hubs and improved hub sealings have been incorporated. Additional mud clearance has been provided to align with the increased tire sizes that have come into use over the years. Standard duty and heavy duty axles are provided to meet local requirements. In the rear axles, strengthening improvements have been made in the crown wheel and pinion of the 135 and a completely new rear axle epicyclic has been designed for the MF 165 and 175. These have been designed with 3 pinion gears instead of 2, to provide the additional strength in the driving torque to the wheels, and have been designed with straddle mounted pinion carriers in contrast to the old carrier mounted units to provide additional strength when operating with wide wheel tracks or heavy vertical axle loads. A sealed brake option is provided on the MF 135 that will be most effective in difficult operating conditions such as paddy fields. Hand brakes have been incorporated on all machines to align with new safety standards. Along this line, we might comment that the legal requirements of all countries have been carefully surveyed and the design arranged to accommodate the many and varied requirements. The transmission has been improved and strengthened where required and can be obtained with either the 6-speed box or the now well-proven and accepted Multipower 12-speed transmission. Changes have been made in the fuel tanks to provide increased fuel capacity in the larger machines and the fuel tank of the MF 135 has been redesigned to facilitate the removal of injectors without detaching the tank.

The fuel tanks of the MF 1100 are a completely new concept; saddle tanks that have many advantages such as removed from engine heat, large capacity, conveniently positioned at ground level for filling without risk of spillage across a hot engine.

Particular care has been taken to provide the optimum operator comfort; improved seats are provided as standard on all machines and spring suspension seats are available as optional equipment on all machines. The relationship of seat to foot pedals, steering wheel and step plates have all been carefully considered and arranged to provide the optimum distances for the average human being. The instrument panel has been arranged for ease of observation by the operator, with provision for additional gauges as standard equipment. In addition to the tractormeter, oil pressure gauge and ammeter, standard on the old machines, we now add water temperature gauge, fuel contents gauge and a speedometer where required. Electroluminescent lighting is provided in the design for operation at night and even a cigarette lighter is available as optional equipment.

The heart of a modern tractor is its hydraulic system and in this area we have incorporated improvements and new features that extend the use of the Ferguson System far beyond that available on the old machines and beyond that available on any competitive machine. This is a new breakthrough in the Ferguson concept and makes it possible to extend the advantages of the Ferguson system to pull type implements and equipment. The internals of the hydraulic system have been improved by increasing the pump capacity and increasing the lift capacity on all machines. The lift capacity of MF 135 and 165 has been increased by approximately 10% and the lift capacity of the MF 175 has been increased by 30% as compared to the old MF 65. The internals of the hydraulic system have been redesigned to provide improvement over the old draft control. Better response speed is provided and improved stability is built in. A completely new feature designed into the system and available as optional equipment is Pressure Control. Under this system, drawbar equipment such as plows, harrows, trailers, and balers can be operated with the advantages of weight transfer that permits a lightweight tractor to do a heavy duty job. It is not necessary to build in dead weight as our competitors must in order to provide the traction. We obtain the traction when we need it by this sophisticated means of transferring a portion of the implement weight of just the amount required to provide proper traction. This is a big plus and in many cases makes a small machine the equal in lugging ability of the next larger category. The operator's controls have been rearranged for convenience of operation. A separate draft control lever has been provided, incorporating all the draft control functions, so that in this mode of operation only one lever is needed. A separate response control lever is provided so that once set, this need not be disturbed until the implement is changed. The position control lever, separated from the draft control lever, provides this useful function and when a tractor is equipped with Pressure Control, this same lever in another sector provides the adustment for the graduated response required for optimum weight transfer from trailed equipment. Subtle improvements in operation are being developed in using the Pressure Control system in conjunction with the draft control system. These new hydraulics are a "first" in the industry and can have the same impact in increasing the utility of a tractor as did the introduction of the original Ferguson System.

Additional auxiliary hydraulic services are provided as optional equipment.

A separate pump can be factory-installed in conjunction with valves to permit us to utilize this hydraulic power, either separately or simultaneously, with the use of the standard hydraulics, or the flow of the auxiliary and main pumps can be combined to provide for the high speed, high power requirements of auxiliary equipment. When the auxiliary pump is supplied, an oil cooler is provided to ensure controlled temperature of the hydraulic system. This same pump can provide the power for the hydraulic clutch of the Multi-Power transmission when this is fitted to the tractor.

Auxiliary equipment to fully utilize these features is now under development and will be available with the machines. Special hitches for the utilization of Pressure Control, automatic trailer hitches and quick hitches for mounted equipment are all under development.

The sheet metal, of course, is new and and this has been styled to present a pleasing, massive appearance with a family similarity throughout the entire line. But there is more than style built into this sheet metal. It has been arranged for accessibility when servicing, for simplicity and ease when fuelling the tractor, or servicing the battery or cooling system. The application of cabs and roll-over bars has been carefully considered for ease of installation, so that a tractor may be serviced without removal of the cab. Ready access for the air cleaner is provided and a free, unobstructed flow of air through the radiator and across the engine has been considered in the design.

With the new line of tractors the organization, for the first time in its long history, had machines to offer that covered the higher-powered tractor market as well as the market for smaller tractors. In addition, the company continued to be technologically competitive in the market, as it had been since the merger with Harry Ferguson.

The experience of the company in developing combines after the 1953 merger was much less successful than in tractors. It is true that Massey-Harris had been very successful in maintaining its position in combines by gradually improving its basic self-propelled model. Initial work on a new line of combines began in 1954, and prototypes appeared in 1955. But, with the merger, much engineering effort was directed towards developing two lines of tractors and the bias of Ferguson engineers was in any case toward tractors. Moreover, the activity of the company in forward combine design was divided. There was the traditional Massey-Harris approach and also the Ferguson approach, the latter involving mounted and semi-mounted machines. For these reasons the pace of combines development slowed dangerously.

Meanwhile the existing models began to suffer by comparison with competing machines, particularly in Europe, despite the introduction in 1957 of the #82 and #92 self-propelled combines in North America, and improved versions of the #80 and #90. All of this explains the deteriorating position of the company in the combine market after 1959 (see Table 12).

Progress in the design of the new line, the TX line as it was referred to internally, was further reduced by the decision, based on a misjudgment by the Marketing Division, to concentrate heavily on the development and production of the small MF-35 self-propelled combine for North America. The combine appeared in the market in 1958, but did not meet sales expectations.

Finally the TX line development was impaired by the failure of the company's engineers to solve quickly several difficult design problems. The objectives to be achieved were: increased capacity, more even weight distribution, more accessible location for the engine and one that would reduce the fire hazard, and many modern conveniences. Normal cost objectives imposed by competitive conditions, and width specifications imposed by highway regulations, had also to be observed.

The early models of the TX machine suffered because they were too high, a development which was regarded as retrograde in view of the lower profiles achieved in the #82 and #92 machines (see illustration on p. 415). So redesigning became necessary. Work had also been wasted in a failure to recognize early what the crucial limiting factors on capacity really were. This failure was in part a reflection of the approach to combine design that had prevailed. It was an approach that depended heavily on the instinct and experience of self-taught engineers. That approach, as we have seen, achieved marvellous results in Massey-Harris in the past but, with increasingly complex machines, highly sophisticated testing and design methods had become necessary. Actually, we shall see later that the company probably went too far in depending on the "scientific" approach in combine design.

A significant step forward in the design of Massey-Ferguson combines was taken in late 1957 when the idea of using two "saddle" grain tanks, instead of one grain tank, on the combines gained acceptance. Herman Klemm described the development in a letter dated January 22, 1958, in this way:

As you know, we have established with the #82 and #92 combines a very low silhouette machine. The #92 combine is approximately 92 inches high to the top of the grain tank and approximately 108 inches high to the top of the grain elevator. The whole profile of this machine has been so well accepted that we cannot produce a new machine in excess of the heights we have achieved on our present production machines, as otherwise we are taking a very serious risk. We have searched for the last few years ways and means to reduce the height of our new TX 3 and TX 2 combines and I believe we have now found the answer to this very difficult problem. The new TX 3 machine, of which we are planning to build two prototypes in Kilmarnock which are being designed in Toronto, will have a height to the grain tank of 92½ inches with no protruding grain elevator above that. The engine is mounted on top of the combine and just above the cylinder. The combine is fitted with two saddle tanks, one on each side; the inner side of the saddle tank is formed by the structural side of the basic combine. This should favourably affect the weight and should also vastly aid the manufacture of grain tanks because of their simplicity. The total width of the machine, when the header and the step ladder are removed, is 95 inches; in other words, within highway limitation. The centre of gravity is very low and should be extremely favourable for

TABLE 11

Massey-Ferguson Limited/Tractor Production (1955-1966)

	1955	1956	1957	1958	1959	1960	1961	1962	1963	1964	1965	1966
NORTH AMERICA												
WOODSTOCK, Canada												
16 Pacer	1,159											
Pony			122									
ST. THOMAS, Canada												
MF-470 Tractor Loader										49	282	350
RACINE, United States												
23	220	60										
33	3,359											
333		2,644	100									
44	3,616											
244	2	273	427									
444		3,961	1,889	394								
404		50	68									
55	1,038	140										
555		1,087	893	854								
303			76	156	796							
Total Racine	8,235	8,291	3,533	2,044								
DETROIT, United States												
TO-35	27,623	3,659	5,951	9,839	14,670	6,471						
MF-35						1,204	13,258	13,386	12,526	10,094		
MF-135										1,904	12,958	9,843
F-40		5,038	4,059									
MH-50		9,250	4,658									
MF-50			292	6,183	7,307	256	1,402	3,602	2,892	1,705		
MF-150										246	3,281	2,021
MF-65				10,269	9,986	8,528	6,584	7,670	9,630	8,260	108	
MF-85					4,350	3,400	813					
MF-88					1,450	700	345					
MF-90								4,035	4,452	4,315	1,776	
MF-95*				450	650	700	700	525				
MF-97*								600	1,756	1,439	305	
MF-98*					25	475						
MF-165											6,105	8,687
MF-175											1,198	1,852
MF-180											2,042	2,714
MF-202			837	1,920	1,624	1,527	757	1,055	1,271	869	1,009	974
MF-203							680	659	370	291	566	533
MF-204					1,100	790	645	555	614	530	715	854
MF-205							375	310	230	186	312	230
MF-244G			427									
MF-356							125	380	481	356	308	460
MF-406					436							
1001					500							
MF-302									95	314	433	428
MF-304									66	311	398	460
MF-1100/1130										4	867	6,170
MF-2135										82	1,702	1,785
MF-2500												327
MF-3165											946	1,496
Total Detroit	27,623	17,947	16,224	28,661	42,098	24,051	25,684	32,777	34,383	30,906	35,029	38,834
TOTAL N.A.	37,017	26,238	19,879	30,705	42,098	24,051	25,684	32,777	34,383	30,955	35,311	39,184

* Tractors purchased from suppliers.

TABLE 11 (continued)

Massey-Ferguson Limited/Tractor Production (1955-1966)

	1955	1956	1957	1958	1959	1960	1961	1962	1963	1964	1965	1966
UNITED KINGDOM												
KILMARNOCK, Scotland												
745	2,965	3,126	1,245	889								
COVENTRY, England												
TE-20	64,342	41,049										
FE-35			65,219	51,379	48,862	53,900	49,656	43,081	47,215	44,469	582	
MF-203							402	1,382	1,111	1,189	1,074	976
MF-205							200	567	606	695	844	792
MF-702					63	1,167	1,370					
MF-765				8,585	9,887	12,700	17,321	20,329	23,288	25,554	2,861	
MF-135											33,802	44,246
MF-165											13,148	20,640
MF-175											5,651	8,658
MF-2135											707	701
MF-3165											1,136	2,365
MF-3303/5												175
Total Coventry	64,342	41,049	65,219	59,964	58,812	67,767	68,949	65,359	72,220	71,907	59,805	78,553
TOTAL U.K.	67,307	44,175	66,464	60,853	58,812	67,767	68,949	65,359	72,220	71,907	59,805	78,553
FRANCE												
MARQUETTE												
812	11,536	13,101	1,894									
820			11,953	14,297	4,033	948	46					
821					4,036	5,004	355					
20-25-8									810	645	25	
Total Marquette	11,536	13,101	13,847	14,297	8,069	5,952	401		810	645	25	
BEAUVAIS												
MF-825 & 30-8							10,171	12,725	9,341	9,037	707	
MF-35-8 & 35X						10,702	6,178	4,473	1,481	1,426	316	
MF-35-8 & 37-8								4	3,363	5,980	987	
MF-42-8								3,161	3,210	3,195	370	
MF-802						200	300	120				
MF-865						2,104	2,673	1,729	643			
MF-65-8 MK II									1,292	3,154	739	
MF-122											590	590
MF-130											8,439	7,510
MF-2130											175	47
MF-135											2,115	3,575
MF-140											5,898	8,459
MF-145											2,230	3,300
MF-165											4,742	5,686
Total Beauvais						13,006	19,322	22,212	19,330	23,152	27,308	29,167
ST. DENIS												
TO-20	7,910	9,286	3,145									
TEF-20	3,985	6,545										
FF-30			15,622	13,953								
MF-835				3,408	19,453							
MF-802					62							
Total St. Denis	11,895	15,831	18,767	17,361	19,515							
TOTAL FRANCE	23,431	28,932	32,614	31,658	27,584	18,958	19,723	22,212	20,140	23,797	27,333	29,167
BRAZIL												
MF-11-50								1,175	2,982	4,209	2,859	3,631
MF-11-65											79	420
TOTAL BRAZIL								1,175	2,982	4,209	2,938	4,051

TABLE 11 (continued)
Massey-Ferguson Limited/Tractor Production (1955-1966)

	1955	1956	1957	1958	1959	1960	1961	1962	1963	1964	1965	1966
ITALY												
FABBRICO												
L-25	1,650	1,410	240									
C-25, R-25		50	950	947	853							
L-30			600	860	274							
C-35					367	688	615	438				
C-1-35							59					
L-35	501	502			325							
MF-44									5	140	186	247
L-45	165		30	30	21							
L-55B & L-55	200	200	10	26	116							
R-50 & DT-50					215	915	347					
R-50J							404	955	446	665	400	
MF-244									248	289	536	768
R-3000						2	493	828	879	759	1,009	835
C-4000								296	1,105	341	355	813
C-4500										788	121	
C-1-4000								78	306	169	113	
DT-4500										190	54	
R-4000						1,184	1,907	1,574	1,429			
R-4500										1,161	551	
R-6000							454	950	1			
R-7000									995	1,216	662	266
DT-7000										270	248	46
C-1-8000										14	79	12
MF-3366											5	97
C-1-5000												162
C-5000											857	713
DT-5000											166	235
R-5000											885	1,285
DT-8000												26
R-8000												499
Total Fabbrico	2,516	2,162	1,830	1,866	2,171	2,789	4,279	5,119	5,414	6,002	6,227	6,004
COMO												
L-35/8						141	30					
L-44/M			182	290	115	36	68					
Total Como			182	290	115	177	98					
TOTAL ITALY	2,516	2,162	2,012	2,156	2,286	2,966	4,377	5,119	5,414	6,002	6,227	6,004
Grand Total	130,271	101,507	120,969	125,372	130,780	113,742	118,733	126,642	135,139	136,870	131,614	156,959

adaptation to the Hillside combine. I believe we have now attained all the basic objectives for a sound design and favourable production costs and appearance-wise, we have achieved a low silhouette and a fine, streamlined machine. The saddle tanks help a great deal to hide the belt drives; in fact, the belt drives have also been simplified. I firmly believe that with the TX 3 design, which will be followed by the TX 2, we are moving into an advanced design of combines similar to what we have enjoyed in tractors. . . .

The new concepts were accepted by the company which then also abandoned the Ferguson mounted combine designs, finally concentrating

TABLE 12
Massey-Ferguson Limited/Combine Production (1955-1966)

	1955	1956	1957	1958	1959	1960	1961	1962	1963	1964	1965	1966
SELF-PROPELLED												
TORONTO, Canada												
60	1,365	1,116	1,014	694								
35				404	3,919	2,222		475	550	600		
72					1,375	1,193	282	300	500	407	53	
80	2,293	2,728										
82			2,179	4,297	2,369	1,349	1,667	1,360	205			
90	4,231	3,722										
92			3,438	4,503	6,033							
92 (Super)						4,104	4,304	3,810	3,706	180		
92 (Super) Hillside							110					
MF-300								50	2,707	3,338	2,517	3,609
MF-410										4,200	4,140	3,410
MF-510										900	2,858	2,270
MF-205												201
Total	7,889	7,566	6,631	9,898	13,696	8,868	6,363	5,995	7,668	9,625	9,568	9,490
KILMARNOCK, Scotland												
780	4,227	3,662	4,495	3,505	2,727	1,709	1,903	348				
500-7								1,296	932	1,659	1,244	570
400-7									1,017	1,356	1,614	956
788									1,126	820	792	361
735		50	1,915	1,155	1,211	83	386					
410-7												441
510-7												570
Total	4,227	3,712	6,410	4,660	3,938	1,792	2,289	1,644	3,075	3,835	3,650	2,898
MARQUETTE, France												
830				796	1,397	757	810	404	302	381	300	
890	1,200	1,201	1,527	1,900	1,297	50						
892					100	1,448	1,568	788	572	847		
99-8										5	806	558
510												672
Total	1,200	1,201	1,527	2,696	2,794	2,255	2,378	1,192	874	1,233	1,106	1,230
ESCHWEGE, Germany												
630	850	3,910	5,185	4,498	2,749	2,537	3,011	2,014				
685			9	353	1,375	1,096	2,460	2,235	616			
31-6								982	785	1,581	1,636	1,620
86-6								148	851	800	960	1,020
30-6									1,000	781	980	1,080
87-6									275	1,424	1,845	1,905
95-6										10		
Total	850	3,910	5,194	4,851	4,124	3,633	5,471	5,379	3,527	4,596	5,421	5,625
MELBOURNE, Australia												
585 (Total)			11	190	336	342	392	337	451	543	496	475
TOTAL SELF-PROPELLED	14,166	16,389	19,773	22,295	24,888	16,890	16,893	14,547	15,595	19,832	20,241	19,718

TABLE 12 (continued)

Massey-Ferguson Limited/Combine Production (1955-1966)

	1955	1956	1957	1958	1959	1960	1961	1962	1963	1964	1965	1966
PULL-TYPE												
TORONTO, Canada												
60	1,310	1,202	881	1,145								
50				1,807	418							
35					300	1,201	375	450		100		
72					1,000	200	300		300	160		
405												485
Total	1,310	1,202	881	2,952	1,718	1,401	675	450	300	260		485
BATAVIA, United States												
50	3,000	2,221	1,293	256								
KILMARNOCK, Scotland												
750	250											
MELBOURNE, Australia												
6	991	459										
506			333	426	201	176	175	300	160		110	
508	1,286	525	474	386	718	250						
585					1	786	1,140	1,020	949	1,010	977	905
Total	2,277	984	807	812	920	1,212	1,315	1,320	1,109	1,010	1,087	905
TOTAL PULL-TYPE	6,837	4,407	2,981	4,020	2,638	2,613	1,990	1,770	1,409	1,270	1,087	1,390
Grand Total	21,003	20,796	22,754	26,315	27,526	19,503	18,883	16,317	17,004	21,102	21,328	21,108

its engineering resources on the TX line. It was hoped that the two models proposed by Klemm would be available for the 1962 sales season in Europe where competition was playing havoc with the company's old models. But delays occurred at Kilmarnock so only the new MF-500 combines were produced there in 1962, the MF-400 following in 1963. In 1962 also the new small MF-300 combine went into production at the Toronto plant and in late 1963, after the opening of the combine plant at Brantford, MF-410 and MF-510 combines, machines of greater capacity than the MF-400 and MF-500, began to come off the new assembly line.

These combines, besides meeting the objectives that were set when design work began and by using innovations such as the "saddle" tanks, also supplanted the traditional body frame with an all-welded sheet metal body. Strength with minimum weight was thereby achieved. The first models were not nearly as successful under all crop conditions as had been hoped and the company's competitive position suffered as a result. An unexpectedly long period of design modification ensued. In retrospect it seemed that the company had over-estimated the extent to which they could depend on scientific laboratory testing in designing combines – the approach the Ferguson engineers used so successfully in tractor design work; and they had also under-estimated the extent to which they would still have to rely on trial-and-error – the Massey-Harris approach. It had been a long haul, but in the end the new line gave the company a base for strengthening its position in the combine market after a period when its share of the market had sadly declined.

This experience with combines forcefully demonstrated to the company the vital need for forward product planning and designing on a continuing basis. It had demonstrated that in an industry with rapid technological change the most devastating form of competition is the pace of product improvement, and that any slackening in that pace will, before long, appear in sales and profit figures.

In 1964 the company extended its product line by purchasing Badger Northland Incorporated of Kaukauna, Wisconsin. In this way it acquired facilities for manufacturing and marketing a line of forage harvesting machinery and farmstead equipment: barn cleaners, silo unloaders, tube and bunk auger feeders, chain conveyers, dump carts, handling systems, cleaning systems, mixer mills, wagons, forage harvesters, corn choppers, rotary mowers, rotary tillers, garden and lawn equipment and two models of garden tractors.

The company's steel office furniture and equipment business, centred at Waterloo, Ontario, and inherited from the H. V. McKay company in 1955, had always been a small sideline. Thought was frequently given to disposing of it, but no attractive opportunity for doing so appeared. Under J. W. Vingoe the profits of the operation improved substantially and it was felt that extension of its line into wood office furniture and equipment might be desirable. This was done through the acquisition of Art Woodwork Limited, Montreal, in 1964. That company manufactured and sold wood furniture for offices, laboratories, libraries and other institutions, and engaged in an active export business, particularly of phonograph cabinets to the United States.

An important but somewhat different product extension occurred when the company decided to become more active in the financing of its dealers, farmers, and industrial equipment customers than previously. In the early 1950s it began to introduce financing plans to assist dealers and farmers, the former with a limited period consignment-type purchase plan, the latter with a periodic time payment service. Both have changed over the years, the latter acquiring life and physical damage insurance features in 1957 and "payment shipping" features in 1960. Financing of dealers now includes not only new and used machinery but dealer shop equipment, service tools, and display facilities as well.

The legal basis for this growing activity was established in 1960 with the formation of Massey-Ferguson Finance Company of Canada Limited, and Massey-Ferguson Finance Corporation, of Springfield, Illinois (now Massey-Ferguson Credit Corporation). Within the North American Operations Unit, this financing activity is the responsibility of the company's North American Finance Operations and it has offices in Toronto and Des Moines.

Underlying this activity was the feeling within the company that existing financial facilities were not completely adequate, that certain financial arrangements – such as those inducing off-season purchases –

would enable it to spread its production more evenly over the year without acquiring a burdensome inventory, and that both of these factors provided opportunities for profitable financing operations. Experience has justified this view. The company's dealer and farmer receivables in 1960 amounted to $153.1 million, while in 1966 they totalled $330.0 million. Its finance company subsidiaries have undoubtedly become a permanent feature of its operations.

It is apparent that the period 1956 to 1967 saw many changes in the company's North American operations and much improvement. More changes are certain to come and greater improvement is necessary. Both are explicitly recognized by management.

Sales in North American operations increased from $143 million in 1956 to $387 million in 1966, and profits were made every year after 1957. The company's share of the market has increased, although the increase after 1959 has been slight (see Chart 11), and until 1964 its profits in North American operation were not as great as anticipated (see Chart 6). North American operations suffered from frequent senior executive changes which prolonged the feeling of uncertainty initially created by the 1956 change in management. Increased stability was introduced when J. G. Staiger moved from corporate group to North American operations in March 1964 to become General Manager there, and when, upon Staiger's move to Group Vice President in 1966, he was succeeded by another corporate executive, J. J. Chluski. However, the latter left the company over a policy dispute in late 1967 and uncertainty returned. Reorganization of the company's marketing and manufacturing facilities in North America took longer than had originally been expected, and not until 1965 did it have both a new line of tractors (including large tractors) and of combines; its purchase of a line of forage harvesting machinery also did not occur until 1965. In a more general sense, however, the major obstacle in the way of more rapid expansion in North America was the company's failure to penetrate more deeply into the mid-west corn-belt market of the United States. There, combined with its emerging industrial and construction machinery business, lie its major opportunities for future expansion.

CHAPTER 12

Expansion in Europe

The Chairman and Chief Executive Officer of Massey-Ferguson Limited, Lieutenant-Colonel W. E. Phillips, found the Eastern Hemisphere Division in a financially strong position when he visited it in July 1956. Sales were rising in the United Kingdom, France, and Germany, although some were running behind budget estimates for the year. In 1956 those three operations generated $9.0 million profits, a year in which North American operations suffered a loss of $7.4 million. The German company was just bringing new combine facilities into operation, and in France exceedingly rapid increases in sales seemed to call for further expansion of facilities. No such pressure existed on the company's facilities in the United Kingdom except on the facilities of the Standard Motor Company, the principal source of the company's Ferguson-system tractors. While parent company management spent most of its time over the next several years solving its North American problems, the sheer size and importance of its European operations made it certain that further development there could not be postponed.

That development began to be influenced by a new force – the negotiations leading to the formation of the European Economic Community. Six European countries had in 1955 agreed to work for the establishment of a customs union and before long the Common Market Treaty was signed. The major question for Massey-Ferguson was how should it deploy its manufacturing facilities in Europe in view of the imminent development of the European Common Market? Under free trade arrangements it would be appropriate to plan for better co-ordination and closer relations between the German, French, and United Kingdom operations than had previously been necessary

To recapitulate for a moment, in 1956 the company had factories in

David Forbert

A farmer in Greece pauses from his tractor to take lunch beneath an olive tree.

all three of the aforementioned countries. In the United Kingdom it had a factory at Manchester that manufactured implements, one at Kilmarnock that produced the MF-780 self-propelled combine, the MF-1 baler, the MF-745 diesel Massey-Harris tractor, and it was just beginning to produce the small MF-735 self-propelled combine. The Standard Motor Company at Coventry was in the process of shifting from the production of the TE-20 Ferguson tractor to the new FE-35.

In France the company's factory at Marquette was producing the large MF-890 combine, the MH-812 (Pony) tractor, as well as a number of implements, and it was preparing to introduce a slightly modified Pony tractor, the MH-820. At St. Denis, just outside of Paris, Société Standard-Hotchkiss, in which Massey-Ferguson had a 25 per cent equity interest, Standard Motors a 50 per cent interest, and Société Hotchkiss-Delahaye (later Société Hotchkiss-Brandt) the remainder, was assembling the Ferguson TE-20 tractor with gasoline and diesel engines in the vineyard and narrow models. Many components for the tractor, including the gasoline engines, were being imported from the United Kingdom, but the diesel engine was purchased from Société Hotchkiss-Delahaye, a licensee of the Standard Motor Company. In effect therefore, in France the company was pursuing a two-line policy with respect to tractors, even though it was developing a single outlet distribution system.

Finally, in Germany the company had its old factory at Westhoven and a new one at Eschwege, and it had begun to produce the small MF-630 combine which it had designed in addition to straw presses and roller chain.

The company's attempts to appraise the effects of the Common Market on its operations were neither comprehensive nor conclusive. Large areas of uncertainty and the absence of adequate statistics placed limitations on conclusions that could be drawn. Would the United Kingdom join the Common Market? What were the long-run relative costs in its various European operations units considering possible changes in relative prices, exchange rates and wages? In any case, how could such costs adequately be measured when the company's plants in Europe differed so much in character? No firm and confident conclusions were drawn.

Several simple facts seemed to be indisputable. It seemed that the company would, as a minimum, find it economical to have tractor and combine *assembly* operations on the continent, as well as in the United Kingdom, and that inter-company shipment of components might well become a permanent feature of its operations. Also there was no evidence that relative cost differences (actually relative variable cost differences since facilities already existed) would necessitate closing down operations in either Germany or France. The question was, where should expansion take place?

German combines of competing companies seemed to be providing strong export competition. Consideration was therefore given to expanding Eschwege into a central source of combines for Europe. A new small tractor was required and again it was thought that Germany might be the appropriate location for its production. But, on the other hand, the French domestic market was so buoyant and the company's position in it so strong that expansion there seemed necessary for satisfying the local demand for combines and tractors – particularly since any substantial reductions in tariff would not take place for a considerable time. These latter considerations, rather than any strong view on the implications of the Common Market, and the feeling that the company should not manufacture tractors in additional new locations, finally became decisive in its planning.

A new small tractor would be produced in France, combine production would be expanded there, and the MHF-630 combine would be produced there as well as in Germany (called the MF-830 in France). At the same time production of the MF-630 would be expanded in Germany and a medium-sized combine, the MF-685, would be added. In the United Kingdom it was also planned to begin producing the small combine, there called the MF-735; tractor production at Kilmarnock would end with the phasing out of the #745 Massey-Harris tractor. Tractor production plans in the United Kingdom were confined to the FE-35 and to the planned introduction in 1958 of the MF-65.

In 1956 and 1957 the European operations of the company, particularly in the United Kingdom, did not observe the company's policy of increasing the interchangeability of components between machines produced in different countries. Production of the new small combine in the United Kingdom, France, and Germany gave those centres of operation a valuable opportunity for increasing interchangeability. But in the United Kingdom operation, apparently with the full knowledge and co-operation of its engineering, marketing, and top management executives, the design of the small combine was changed. This delayed the development of interchangeability in the European combines programme. It also was an important factor in making the parent company decide that it should retain line responsibility in matters of engineering design. Interchangeability of major components for the MF-35 and MF-65 tractors being introduced at Coventry was, however, achieved from the beginning. In time, interchangeability in combines and other implements was also attained, and this began to give the company flexibility in its European manufacturing operations, as elsewhere, even after the plants had been built.

Expansion of Massey-Ferguson in France after 1956 arose from its desire to acquire complete control over its local source of tractors and engines, and it also decided to build a combine plant at Marquette. Besides these projects the company built a central repair parts warehouse together with an inventory control centre at Athis-Mons (Juvisy) near Paris in 1958. A product education and service training centre for its dealers and employees was established at Gif-sur-Yvette also near Paris. In late 1958 the Head Office departments of the French Massey-Ferguson company began to move from Marquette to Paris and official transfer of the registered head office took place on July 1, 1960.

Land for expansion of combine facilities was purchased in 1955 at Marquette, just across railroad tracks from existing facilities. Loans to finance the expansion were obtained from the Crédit National in 1957 which, together with retained earnings of the French company, was sufficient to finance the project. Soon after these combine facilities were completed, the market and the company's share of it began to decline, import competition increased, and costly excess capacity appeared at Marquette. It remained for a long time. In 1964 the decision to introduce the MF-510 combine into production at Marquette was taken because excess capacity did not exist at Kilmarnock and perennial labour problems there discouraged expansion, and because the United Kingdom seemed not to be entering the Common Market.

While this combine plant was being planned and built, initial decisions were also being made to increase the company's tractor assembly and manufacturing facilities in France. Rapid expansion of sales in France prompted this move, as did the view that import competition arising from the development of the Common Market would not be too important for about ten years. It was also regarded as desirable to develop a source of tractor components in France in addition to the United Kingdom source and to source some export markets with tractors produced in France.

An increase in capacity alone was not enough. The company in Europe was faced with increased technological competition in tractors and with a market shift to diesel engines intensified in 1956 by the introduction of low-cost diesel fuel for farm use. Plainly, the old Pony (MH-812) tractor with its gasoline engine would have to be sacrificed in favour of a small tractor with a diesel engine. The Ferguson TE-20 produced at St. Denis by Standard-Hotchkiss would also have to be replaced by the more modern MF-35 that was already in production at Detroit and (beginning in 1957) at Coventry.

Stop-gap measures were resorted to. An improved Pony, the MH-820, with a five- instead of three-speed gear box and improved hydraulics – but not of the Ferguson-system kind – powered by a Hanomag diesel engine was put into production in 1957 at Marquette. At St. Denis, Standard-Hotchkiss replaced the TE-20 with the FF-30, but apart from colour scheme the two tractors were essentially the same. Still, it filled a gap until the MF-35 could be introduced in 1958. To relieve its over-crowded facilities at St. Denis, Standard-Hotchkiss purchased 45 acres of land at Beauvais in late 1956, on which it planned to build a small plant for manufacturing components for its tractor. The site at Beauvais was chosen because the French government's policy of industrial de-centralization made it impossible to expand near Paris and because local authorities at Beauvais co-operated in providing land and utility facilities. Situated north of Paris on the main road to Calais, about forty miles from St. Denis, Beauvais seemed to be fairly conveniently located. The small plant was officially opened in March 1958.

But Massey-Ferguson still did not have a new small tractor in France to replace the Pony, and it did not control either its source of "Ferguson" tractors in France or its source of engines. In July 1958 the company succeeded in acquiring a 25 per cent equity in Standard-Hotchkiss owned by Hotchkiss-Brandt; in return the company agreed to consult with Hotchkiss-Brandt over the next ten years in purchasing engines in France. This gave Massey-Ferguson the same equity interest in Standard-Hotchkiss as that held by Standard Motors, and in 1959 it acquired complete control when it purchased all of the Standard Motor Company's tractor facilities. Its objective of controlling its source of tractors in France had been achieved, but control of its source of engines did not come until later.

Initial plans for developing a new small tractor to replace the MH-820 were implemented in 1958, but since it was obvious that those plans would take time to develop, an improved model of the Pony, the MH-821, was introduced in January 1959. When tractor sales declined sharply in 1959, it was interpreted partly as an indication that the company was in danger of suffering severely from not having modernized its small machine. Since market surveys at the time suggested that a small tractor would continue to form almost one-third of the total market for tractors, there was increasing need for completing the design of a new small tractor, subsequently known as the MF-825. Soon the MF-825 project became the dominating influence in the life of the French company, causing it and parent company executives very considerable anxiety indeed.

Design work had begun at Marquette, but in mid-1958 it was discovered that a completely new gear box would have to be designed, which would mean several years' delay. It was the first of many setbacks. The whole task seemed much greater than had earlier been estimated, and design of the tractor was shifted to the Engineering Department at Detroit. More difficulties followed, and these can be traced to initial market surveys and cost appraisals. Both appraisals and surveys had assumed a relatively small, simple, low-cost tractor. It soon appeared that cost estimates would have to be revised upward because original estimates had been crude; and also because the design of the tractor, in response to the demands of local marketing executives, was gradually becoming more sophisticated. The proposed tractor would also have to be somewhat larger than had originally been contemplated. Curiously, as the design began subtly to change and cost of producing it began to rise, its market potential was not adequately reappraised.

The company did, however, begin to investigate how costs of producing the tractor could be reduced. A report on the manufacture of the tractor was completed in early 1959. The question was whether a new tractor plant should be built at Beauvais to assemble all the company's tractors in France, or whether the MF-825 should be assembled at Marquette and the MF-835 and MF-865, as well as the industrial #802 tractor, at St. Denis. The report concluded that the tractor could be competitive only if the company used a modern assembly line and if it made many of the castings, gears and hydraulics itself. This was an argument in favour of the Beauvais project and one that did not hinge entirely on inadequate existing capacity, but rather on inadequate *efficient* capacity.

It was initially estimated that the Beauvais project would cost $1.4 million, and that preparing Marquette and St. Denis for producing the new tractor would cost $.6 million. Since calculations showed that costs of production at Beauvais would annually be $.5 million less than they would be if production were continued at Marquette and St. Denis, the argument in favour of the Beauvais project seemed decisive. When Massey-Ferguson acquired in mid-1959 all of the equity in Standard-Hotchkiss a further obstacle was removed. Although the total cost of the project had to be revised upward very substantially, the Board approved the project and construction was under way before the end of the year. When completed the Beauvais project was continental Europe's largest and most modern tractor plant, one that could produce all the company's tractors in France at a rate, if necessary and with additional equipment, of 250 tractors per day.

The Beauvais project did not solve the difficult problem of the company lacking its own source of engines in France. Hotchkiss-Brandt had been supplying engines for the Ferguson-system tractors and by agreement would continue to do so if the price offered by it was acceptable. True, Massey-Ferguson had acquired F. Perkins Limited but the latter only owned 49 per cent of the French Perkins company, Société Française des Moteurs Perkins, and in any case Perkins did not manufacture in France: it had granted Ateliers Guillemin Sergot et Pegard (G.S.P.) of Courbevoir a license to produce its P series engines.

The need for a new policy concerning a source of engines in France arose not only because the new MF-825 tractor required an engine, but also because the company had experienced severe quality problems with the 23C Standard Motor engine supplied to it by Hotchkiss-Brandt for the MF-835 tractor. High oil consumption, loss of performance, oil leakages and poor starting had plagued it.

In late 1959 the company decided that it should equip its St. Denis facilities (the former Standard-Hotchkiss plant) to produce the Perkins four-99 engine for the MF-825 tractor. This it could legally do because the Perkins licensing agreement with G.S.P. included only P series engines, and the arrangement with Hotchkiss-Brandt excluded engines made within Massey-Ferguson facilities. But the cost of all this was going to be high. The cost of building up the facilities at St. Denis and constructing the Beauvais plant demanded that substantial new capital be found. A long-term loan of N.F. 20,000,000 was obtained from the Crédit National, and French banks provided short-term capital as usual. But the company was required to increase its capital by N.F. 20,000,000. This meant that the parent company had to invest a sizable amount of money in its French subsidiary – for the first time in the post-war period.

Massey-Ferguson then negotiated termination of its agreement with Hotchkiss-Brandt, although it continued to purchase engines from that company for a few years. It then obtained control of the Perkins company in France and turned over responsibility of operating the St. Denis engine plant to Perkins. And in 1963 Perkins negotiated termination of its licensing agreement with G.S.P. In this way, after many years of protracted negotiations over sources of engines in France, Massey-Ferguson had completely clarified its position and gained the control that seemed to be required for flexibility in tractor planning and production.

Unfortunately that was not the end of the company's difficulties in France. Production of the long-awaited MF-825 tractor in the new Beauvais plant began in late 1960, although regular production did not begin until 1961. It was immediately discovered that costs were much higher than expected. Even worse, a disturbingly large number of quality problems became evident when the MF-825 tractors reached the market, and these problems included difficulties with the engine produced in the plant at St. Denis. Sales, which had in any case levelled off in the industry, suffered further because of the unfortunate experience with the MF-825 tractor and the declining competitive position of the company's old combine models. Overhead expenses of the French company became burdensome. The decline in production, the increased capacity that had been built up over the preceding years at Marquette, Beauvais, and St. Denis, and the cost of terminating contracts for engines, involved a large increase in unit overhead costs.

There was no doubt that all phases of planning surrounding the MF-825 project, and most certainly initial experiences in manufacturing that tractor, reflected serious deficiencies in the company's management in France. Those difficulties also emphasized that supervision and control of parent company management had not been successful. The cost of these inadequacies was high. Losses in 1960 totalled $2.7 million, in 1961 $4.5 million, and in 1962 $4.7 million. Obviously something had to be done.

A change in management was the starting point. W. Lattman, who was well known to the French authorities, became President-Directeur Général in late 1961. J. J. Chluski was sent from the parent company to reorganize the French operations unit and place it in a profit position. His first task was to regain control over manufacturing operations in terms of quality and cost, and this involved changing both personnel and procedures. Much tighter quality control, prior to assembly, of components supplied by subcontractors was instituted. The degree of manufacturing within the company was increased. In marketing, it was necessary to improve the morale of dealers who had been discouraged by quality problems. Fortunately the dealers were inherently strong and experienced, and they provided great support over a crucial period.

By 1963 losses had disappeared, Chluski returned to Toronto, and a new President-Directeur Général was appointed. Following the established policy of the company, the new executive was a French national. In 1964 both sales and profits in France greatly exceeded the amounts that had been planned at the beginning of the year. With that improvement and with the decision to manufacture the MF-510 combine at Marquette, the burdensome problem of excess capacity began to diminish, and the company's capacity to supply an expanding market began to furnish its rewards.

In 1964 it also became apparent that the company's tractor capacity in Europe, including that within the boundaries of the Common Market, would be inadequate within a few years. Expansion at Beauvais was physically quite feasible, and it also seemed economically attractive because it would reduce unit costs of production there bringing them more in line with those at Coventry. It was desirable to build up an additional source for tractor component parts. For these reasons the Beauvais facilities were expanded by 80 per cent to 581,000 square feet to permit increased production of rear axles, gear boxes, and hydraulic systems for the company's tractor plants in the United Kingdom and United States. There was also increased production of finished tractors, particularly the MF-135 and MF-165. At the same time a new administrative building was constructed at Petit-Clamart, six miles south of the centre of Paris, where most of the company's administrative and certain other personnel, previously located in eight different places, were to be concentrated in one building.

As a result of all these developments, the company has become a major factor not only in the French farm machinery market but has also become the largest exporter of farm machinery from France.

Because it had been decided to concentrate continental European production of tractors in France, expansion in Germany after 1956 was in the area of combines, straw presses, roller chain, and a few implements. But besides expanding its facilities for producing this machinery, the German company also embarked on a plan for consolidating its operation.

The German company needed to strengthen its product line and improve its system of dealers, the latter being much weaker than the one in France. It was decided that another combine, the MF-685, would be added to the MF-630 and that the MF-35 and MF-65 tractors would be imported from Coventry. The suggestion that the French Pony tractor be introduced was rejected by the German company because it felt that it would be regarded as obsolete by German farmers. The company decided to wait for the new French MF-825 tractor.

Increased capacity was required at Eschwege to introduce the MF-685 combine. The company went to the government of the Land of Hesse for financial assistance as before. By an agreement reached in June 1956, further credits, more or less on the same basis as before, were made available. Prototypes of the new MF-685 combine were available in 1957 and production began in 1958. Whereas 850 combines had been produced in 1955, by 1958 this had increased to 4,851.

It was neither convenient nor efficient to have the roller chain plant and administrative offices at Westhoven and the combine plant at Eschwege. Nor did it appear that adequate management personnel would be available at Eschwege. Many of the company's employees working at Westhoven would not consent to moving to the small town of Eschwege. The Hessian government, however, was anxious to rebuild Kassel – just forty miles from Eschwege – a city that had virtually been destroyed during the war.

Negotiations between the German company and the Land of Hesse resulted in an agreement in August 1957 for the transfer of the Westhoven roller chain facilities to Eschwege and the administrative offices to Kassel. This was facilitated by financial support from Hesse, without which the move would not have been economically feasible. In August and September of 1958 the move was effected, and when the Company sold its Westhoven property and its old "Industriestätte" property in Berlin the process of consolidation was complete. The new expansion at Eschwege increased the facilities there to 340,000 square feet, and this was eventually increased to 596,066 square feet in 1963. This was about one-third the size of facilities in France at that time. A new engineering building was erected in Kassel in 1962, where the company not only continued to design combines but also to adapt imported Massey-Ferguson products to the specifications of the highly competitive West German market.

In 1959, the German company began to reorganize its retail distribution system changing it from a combination of agents and distributors to company-owned branches and distributors, each with its own network of dealers and co-operatives. At the same time a marketing function, along North American lines, was developed. There was noticeable scepticism among company personnel concerning the suitability of "North American" programmes such as dealer conventions with lively entertainment. To their surprise these programmes were successful.

After 1961 sales of Massey-Ferguson in Germany were adversely affected by the quality problems of the MF-825 tractor that had been imported from the French Massey-Ferguson company. Still, the company had come a long way from earlier years when its fate was in doubt. Even in 1956 Massey-Ferguson sales had amounted to only $7.0 million. By 1960 they had increased to $20.3 million, and in the succeeding year to $34.2 million. In 1966 they totalled $46.6 million. Profits were earned by the German Massey-Ferguson company in every one of those years, although not always sufficient to provide a satisfactory rate of return. The organization had become reasonably well established, but it had now to make further progress in the highly competitive German market created by international competitors such as Deere-Lanz, Ford, Fiat and Renault. Increased strength in its retail distribution system was mandatory.

Reorganization and expansion of operations in the United Kingdom after 1956 materially changed the character of Massey-Ferguson. The acquisition of tractor and diesel engine facilities that dominated these years of expansion will be the subject of the two succeeding chapters. How the United Kingdom operations unit was used as a "guinea pig" for the introduction of the system of Integrated Planning and Control we have already noted. An equally important change in the United Kingdom related to reorganizing the company's distribution system. The problem was similar to the one in North America: two separate dealership systems, one for Ferguson, the other for Massey-Harris.

Until 1959, the period of the parent company's preoccupation with North American operations, the two dealership system had not caused great concern, for sales in the United Kingdom expanded at an exceedingly high rate. In 1956 they had amounted to $38.5 million, and in 1959 to $69.3 million. Towards the end of this period, however, the company noticed that its share of the tractor market had begun to decline. J. H. Shiner, Vice President Marketing of the parent company, investigated. He was soon convinced that the United Kingdom marketing structure had to be reorganized.

These marketing difficulties came to light about the same time that the company acquired the Coventry tractor plant and the Perkins engine plant at Peterborough. As a result the whole operation in the United Kingdom, not just marketing, suddenly seemed to require immediate concentrated attention. In November 1959 a single management group was established under the title "United Kingdom Operations," although Perkins continued as a separate entity. The managing director of the United Kingdom operations unit was relieved of his responsibilities and was given a special assignment, but he soon left the company. Thornbrough then assumed executive control of all United Kingdom operations, H. A. Wallace was appointed Director Manufacturing, and J. G. Staiger became Comptroller. For the next year the United Kingdom operation was in effect under the direct control of parent company executives, the latter only just having succeeded in relinquishing line responsibilities in the North American operations unit. They located themselves in the United Kingdom and for a period managed both United Kingdom operations and parent company affairs from there.

Internal re-arrangement of facilities and staff relocation arising from the acquisition of the Standard Motors tractor facilities, Perkins engines, and the decision to concentrate all executives at Coventry (the Massey-Harris head office had been at Manchester) was a major task but went reasonably smoothly. A much greater problem existed in improving the company's distribution system in the United Kingdom. There were several major reasons why reorganization in distribution seemed necessary. The concept of a complete marketing function of the kind introduced into North American operations had not yet been developed in the United Kingdom. Also, because the Ferguson and Massey-Harris distribution systems had never merged, combines and tractors in numerous instances were being sold in a given sales area by different dealers. Furthermore, United Kingdom sales organization had been strongly influenced by the approach to distribution that had earlier been taken by the Ferguson organization. That approach was essentially distributor-oriented, in this case "distributor" being defined as an independent businessman selling directly to farmers and indirectly to them through dealers, subdealers, and associate dealers. One consequence of this was that there were too many retail outlets and often too many intermediaries between manufacturer and farmer.

Another consequence was that relations between the Ferguson sales personnel and these distributors had become very personal. The two had even co-operated in establishing a national distributor council and area associations for the purpose of communicating company plans and policies and for exchanging information. It appeared that through this exceedingly close relationship the distributors had become rather influential in the company's decisions relating to the introduction, pricing, and methods of distribution of its products. It was a different approach from the one the company was pursuing in North America and, because of declining sales, it did not appear to be appropriate for future operations.

Thornbrough had his first meeting with the distributors at the Smithfield Agricultural Show in London in late 1959. At that meeting he spoke freely about the company's difficulties in the United Kingdom and in a general way about its plans for overcoming them. Perkins would provide better engines, a centralized parts operation at Urmston would solve problems in that area, and so on over the whole range of difficulties. The meeting went quite well.

J. H. Shiner, the corporate senior marketing executive, was brought over to examine the problems of the distribution system. He temporarily assumed the position of Acting Director Marketing and began to organize a programme involving travelling to the distributors and explaining to them how Massey-Ferguson planned to increase the strength of the United Kingdom operation and improve its penetration in the market. When it became clear that the proposal would drastically change the distribution system, the Director Marketing, a senior member of the former Ferguson organization, resigned his position.

In order to effect all this Shiner required a senior marketing man who understood the company's over-all plans and procedures. He and Thornbrough discussed this and decided that L. H. Pomeroy, another North American executive, temporarily in Australia to reorganize the marketing structure there, should undertake the task. Shiner flew to Australia, talked to Pomeroy about it, and with little delay Pomeroy was on his way to his new appointment as Director Marketing, United Kingdom operations.

A comprehensive programme for meeting with distributors was organized, and all the necessary material outlining the company's plans and policies was printed. Then, just a few hours before the beginning of the trip to the distributors, the sales manager, also a former Ferguson employee, presented his resignation. This in effect meant that the presentation would have to be made entirely by North American executives which, in the existing environment, was not likely to go down too well. The company's programme was, in fact, non-committally received by the distributor organization.

While this was happening, the difficult problem of dealing with a certain distributor, who controlled about 10 per cent of the company's total sales in the United Kingdom, had to be solved. The company and that distributor could not agree on policies to be pursued, and yet because of the large volume of business in his territory something needed to be done to reconcile differences. This became more urgent when it was realized that a large portion of the company's decline in sales had occurred in the distributor's territory. Management attempted to achieve various arrangements with him but were unsuccessful. They then terminated his franchise. A law-suit threatened and after months of negotiation the matter was settled at a substantial cost to Massey-Ferguson. Still, from the viewpoint of reorganizing marketing operations in the United Kingdom, it represented a definite turning point. The impression seems to have been created that if the company was prepared to risk about 10 per cent of its business by terminating a distributor, it obviously was determined to implement its new plans. Gradually the number of dealers was reduced, the national distributor council and area associations were abolished, and a marketing organization with regional managers, a general sales manager, a director of marketing, and supporting staff positions was established.

A curious change began to occur in the attitude of distributors towards the new marketing organization. In part this was because of the personality and temperament of Pomeroy, and in part because some of them began to be convinced that the new marketing plans would assist them in operating under increasingly competitive conditions. Pomeroy was able to return to Toronto in February 1962, and his place was taken by P. J. Wright, who had come into the company a year before and who continued the work of reorganizing the distribution system. In the end the company's distribution system in the United Kingdom was probably the best one it had among all its operations units. Sales in 1960 had declined to $65.6 million from $69.3 million in 1959, but in 1961 they rose to $76.3 million, and in 1966 they totalled $103.3 million. The company's share of the market in the United Kingdom in 1960 was about 30 per cent, in 1965 it was 35 per cent.

Expansion of manufacturing operations, particularly in the United Kingdom but also in France and Germany, greatly increased the relative importance to the company of its operations in Europe. Chart 10, for example, shows that since 1959 approximately 60 per cent of the company's employees have been located in Europe, and after 1960 about 40 per cent of the company's assets employed were accounted for by its European operations. These figures include the new Italian operation.

The company's sales in Europe in 1966 represented 38 per cent of its total sales, a year in which sales in North America were 42 per cent of the total, and in individual years European sales were larger than sales in North America. By any measure the company had become at least as European as it was North American, and the ties between the two markets had become exceedingly close because of the shipment of component parts from United Kingdom operations to North American operations.

CHART 10
Massey-Ferguson Limited/European operations
Ratios to world-wide operations

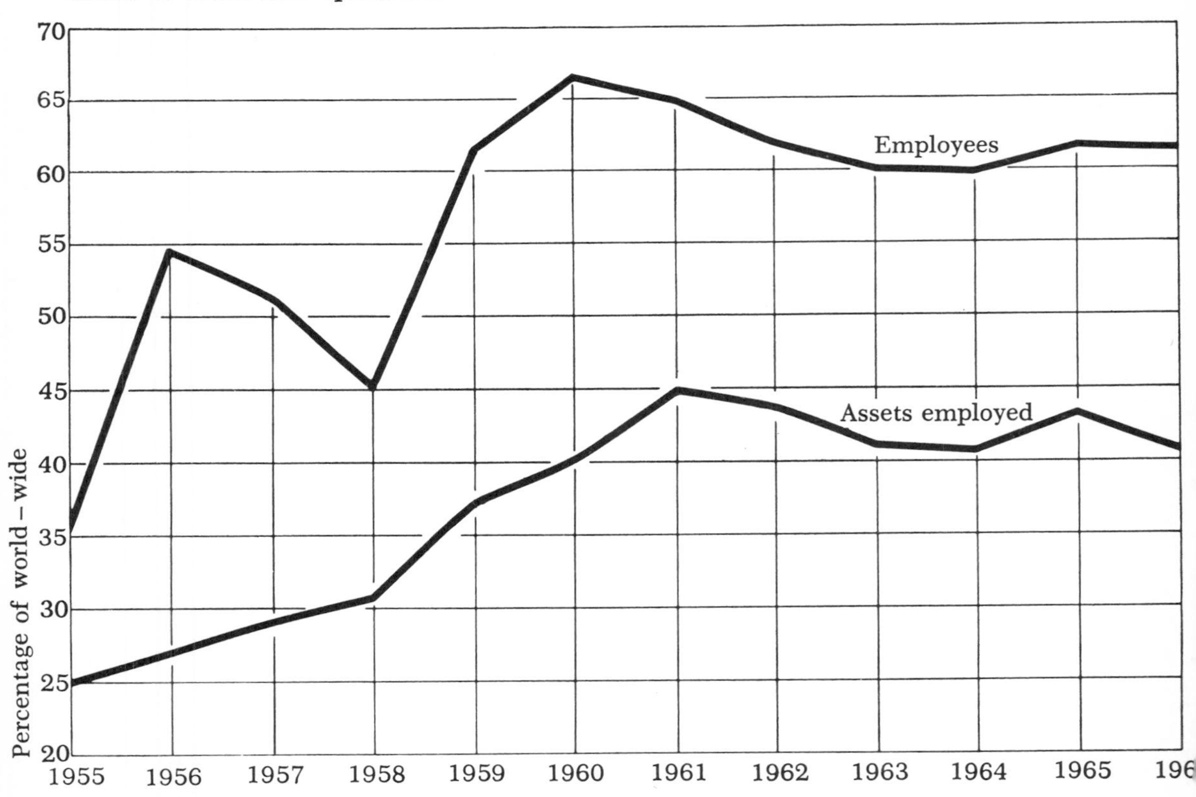

CHAPTER 13

Negotiations with Standard Motor Company

Acquisition by Massey-Ferguson in 1959 of the tractor manufacturing facilities of the Standard Motor Company, particularly its Banner Lane, Coventry, plant, gave the company greatly increased control over the manufacture, world-wide, of the most important product in its product line. This was a major event in the development of the company. But the history of the relations between Massey-Ferguson and Standard is also interesting for more general reasons. Their association appears to be a classic example of a business arrangement where contractor and buyer are heavily dependent on each other because of the relative importance to each of the trade passing between them and because of the great difficulties facing both in making substitute arrangements with other companies. The experience of these two companies suggests that it is an inherently unstable arrangement.

It all began in 1945 when Harry Ferguson convinced the Standard Motor Company that it should manufacture his TE-20 tractor for him. A ten-year agreement was signed on August 20, 1946. Relations between Standard and Ferguson were never particularly friendly, and sometimes most unfriendly. However, in spite of that, Ferguson never wished to become involved directly in manufacturing, even if he had had sufficient capital to do so, and at the time of the merger of Massey-Harris and Ferguson the Standard Motor Company was still the sole supplier of TE-20 tractors to Ferguson in the Eastern Hemisphere. Standard had first option to undertake the manufacture of Ferguson tractors anywhere in the Eastern Hemisphere, but only in France had the company actually begun planning the manufacture of the TE-20. A French company, Société Standard-Hotchkiss, was formed for that purpose in April 1953, with the Standard Motor Company holding 50 per cent of the equity and French shareholders the remainder.

On November 24, 1953, shortly after the merger, Massey-Ferguson entered into a new twelve-year contract with the Standard Motor Company. This agreement fixed the legal basis for their commercial relations until Massey-Ferguson acquired the tractor manufacturing facilities. It seems that in their initial encounters with the executives of Standard the former Massey-Harris executives revealed a certain arrogance that was not likely to generate great friendship. They seem to have ignored all the lessons that Ferguson management had learned in living with an inherently unstable business arrangement.

Massey-Ferguson first entered into a shareholding relationship with Standard through their mutual interest in the success of the tractor project in France. Société Standard-Hotchkiss appeared not to be making adequate progress in planning for the manufacture of the TE-20 tractor in that country; yet Massey-Ferguson, fearing the increasing obsolescence of its Pony tractor, was anxious to have it available for the French market. When Standard-Hotchkiss required more capital and French shareholders were disinclined to provide it, an opportunity was thereby provided for Massey-Ferguson to acquire an interest in the company; both companies thought it desirable for Massey-Ferguson to take up the equity issue so as to increase its influence in managerial matters in Standard-Hotchkiss. This was done and Massey-Ferguson acquired 25 per cent of the equity in the French company, with Standard still holding 50 per cent, and the French Hotchkiss group the remainder.

But Massey-Ferguson had no equity interest in, and therefore no managerial influence on, the Standard Motor Company itself and its principal tractor operations at Coventry. And it was evident soon after the merger that the many quarrels which Harry Ferguson had had with Standard over prices paid for tractors and components produced at Coventry had not ended with the merger. In early 1954 Massey-Ferguson undertook an investigation into Standard's tractor costs although on rather limited information. The President, J. S. Duncan, was alarmed at the costs revealed, was convinced that gross inefficiency existed at Coventry, and recommended, in March 1954, that the company take a firm stand with Standard. There was also a strong suspicion among Massey-Ferguson management, a suspicion that remained over the next several years, that the Standard tractor manufacturing operations were absorbing the overhead of the car side of the Standard Motor Company.

But what kind of a stand can a major buyer take when he is fully dependent on the subcontractor? How far can he go in demanding and receiving confidential cost data from the subcontractor? How far can he go without having managerial responsibility in demanding that the subcontractor initiate moves to increase efficiency? How can he say what profit margin the subcontractor should enjoy? The subcontractor is faced with the dilemma that suggestions coming from the major buyer may be in the interests of that buyer's shareholders but not of his own shareholders. Certainly objections by the buyer to prices charged is a normal buyer's reaction. Yet to go further and give the buyer confidential cost information to enable him to judge efficiency and make recommendations for operational improvement really involves abdication of managerial responsibility. On the other hand, not to give detailed cost information leaves the buyer free to generate volumes of suspicion, with or without justification. If personalities happen at the same time to conflict, then a solution to the multi-dimensional dilemma is even more difficult to attain.

Over the next several years there was endless discussion on costs and prices but seldom agreement, and Massey-Ferguson continued to feel that it had not received full information. All these difficulties led to five years of troublesome relations, and frequent contemplation by Massey-Ferguson, one way or another, to become more directly involved in the operations of the Standard Motor Company.

There was, of course, one solution. That was for Massey-Ferguson to acquire the tractor facilities, if possible. The President, however, according to notes made for the Executive Committee Meeting of May 27, 1954, did not favour acquisition, partly because it seemed at that moment that friendly relations with Standard might be possible, but more so because of his long-standing view that scarcity of capital prevented it. He explained:

> The prospect of owning our own tractor Plant in the United Kingdom is naturally alluring.
>
> In a market which will become heavily competitive, the possibility of controlling costs and saving an additional profit is attractive, and, indeed, the failure to be in this position might well prove a stumbling block to our success in the years to come.
>
> My recommendation is, however, to leave matters as they stand for the present. This is based upon the following factors:
>
> 1. Standard will, I believe, undoubtedly receive the financial support required to finance the TO-35 capital expenditure and generally to look after their immediate requirements. . . .
> 2. Our 12-year contract insures us against their making unduly high profits, and our very friendly and close relationship with them now gives us ample opportunity for guidance and reasonable control.
> 3. Their position competitively is hazardous. . . .
> 4. Some people argue that Standard should abandon the light car field . . . but to do this would entail radical changes in design and very large capital expenditures.
> 5. Should Standard get into financial difficulties this does not mean that their tractor production would require to be suspended. We or other interests would step in in the event of a crisis, and it might well be that, in such an eventuality, the business could be acquired cheaper than to-day.
> 6. An investment of £1,000,000 would give us virtual control, because the holdings are widely spread. . . .
>
> If capital were of less importance to us than it is at present, I would be in-

clined to recommend that we obtained control of the company as a safeguard against high manufacturing cost, and perhaps even a safeguard against an attempt to control by Ferguson, but all things being as they are, my recommendation is that we wait and see how they handle their affairs. . . .

The reference to "very friendly and close relationship" was not an appropriate description for long. Fundamental difficulties in their relationship were discussed by Standard and Massey-Ferguson even in 1954.

E. W. Young, the Managing Director of Massey-Ferguson's Eastern Hemisphere Division in London, reported to Duncan on a visit with officials of Standard in a letter dated October 12, 1954:

Both Lord Tedder and Mr. Dick advanced the view that our two organizations cannot continue to work together indefinitely in the position of sole supplier and sole purchaser of products which to them represent a large percentage of their manufacturing turnover and to us, an important part of our sales turnover. They both appeared fully conscious of the fact that they are going to be faced in the future with an intense competitive situation in the automobile field. . . .

They appreciate . . . that the Tractor side of their manufacturing business, particularly with the American content and the possible future expansion of American requirements, may ultimately become the major side of their over-all activities. They suggest, therefore, that in their view, the time is fast becoming opportune for them and us to sit down together, to see whether some identity of aims cannot be established which might have for their objective a closer fusion of our joint interests at some ultimate time. . . .

Nothing came of those sentiments of co-operation, and the difficulties that remained were the ones Young outlined in the same letter:

On the question of our relationship at the present time, both Lord Tedder and Mr. Dick regretted that there appeared to be some fresh atmosphere of suspicion creeping in between us. I have, of course, in routine discussions, had cause to offer to the Standard Motor Company, criticisms on certain questions relating to costs but I have, naturally, not forgotten to remind them that we consider them to be a high-cost producer. . . . As a general reply to my arguments, he pointed out that . . . we have at all times the right to full access to their Books, to their cost structures and to their procurement prices, and indeed, to any relevant factors which influence the price of the Tractor to us. . . . He pointed out to me that the problem of reducing manufacturing costs was for them a day-to-day operation and that so far as the Tractor was concerned, they had not been wholly unsuccessful. . . .

I think if you are prepared to accept the good will of their intentions and agree that at least something has already been done in the right direction, the crux of the question is some reconciliation about the profit margin to which the Standard Motor Company is entitled. . . . Their view was that a manufacturing margin in the neighbourhood of 7½% was one to which they were entitled.

In May 1955 Standard informed Massey-Ferguson that basic tractor prices would have to be raised in view of material costs and the need for additional profit. Duncan cabled: "I . . . consider it essential we answer with unequivocal refusal to consider anything but proven cost increases . . . since last cost review. . . ." Again it was the proposed profit margin on which no agreement was in sight.

Massey-Ferguson once more began to think of purchasing the tractor division. A. A. Thornbrough wrote a memorandum on the subject on July 13, 1955, in which he favoured investigating the effects of acquiring those facilities, and co-operating with Standard. He did not favour precipitate action unless forced. He concluded:

Purchase of the Tractor Division by M-H-F would from a cost point of view at best leave the situation pretty much as it is now. It could be worse. Unless there is a situation necessitating very prompt action, I would suggest immediate action on our part to determine what material savings can be achieved in Ferguson tractor manufacture, what is available in the way of existing facilities such as foundries etc., and the investment required to accomplish the savings. Mr. Alick Dick, in our last meeting before his hospitalization, raised again the joint solution of some of these problems by both Standard and M-H-F. In the absence of any immediate action on Standard, I believe we would have everything to gain by joining hands with Dick on this matter. It certainly would enable us better to evaluate how well M-H-F could do by operating the Tractor Division separately.

In any event things went on as they had with no decisive moves by Massey-Ferguson. To strengthen their position in cars Standard had discussions with the Rootes group and the Rover company in 1954 and 1955. Massey-Ferguson heard about this indirectly, and was irked that Standard had not discussed it frankly. At the same time, however, Massey-Ferguson was corresponding with a French company over acquiring a partial or complete interest in it. The company in question was Hotchkiss-Delahaye S.A., a licensee of Standard for the manufacture of engines for the tractor assembled by Standard-Hotchkiss S.A. at St. Denis. None of these "secret" negotiations came off.

But costs and profits of tractor manufacturing at Coventry were not the only issues causing difficulties between Standard and Massey-Ferguson. The two also differed on matters relating to design changes. Massey-Ferguson felt strongly that new tractors had hurriedly to be brought forward, while the former feared the expenses involved. For example, in November 1955, Standard felt it could not manufacture the proposed new large tractor excepting only components that were common to the FE-35, and the diesel engine. Discussions on these matters made it increasingly evident that a major disadvantage of the basic relationship between the two companies was that it prevented Massey-Ferguson from engaging confidently in long-run planning of its tractor and engine requirements. The subcontractor, after all, could suddenly refuse to make capital expenditures required for fulfilling such plans, with disastrous consequences for Massey-Ferguson.

A new source of irritation appeared in late 1955, as far as Massey-Ferguson was concerned. In the course of discussing with Alick Dick of Standard problems of switching production from the TE-20 to the FE-35 tractor on October 31, 1955, A. A. Thornbrough learned that Standard was interested in amalgamating their company in France, Standard-Hotchkiss (in which Massey-Ferguson had a 25 per cent interest) with their engine licensee in France, Hotchkiss-Delahaye. When Massey-Ferguson executives discovered later that negotiations had begun without their being a party to them, they were not pleased.

W. Lattman, a Vice President of Massey-Ferguson parent company at Toronto, wrote to E. W. Young, of the Eastern Hemisphere Division, on January 17, 1956, about this development. He also discussed Massey-Ferguson's recent (and in retrospect incomprehensible) decision to sell its interest in Standard-Hotchkiss:

> The recent actions of Standard Motors have shown a complete lack of consideration for our Company, both as a stockholder in Standard-Hotchkiss and as Standard Motors' major customer. We have, indeed, been shocked at Mr. Dick's discourteous action in entering into negotiations with Hotchkiss-Delahaye involving a company in which we have a 25% interest without prior discussion with us, especially since Mr. Dick knew that we were opposed in principle, to the contemplated move. . . .
>
> In view of Standard Motors' recent actions and attitude, we can see no future in this company having a participation subordinate to that of Standard Motors, and while we are being forced most reluctantly, to subordinate certain of our programmes to programmes involving Standard Motors' interests, we have authorized you to negotiate the settlement with Standard Motors [for sale to them of Massey-Harris-Ferguson shares in Standard-Hotchkiss].

Standard's view was that it had been given responsibility for manufacturing the Ferguson tractor in France and that the proposed amalgamation between Standard-Hotchkiss and Hotchkiss-Delahaye was part of its long-run programme of moving from partial to complete manufacturing. Failure to do so, it felt, would jeopardize the continuity of tractor production in France. This was a strong argument. Massey-Ferguson's objection to it seems to have been based essentially on matters of procedure and protocol. It is significant that Massey-Ferguson management decided, in the heat of the moment, to relinquish what control it had over the source of supply of the Ferguson tractor in France. Did this reflect the absence of any plan or policy for arriving at a permanent solution to the unstable arrangement the company had depended upon for a large portion of its supply of tractors and engines? Probably it did, but it is also true that in January 1956, being exasperated over their relations with Standard, the Executive Committee decided to begin accumulating stock in Standard Motors itself, and gave one of its Directors, J. A. McDougald, authority to proceed. The company, in the end, did not relinquish its influence in France because Standard would not agree to purchase Massey-Ferguson stock in Standard-Hotchkiss on the terms offered.

On February 3, 1956, E. W. Young of the Massey-Ferguson Eastern

Hemisphere Division in London received a message from Alick Dick saying that he wanted to see him on an important matter. They got together and Young reported on the meeting to Duncan:

After some introductory remarks, Dick opened up by saying that he and his co-directors were concerned by what they felt to be a deterioration over the last few months in the previous goodwill and understanding that had existed between our two organizations. Other than a passing reference to past events in relation to the Standard-Hotchkiss company and Hotchkiss-Delahaye, he was not specific. . . .

As the conversation developed, he went on to say that he and his colleagues felt there was some merit in considering the possibility of a direct and definite merger between the U.K. companies of the M-H-F group and the Standard Motor Company. . . . He felt strongly that no half measures would ever achieve the objectives which he felt to be essential in our mutual interest. . . .

Recognising our previous reactions about being directly involved in the car business, and also the fact that such a merger would mean a great deal less facility in the handling of inter-company pricing, dividends, engineering changes, organization, etc., we were rather non-commital in our reaction. . . .

The reaction of the President to this approach was not a helpful one. He answered Young's letter this way:

He [Alick Dick] . . . is right that our relationships have deteriorated somewhat. This is regrettable and we should continue to do our best to keep them on the friendliest possible basis – but this deterioration has resulted from his actions. May I recall just a few of these: –

Our disappointment at:

a) Standard Management's defective planning which has cost us millions of dollars in the U.S.A.
b) Their periodical and unpredictable change of policy with regard to the building of the large tractor and the diesel engine.
c) Their having invited us into the Standard-Hotchkiss situation without revealing their prior commitments or their future intentions.
d) Their inexcusable action in offering to amalgamate Standard-Hotchkiss, in which we hold a 25% position, with Hotchkiss-Delahaye, without consulting us, and with the full knowledge that we were opposed to any such move.
e) Last but not least, after agreeing to their suggestion that we establish a top policy committee to discuss all major problems, that they should have entered into negotiations with Rootes without advising us. . . .

For your information we are not interested in selling out any part of our business to anyone and following Mr. Dick's suggestion that "we would be strongly represented on his board" it is quite clear that he fancies himself as Managing Director of a company who would control Standard and Massey-Harris U.K. and U.K. export interests. That day hasn't dawned, nor will it! . . . .

You might . . . say [to them] . . . that before talking larger issues, I should like to see the present smaller ones settled in the normal business manner which one is entitled to expect from corporations of the importance of ours, and that when these matters are behind us I shall be most happy to sit down with Lord Tedder and himself, on a subsequent trip to England, and discuss the whole situation along the broadest lines.

In spite of the apparently confident posture of Massey-Ferguson executives, their principal difficulty was that they had not yet offered any positive plan for resolving permanently their difficulties with Standard.

Young wrote to both Phillips and Thornbrough in August 1956 about the deteriorating car business at Standard: "To-day, it has been published that the automobile division of the Standard Motor Company is going to operate immediately on a 3-day week, and over and above this 1,000 additional workers have been declared redundant, and will be released at the end of August. . . . The situation is one which cannot but be of great concern to us in its implications on the tractor side. . . . " These implications had increased greatly with the advance of the company's policy of using components from Coventry for its assembly operations at Detroit, and of planning for Standard to play a dominant role in supplying tractors and engines. If all the company's tractor activity was tied to what it regarded as an inefficient producer, over whom it could exercise no effective management control, and one who might even become insolvent through weakness in another manufacturing line and might sell out to an unfriendly party or even a competitor, there could be the most serious financial consequences for Massey-Ferguson.

The Board had earlier decided that the company should purchase a substantial block of shares of the Standard Motor Company on the open market. Secrecy was vitally important to maintain stability in the price of Standard stock, and the company chose to deal not through Wood, Gundy and Company, whom everyone knew to be its underwriter, but rather through Dominion Securities Corporation Limited, another Canadian investment dealer. On October 3, the Executive Committee was informed that the company had acquired about 5 million shares of Standard, approximately 18 per cent of the common shares outstanding. Strict secrecy had been maintained and the market had not been able to identify the buyer. A resolution of the Executive Committee authorized the Chairman, W. E. Phillips, to conduct negotiations with Standard to acquire up to 5 million additional shares at not more than 8 shillings and 6 pence per share, but few additional shares were immediately purchased.

Phillips first attempted to bring Standard and Rootes together, for he thought that a merger between them would strengthen the automobile side of Standard; but no positive results were achieved. Another approach seemed to be necessary. In June 1957, before Phillips went to England, he asked the company's auditor in the United Kingdom to provide him with some financial information on the Standard Motor Company, and in the same letter he revealed that a new stage in his company's approach to Standard was developing:

I think you understand my position in this matter and, after consulting with my colleagues on the subject matter of your report, we are working our way towards the conclusion that we must, sooner or later – and the sooner the better – make an attempt to secure the Standard shares by offering our own shares.

This will require considerably more study than I have been able to give it, but I am convinced that we must face up to the fact that this is the only practical proposal that we can entertain. Any other approach gives us less than the desired results and I think you may assume that this is the proposal we will attempt to develop with you when I arrive in England early in June.

It may well involve two stages. The first that I have in mind would involve a proposal that we purchase at the best possible price some five million shares from the presently unissued treasury shares. This would be for cash and would of course involve a general agreement on our having a substantial interest in the control of operations which would permit us to study, on our own account, some of the problems which might arise as between the motor car and the tractor operation.

This approach to acquiring control of Standard, that is through an exchange of shares, did not mean that the company was interested in entering the car business. A memorandum of June 13, 1957, stated "The policy of the Board is, so far as may be practicable, to secure control of the manufacture of its tractor without entering the motor car trade." Separation and sale of the car business would be possible after control of the company was acquired. Schemes involving forming a tractor company with Standard were discussed internally and were frequently a feature of later negotiations. But for the moment it was thought that the best strategy was to acquire control through an exchange of shares.

It should not be forgotten that apart from the temptation to acquire the tractor manufacturing facilities as a way of overcoming the difficulties inherent in the existing subcontracting relationship, there was also the very positive attraction of the Coventry facilities themselves. While Phillips and H. A. Wallace, Vice President Manufacturing, were in London in mid-1957 to negotiate with Standard, Wallace gave his opinion of the facilities and of the company's need to acquire them:

This is a fine facility. While agreed that the transfer equipment in the Banner Lane plant has been overdone, the fact that it is there makes the facility an exceptionally fine one. It is true that there is a very limited flexibility, except at a high price, at Banner Lane, but full use of the facility can be made for some time. . . .

My specific observations are as follows:

1. This is an excellent facility and unless it is acquired, Massey-Harris-Ferguson will be in almost as vulnerable a position as Harry Ferguson was with Ford.
2. In my opinion, the acquisition of Standard Motors must include the engine manufacturing facility because Massey-Harris-Ferguson is just as vulnerable in this area as they are with tractors.

The reference to the importance of the engine manufacturing facilities is significant because later, when Massey-Ferguson acquired the diesel engine facilities of F. Perkins Limited, its dependence on Standard diminished and the bargaining power of the latter was considerably weakened.

It was on July 17, 1957, that the world heard through a press release that Massey-Ferguson had sent an offer to Lord Tedder, Chairman of the Standard Motor Company, to purchase all the ordinary shares of Standard. With that offer the fireworks started. The offer involved one share of Massey-Ferguson common stock plus twelve shillings, for each unit of eight shares of Standard. Massey-Ferguson shares had traded at over $7.00 on the Toronto stock exchange which meant that at current stock prices and current exchange rate, Massey-Ferguson was offering about $25 million for Standard. It worked out at about 9 shillings and 6 pence per Standard share. Lord Tedder recommended to the shareholders that they should accept the offer.

Whether or not the offer was financially attractive depended entirely on the extent to which current prices for the stock of the two companies reflected earning prospects. To assist in making the offer successful, such prospects ideally should reveal weakness in earnings of Standard and strength in Massey-Ferguson earnings. But no new earnings information was made public. It is now known, of course, that non-recurring factors were producing a very large loss in the Massey-Ferguson organization which, while transitory, would probably not have been accepted as such by Standard shareholders if they were to hear about them. E. P. Taylor sensed this at the time and wrote to Phillips about it the day after the offer was sent to Lord Tedder:

> I am convinced that a very difficult task lies before us to make the offer successful. This will be particularly true if we are required to have an audit which will put the spotlight on the unsatisfactory figures we are experiencing in North America this year. I cannot emphasize this too strongly that every effort should be made to avoid an audit.
>
> It is also very important that Standard must make a poor showing this year. In that respect, one sentence in Tedder's remarks was unfortunate, which as you know is to the effect that substantial profits are currently being earned by Standard. I also think that prior to the forwarding of the offer and prospectus to Standard shareholders, the Canadian market must be groomed and informed that world-wide profits this year will be nominal, but that the expectations for next year are for quite substantial profits.

In the United Kingdom the press at first was divided on whether the offer should be accepted or not. The absence of earnings information was referred to. There was also much discussion of loss of control of Standard to another country, and a Member of Parliament raised that issue in the British House of Commons. A well-known Cardiff accountant announced immediately that he would write to the shareholders recommending that they refuse the offer and demand a better one. He remained in vocal opposition to the offer over succeeding weeks. This view soon received increasing support and Massey-Ferguson experienced a "bad" press. The *Stock Exchange Gazette* of London told its readers on July 26, 1957:

Nearly everyone knows that the price of Standard Motor shares is no indication of the value of the underlying assets, and nearly everyone knows that if M-H-F is successful in this bid it will be obtaining an absolute bargain. The directors are there to prevent that, and it is hoped that the offer will never be made formally.

The terms had been approved, it should be noted, not only by the Board of the Standard Motor Company but also by two leading merchant banks, J. Henry Schroder, and Helbert Wagg and Company.

While the view was beginning to be general that the offer was inadequate, it also became clear that the formal offer as such could not be made until about the middle of September because of the numerous formalities required by Canadian law. The United Kingdom government meanwhile gave permission to Standard shareholders to accept the offer. On August 21, Phillips submitted a proposal to the shareholders of Massey-Ferguson to increase the number of common shares by 7½ million in order to acquire the Standard facilities. He also told them that if the offer was not accepted by Standard shareholders, the company would be forced to build its own plant in North America or Germany. The United Kingdom press reaction was epitomized by the *Daily Express* article heading of August 23, 1957, "Massey Man Gets Tough Over Bid." The term "blackmail" appeared in the press.

Quite apart from the appropriateness of the initial offer, developments were now beginning to appear that gave Massey-Ferguson no choice but to beat a hasty retreat. The stock market in general entered a decline that lasted six months and the market became increasingly aware that Massey-Ferguson would have a poor showing in 1957. Devaluation of the French franc cast some uncertainty over the valuation of the company's assets in that country. The company's stock declined over one dollar per share in price and on September 8 Massey-Ferguson issued a press release withdrawing their informal offer. No formal offer was ever received by the shareholders of Standard. The press release ended with this reassurance:

Massey-Harris-Ferguson and Standards will, of course, continue to co-operate closely as in the past and will explore all areas in which the two companies are interested with a view to developing a closer and more profitable association wherever possible.

These were noble sentiments, but the fundamentals of the situation cast grave doubts on their sincerity. The end to unrelenting press discussion of the issue might have suggested that something fundamental had indeed been settled. This could not have been further from the truth. In fact, something of a crisis was emerging regarding Standard's work on the new MF-65 tractor programme. Thornbrough wrote a letter to Alick Dick on September 30, 1957, that left no doubt about the urgency of the situation:

To say that I am disturbed about the quality control programme at Standard

> Motor Company, particularly regarding the MF-65 programme, would indeed be an understatement.
>
> The information I have received is not only disturbing but comes as a surprise as well. As long as two years ago I personally discussed the importance of a sound quality control programme with regard to our component planning. . . . Full agreement was understood.
>
> I should like to itemize some of the things which are unsatisfactory concerning the existing programme:
>
> 1. There has been practically no organizing or administration of a basic inspection programme.
> 2. There is no over-all correlation of the various inspection activities under a sound basic written plan.
> 3. There is no staff activities under the chief inspector by which the current difficulties are analysed. . . .
> 4. There is no system of putting on record the day-to-day inspection information . . . of performance.
> 5. Approximately 70% of all completed tractors presented to Massey-Harris-Ferguson Ltd. for final inspection are rejected for defects. . . .
> 6. There is a dual standard of inspection practice by which parts or assemblies rejected for the North American schedule are used in FE-35 production. . . .
> 7. The Chief Inspector of Massey-Harris-Ferguson Ltd. is not contacted or advised when major decisions as to usage of sub-standard parts are made. . . .
>
> Don't you think . . . that it is a little late to be investigating such a basic and comprehensive programme when we are expecting shipments on the MF 65 programme in November? Why was it necessary to let all the valuable time slip by. . . . ?
>
> Certainly it is not necessary for me to point out to you how important the success of this component programme is to both the Standard Motor Co. and Massey-Harris-Ferguson Ltd. This is probably the largest programme of its type that has ever been attempted. We feel that the stake is so great that we have a right to know that these components are being properly quality controlled. . . . It is not our intention . . . that we accept sub-standard parts.

Before long, too, the company was experiencing very sharp cricitism in the field because of its inability to supply adequate amounts of spare parts for the TE-20 and FE-35 tractors, and investigation pointed to the failure of Standard to supply parts on schedule. Negotiations between the two companies relating to prices for components to be supplied were faltering and in February 1958 Standard spoke of the possibility of arbitration. It should not be assumed that Massey-Ferguson personnel were innocent in the failure to reach an agreement on prices, and in the case of one senior executive, the fault for failure lay with the company. Still, it was against this background of experience that the Executive Committee of Massey-Ferguson again authorized Phillips on February 7 to enter into negotiations with a view to purchasing some or all of the shares of Standard. No immediate steps were taken, in part because tension eased somewhat in mid-February when Thornbrough was more successful than a previous Massey-Ferguson executive had been in negotiating prices. Quality and spare parts problems also seemed to be diminishing.

A ray of hope and good will seemed to be glimmering. It was short-lived.

On May 23, 1958, an announcement appeared in the United Kingdom press that made an open break between the two companies inevitable and permanent. The announcement said that the Standard Motor Company had, by an exchange of shares, offered to acquire Mulliners (Holdings) Limited, owners of Mulliners Limited, makers of car and commercial vehicle bodies. Massey-Ferguson executives in general, and Phillips in particular, were furious. Such an exchange would mean that the Massey-Ferguson share in Standard would sharply be diluted since Mulliner shareholders would receive 3,656,250 new shares of Standard stock. What is more, when Lord Tedder had mentioned the proposal several weeks before, Phillips had expressed his objection to a share exchange and had offered to purchase with cash and direct from Standard sufficient shares to enable Standard to purchase the body plant.

The formal offer by Standard for the shares of Mulliners was dated June 11. To the great surprise of the financial community this was followed on July 3 by an offer by Massey-Ferguson Limited to buy from Mulliners shareholders all the Standard shares they would receive at a price of eight shillings each. This was well above the market price.

There then followed a press release from Standard, which together with a private letter written by Phillips, provide graphic illustration of the depth to which relations had descended. The press release stated:

> The Board of Standard wish to make it clear that the proposed offer announced yesterday by Massey-Ferguson to purchase the Ordinary Shares in Standard to be issued to Stockholders in Mulliners was made without any consultation with them.
>
> The offer is, of course, intended to enable Massey-Ferguson to acquire a further important interest in the Ordinary capital of Standard (beyond their present holding of 20¼%) without the necessity of an offer to the general body of Stockholders and at a price below that proposed in the negotiations of last summer.
>
> The Board of Standard regret this unilateral action which they feel is inconsistent with the mutual confidence which is so essential in the interests of both companies.
>
> Standards are, however, determined to do all in their power to re-establish this mutual confidence.
>
> In the meantime they can only interpret Massey-Ferguson's offer as a step towards progressively obtaining full control of Standard; a very different situation from that envisaged in the proposed merger last year. Such a development the Board could not support in view of the fact that the interests of Massey-Ferguson as the largest customer and largest Stockholder of Standard must inevitably conflict at times with the interests of the other Stockholders.

Massey-Ferguson made no public announcement even though much of the press, including the *Economist*, were exceedingly critical of its actions. But a letter Phillips wrote to W. Lionel Fraser, chairman of the banking firm of Helbert Wagg and Company, explains how he saw the situation. The letter shows that he saw it with some feeling:

When Tedder was last in Toronto in March of this year, he mentioned at lunch one day that Standard was contemplating the purchase of Mulliners (Holdings) Limited, and would probably propose an exchange of shares.

I told him at once that we would regard any such share exchange as unfortunate – if not an actually unfriendly act – and proposed, as an alternative, that we should put them in such a position that they could make a cash offer to Mulliners, which would probably mean a lower price, and that to do that we would purchase up to five million dollars of Standard's unissued Ordinary Shares.

I pointed out that, as on a previous occasion when such a transaction had been discussed in another context, we would readily agree to underwrite an offer of the requisite number of shares which would be made pro rata to all the shareholders of Standard.

Tedder then appeared to receive these suggestions in a constructive mood and said that he would keep me informed.

I enclose a copy of his letter to me of April 3rd last, the last paragraph of which speaks for itself. I told Tedder that I would be in England about May the 15th and we had arranged by exchange of cables a meeting in London for the 17th of May. However, he was not available and I lunched with him at his house on the following Monday.

Up to that moment, I had not had a single word from him regarding the Mulliner project. You can imagine my astonishment when he told me that the matter was now in the hands of yourself and Schroeders and that it was expected an agreed offer would crystallize within the next few days.

At this same meeting he rejected out of hand the several other matters of common concern which we had discussed in Toronto.

I have shared with you for a long time the view that the development of mutual confidence in the growth of understanding and good faith between us was a sine qua non to the orderly and constructive resolution of our common interests. Tedder's failure to provide even a moment for discussion of the whole Mulliner proposal leads me to the conclusion that there is no longer any purpose in pursuing the now illusory objective of co-operation and mutual understanding. Tedder preaches both and practices neither.

I conclude that his failure to carry out the undertaking implicit in the last paragraph of his letter to me of April 3rd is positive evidence of something less than good faith, and I put it to you, though I know you will not agree, that our action in making a blanket offer to the Mulliner shareholders was a perfectly proper answer to an improper action of Tedder's.

I note in the Press Tedder's statement that we did not consult him on our offer to the Mulliner shareholders.

Now that you must agree is not the whole truth. We certainly did not and would not consult him or anyone else in our decision to make this offer to the original Mulliner shareholders but I see nothing in his statement to the Press to indicate that the matter of Mulliners was the subject of a serious proposal from us as early as March.

When Tedder was here, I asked him to sell us sufficient of Standard shareholdings in Standard-Hotchkiss of France to give us each 37½%, and, on that basis of equality of shareholdings, we could then approach the obvious economic objective of manufacturing all our French tractors in the one production facility. This again, he rejected without explanation.

Co-operation is a two-way street and I have been forced to the conclusion that Tedder does not want to co-operate, nor does he desire any better understanding with us.

We have refrained from saying anything at all up to this time but it may well be necessary for us to arrange for the release of certain facts relating to our difficulties with Standard and its management.

This is a fantastically confused situation. We are the largest single shareholders of Standard and we are the largest customer but any unprejudiced outsider might well conclude that we are acting like a crowd of drunken Russians.

We do not seek control of Standard. We seek primarily the ability to achieve the maximum integration of our common manufacturing facilities, so that we may have a reasonable guarantee of lowest possible manufacturing cost, which indeed we must have if we are to continue to sell our tractor products successfully against Ford and others.

You yourself, and Tedder, have several times encouraged us to buy Standard shares and when we do so you shout to high heaven. Perhaps we all need the advice of a psychiatrist.

In closing, may I say that I hope the arguments which arise out of this fantastic relationship will not impair our own personal friendship.

While these latest moves were being discussed on every financial page in a way generally very critical of Massey-Ferguson, the company quietly increased its control over manufacturing in France. Hotchkiss-Brandt S.A., which held 25 per cent in Standard-Hotchkiss, approached Massey-Ferguson with an offer to sell them its interest in that company. Massey-Ferguson accepted and thereby gained 50 per cent ownership of Standard-Hotchkiss stock, the other 50 per cent being held by the Standard Motor Company. So by July 1958 Massey-Ferguson had placed itself on an equal basis in tractor manufacturing in France, even though it had failed to do so in the United Kingdom.

The Mulliners affair proved to be a turning point. Things simply could not go on as they were. Standard Motor management began to accept the principle that Massey-Ferguson would have to control the tractor business and that if satisfactory terms could be reached this should be done by physically separating the car business from the tractor operation. As early as August 1958 senior management of both companies began discussing various plans and on August 15 an official of Massey-Ferguson wrote Phillips that an executive from Standard had " . . . agreed that he would be willing to consider an offer by Massey-Ferguson (a) to buy the whole tractor business for cash, (b) to buy the whole tractor business for cash plus shares in the Toronto Company." He also reported that " . . . it appears that the S.M. Co. have also reached the conclusion that it would not be practicable to form a joint tractor company on a 60% M.F. — 40% S.M. Co. or any other basis." Massey-Ferguson submitted a proposal.

Negotiations, however, did not move very quickly. Before the two companies were even near agreement a representative from F. Perkins Limited had made known to Massey-Ferguson that his company could be purchased; and within eight weeks Massey-Ferguson had made one of its most important acquisitions. This meant that Massey-Ferguson would no longer have to depend on the engine manufacturing facilities of Standard and would find it somewhat easier to establish its own tractor manufacturing facilities if no agreement with Standard could be reached.

Before agreement was finally reached Standard again proposed forming a new tractor company with participation by both, but this Massey-Ferguson flatly refused. Then there was wide disagreement over price. Deadlock appeared certain. Massey-Ferguson even prepared a draft press release announcing both the failure of negotiations and plans for building its own facilities.

But discussions continued, perhaps because of Massey-Ferguson's stronger bargaining position through its acquisition of Perkins and its interest in Standard-Hotchkiss, perhaps because Standard wanted capital for a new line of cars, or perhaps just because of the undeniable fact that existing arrangements were unworkable and had to go. Also, beginning in April 1959, Thornbrough, Staiger, and Wallace temporarily moved to the United Kingdom to reorganize operations there, and Thornbrough personally handled discussions with Dick of Standard. After long, tedious, complicated, and sometimes acrimonious discussions, the issues were clarified and agreement was finally reached.

On July 23, the terms of agreement were announced. The tractor assets in England and France, capable of producing 125,000 units annually were acquired for $32 million net, that is after deducting about $8.3 million which the company was to receive for its holdings of 7,757,938 common shares in Standard. Ownership passed to Massey-Ferguson on August 31. The Banner Lane, Coventry, facility, utilizing more than one million square feet was capable of producing 100,000 tractors annually. In France the company acquired complete ownership of the Saint Denis and Beauvais plants of Société Standard-Hotchkiss. But not included were the engine facilities of Standard, for Massey-Ferguson now owned F. Perkins Limited of Peterborough, England.

Phillips told the press " . . . we shall be masters of our own home and be free to perform more effectively in the increasingly competitive world market for farm tractors and equipment. It is a major leap forward in our policy to manufacture more of our products."

CHAPTER 14

Perkins of Peterborough

Negotiations for the acquisition by Massey-Ferguson of F. Perkins Limited, of Peterborough, England, were more or less complete by December 12, 1958, six days after the President of Massey-Ferguson revealed his company's interest in Perkins. It was probably one of the quickest and smoothest take-overs of its size in recent times.

For Massey-Ferguson it was also one of the most important events in its recent development. It gave the company its own source of diesel engines at a time when the agricultural tractor market was shifting massively away from gasoline engines towards diesel engines; it permitted the company to plan its future tractor line with the assurance that planning for appropriate engines was also under its control; it greatly enhanced the company's ability to enter tractor manufacturing operations in the less developed countries, for by establishing local engine assembly and manufacturing operation (Perkins indeed was already operating in several countries), it made it easier for the company to meet "local content" requirements; and, as we have already seen, it strengthened the company's position in its attempt to obtain the tractor manufacturing facilities of the Standard Motor Company.

But all these advantages were enhanced by the nature of the Perkins organization. The Perkins company had been a pioneer in the development of the light, high-speed diesel engine suitable for trucks, tractors, and boats. It had developed a strong export market. It had begun to expand its assembly and manufacturing operation through subsidiaries and licensing agreements to other countries – Yugoslavia, India, France, Brazil, Spain, Argentina, and Italy. Because of all this the Perkins diesel was a well-known engine and the Perkins organization produced it in very large numbers. Yet, eventually, the Perkins organization decided that its future lay in becoming part of another, and much larger, organization. So while its history reflected the unusual accomplishments of an individual entrepreneur, it also reflected the strong forces of structural change in industrial organization.

The individual entrepreneur in question was Frank A. Perkins. In 1929 Frank Perkins was Vice Chairman and Works Director of Aveling and Porter Limited, of Rochester, England. That company, being caught up in the officially inspired policy of industrial "rationalization" in the 1920s, had become part of a new group called Agricultural and General Engineers Limited (A.G.E.). Among other things, Perkins was concerned at Avelings with the development of a four-cylinder diesel engine of higher speed that would be suitable for a certain make of agricultural tractor. Helping him as Chief Engineer was a gifted design engineer – Charles W. Chapman. The engine was completed and fitted on an International Harvester tractor. From it was developed a six-cylinder engine that could be fitted to a truck. But before this experimental work could proceed further, the Depression ended the life of A.G.E. and with it, Avelings. Frank Perkins was out of a job and so was Charles Chapman. At Peterborough the family engineering firm of Barford and Perkins Limited was also in financial straits.

Perkins apparently had become convinced that the basic idea of a light, high-speed diesel engine was a sound one. If such an engine could be developed it might be sold to truck owners, provided they could be made to realize that converting from gasoline to diesel engines would save them money through reducing their operating expenses. The diesel engine, in practice, is more efficient in its use of fuel than the gasoline, mainly because it can operate at a higher compression ratio (that is, the ratio of the volume of the engine cylinder when the piston is at the bottom of its stroke to the volume when it is at the top of its stroke). In this way the temperature of the air in the cylinder is raised to a level that will cause fuel injected into the cylinder towards the end of the compression to ignite without any external means of ignition. A special injection pump injects fuel into the combustion chamber of each cylinder at the appropriate time, and in a quantity which determines the amount of power to be obtained from the engine.

The high internal engine pressures created by the high compression ratio, as well as the type of fuel pumps and the method of combustion employed, resulted in diesel engines of the early thirties running very slowly and being very heavy. They were much bigger than gasoline engines of the same power output. Therefore they could only be used in vehicles designed specifically for them, and not in vehicles designed for gasoline engines. What was needed to penetrate the high-volume automotive market was a diesel engine which approached gasoline engines in size, weight, and power output.

Power output is dependent on engine speed and torque. To achieve desired increases in engine power it was necessary to increase the speed of the diesel engine to a rate approaching that of the equivalent gasoline engine. This alone would allow the required reduction in engine weight and size. Perkins were the first to do this. Since then, this type of engine, powerful yet small and efficient in fuel consumption, has come to dominate the agricultural tractor engine market as well as a large portion of the land vehicle and marine markets.

But that could not be known when Frank Perkins decided in 1932 to form F. Perkins Limited. The company was registered on June 7, 1932, its address being given as 17 Queen Street, Peterborough. The prospectus read:

> It is proposed to form this Company to develop and manufacture small High-Speed DIESEL ENGINES, and to carry on such engineering work of an allied nature as may from time to time be deemed suitable by the Board of Directors.
>
> The Company is to be a Private Company with an Authorized Capital of £12,000. It is proposed that £10,000 be subscribed in cash in one-pound shares, by G. D. Perks, A. J. M. Richardson, and F. A. Perkins, and that 1,000 £1 shares be assigned to F. A. Perkins, and 350 one-pound shares to C. W. Chapman, without payment. . . .
>
> There is an unlimited field for engines of this class, and up to the present only a small part of the market has been developed. There is a steadily increasing demand for a smaller engine than has yet been produced, and it is such an engine that the Company proposes to concentrate on.

Frank Perkins was Chairman and Managing Director, Charles Chapman was Technical Director and Secretary, and the premises for the new company, these having belonged to the family firm of Barford and Perkins, were rented from Frank's father, J. E. S. Perkins. The family's relationship with Peterborough undoubtedly influenced the choice of a site, but the city also was an engineering centre, had good rail connections, and was close to potential customers in cities such as Coventry. This is how Massey-Ferguson came to have its engine manufacturing organization located at Peterborough.

As soon as equipment could be acquired, the few engineers constituting the company began to develop an engine. As early as 1932 the first engine, a four-cylinder one named the Vixen, was running on the test-bed. There followed innumerable improvements and Perkins began to contact potential customers among the manufacturers of trucks and vans. To obtain more power the cylinders were bored to 80 mm. from 75 mm. and the resulting engine was called the Fox; and to obtain even more power they were bored to 85 mm. and so the Wolf was born – with not much metal left between cylinder bores. In October 1933, Perkins was successful in negotiating an agreement with Commer Cars Limited through its parent, the Humber organization, whereby the latter would show its truck fitted with a Perkins engine at a commercial vehicle exhibition that year and would offer that engine as optional equipment in new vehicles usually equipped with a gasoline engine. At the same time Perkins began a business that has always been important to it – the conversion of vehicles in owners' hands from original engines to Perkins engines. Perkins engine design was greatly influenced by the need to make its engines fit existing vehicles.

Progress was not rapid and the company experienced losses. Developmental work required additional funds and sales were not providing them. On several important occasions over those early years it was only through the strong financial support of Captain A. J. M. Richardson that expenses could be met. Suppliers had at times to be convinced that they should wait for payment. Almost all components were made by suppliers in the early years.

The Leopard, an engine with a 100 mm. bore, and the Leopard II with a 105 mm. bore were hurriedly produced in 1934 to bolster sales. Sales did pick up and some export orders were beginning to appear, but in 1935 a heavy tax on fuel oil for road vehicles almost cost the company its existence. That was one of the occasions when Richardson gave vital financial support. Still, in 1935, sales at £62,571 were more than twice those of 1934, and by crediting the development account with a substantial sum a profit could for the first time be shown.

Not until 1937 was a genuine profit made. Much more important than that, however, was the appearance in that year of the engine that put Perkins on its feet – the famous Perkins P6. Its development, as the others, owed much to the engineering design contribution of Charles Chapman who was with the Perkins company from its beginning until 1942. By the year 1937 the old Wolf was out of date and the Leopard had strong competition. The more efficient P6, a six-cylinder engine, was tested successfully in April 1937, less than half a year after work on it had begun.

The company was on its way. Its development can be seen in the number of engines produced, and in its sales and profits as shown in Table 13. From 1941 to 1945 inclusive, the company sold 9,074 engines. Of these, 50 per cent were for marine application, 26 per cent for vehicles, and 24 per cent for industrial uses. The P6 led the field, accounting for 51 per cent of total sales of the Company, the S6 29 per cent and the P4 18 per cent. Sales had amounted to £157,774 in 1938; by 1945 they totalled £1,304,484, which included 2,278 engines as well as some spare parts.

Additional facilities and therefore additional capital were badly needed. The Eastfield factory was started in 1945 at Newark near Peterborough, and additional capital through issues of common and preferred stock and debentures was obtained with the assistance of the Finance Corporation for Industry. This was to be repeated a number of times because of the almost meteoric expansion of the company. (In 1951 Perkins became a public company.)

To obtain outside funds in the amounts required by this relatively new company was not an easy task. The company's success in doing so, and its decision to expand at great speed after 1947 cannot be separated from

David Forbert

An MF tractor heads for the countryside past the ancient city of Toledo, Spain.

TABLE 13

F. Perkins Limited Production, Sales, and Profits (1933–1958)

	Engines produced (Units)	Sales (£)	Profits (£ Net after taxes)
1933	N/A	7,006	—3,503
1934	N/A	30,274	—521
1935	N/A	62,571	3,386
1936	N/A	99,318	687
1937	N/A	151,262	8,296
1938	N/A	157,774	7,711
1939	N/A	206,320	8,242
1940	N/A	416,781	9,066
1941	1,486	673,758	12,049
1942	1,869	1,019,193	9,724
1943	1,746	1,055,568	5,419
1944	1,695	1,088,205	12,773
1945	2,278	1,304,484	10,016
1946	3,625	1,249,745	10,385
1947	3,895	1,398,623	60,262
1948	6,865	2,370,422	54,953
1949	15,093	4,447,960	136,083
1950	25,218	6,660,177	318,633
1951	34,961	9,348,984	474,962
1952	38,154	11,374,203	324,363
1953	46,168	13,216,796	326,733
1954	61,956	16,802,515	432,478
1955	68,875	19,860,000	418,871
1956	74,838	21,548,000	348,075
1957	56,388	14,239,900	—318,751
1958	77,018	16,006,802	375,265

the man who founded and, in every sense, ran it. Frank Perkins was recognized as a man of integrity. He exuded confidence; even his physical appearance left a firm impression.

The atmosphere of paternalism that he inevitably, and perhaps unknowingly or deliberately (who can know?), generated in the organization seems not to have been resented; since it was accompanied by firmness and fairness, it seems to have created a sense of security among employees. Without the impression of integrity, confidence, and ability that he was able to create, it would have been much more difficult to obtain outside funds for capital expansion.

Just how rapid the company's expansion was is indicated in Table 13. It shows that while the company produced 3,895 engines in 1947, that figure increased to 6,865 the following year, then to 15,093 in 1949, and in 1954 61,956 engines were sold. In value terms sales amounted to £1,249,745 and profits (after provision for taxes) to £10,385 in 1946. In 1954 they were, respectively, £16,802,515 and £432,478.

The company developed important export markets – over half of its output went outside the United Kingdom. Subsidiary companies were established in Australia, Canada, France, and South Africa to service its customers. But the first steps toward manufacturing engines abroad were taken in 1953. In January of that year Perkins granted an exclusive licensing agreement to its distributor in India, Simpson and Company Limited of Madras, for the assembly and progressively the manufacture of the P6 engine for sale in India. Component parts were to come mainly from Peterborough until local manufacturing was possible. This agreement was extended in May 1954 to include all P series engines. In 1954 also a licensing agreement was granted to a government organization in Yugoslavia but it did not include use of the Perkins name. More important, in that year Société Française des Moteurs Perkins was formed in France with F. Perkins Limited holding 49 per cent of the stock, the majority being held by French shareholders; and at the same time F. Perkins Limited granted a ten-year licensing agreement for the manufacture of P series engines in France to Ateliers Guillemin Sergot et Pegard (G.S.P.). The arrangement was that all P series engines manufactured by G.S.P. were to be sold to Moteurs Perkins, which had exclusive selling rights in France.

No further new manufacturing arrangements appeared until 1958 when a licensing agreement was signed with a new company, Perkins Hispania S.A., of Madrid, Spain, in which F. Perkins Limited acquired a minority interest. In that year, too, the company became committed to establishing operations in Brazil. At the end of 1958, Massey-Ferguson entered the life of Perkins.

In all these instances of foreign development Perkins had had to invest little of its own capital, and by confining itself to licensing arrangements and a minority interest, it had not directly become responsible for manufacturing abroad. This was soon to change in the case of Brazil and France.

But we must now return to the main stream of the company's development and experience, centring on its operations at Peterborough. The company's explosive growth was based on a deliberate policy of standardizing on a few engine models, particularly on the P6. In 1951 ninety-five per cent of its effort was devoted to producing that one engine. It was a policy that led to low unit costs of production. The company's market in those years was essentially for vehicle and tractor engines: in 1949 the former took 49 per cent of its output and the latter took 39 per cent, while in 1954 the vehicle market had increased its share to 64 per cent, and the tractor share had declined to 30 per cent. But sales of Perkins engines for vehicles included not only sales to vehicles owners, that is for "conversion" purposes, but also increasingly during the post-war years to large vehicle manufacturers as original equipment.

The vehicle market and also the market for tractor engines grew swiftly because of the market shift away from gasoline engines towards diesel engines – prompted by the success of the concept of diesel engines pursued by Frank Perkins in the early 1930s. Herein lay a hidden danger for Perkins. As long as the diesel engine requirements of vehicle manufacturers were relatively small, it was less expensive for them to purchase the Perkins engine than to manufacture their own. Economies of scale of production were realized in the industry by various manufacturers drawing on Perkins production. But as the requirements of the major manufacturers grew, a point was reached where the output that each required began to approximate the output of an optimum size plant; and at that point the advantages of drawing on the output of a subcontractor began to vanish. Large manufacturers began increasingly to think of producing their own diesel engines.

At this crucial stage, the process was speeded by an unfortunate development in the Perkins Company. It began when in the early 1950s the company discovered that vehicle manufacturers, who had become so important to it, were no longer entirely content with the P6, and that competitors were rapidly catching up, even in some instances overtaking Perkins, in developing improved light-weight diesel engines and in producing them efficiently. It was obvious that the phenomenal success of the P6 had caused the company to devote insufficient time, talent, and capital to design and developmental work.

Hurriedly the R6 was developed. Equally hurriedly it was put into production. It was a failure. But worst of all, its deficiencies appeared after many had been installed in new vehicles. Vehicle manufacturers had relied on it because of their excellent experience over the years with the P6. When disaster struck, the advantages of shifting sooner than had been planned, to producing their own diesel engines, appeared convincing to vehicle manufacturers.

The failure lay in the hurry. Perhaps its past experience with the P6, which it had successfully introduced in a remarkably short period of time before the war, influenced the company. But times had changed. At the time of the introduction of the P6, engines were more simple, year-to-year production was small and greater reliance could be placed on owner-experience – particularly since Perkins emphasized after sales service. Now, however, there was no such convenient opportunity for owner-testing. Mass production meant that the only "safety" device between engineering designs and the quality of the final product was intensive proving within the company. But this was time-consuming. The demand was there and the engine was released for production without adequate testing. As well as design deficiencies, there was the additional problem that testing a prototype engine is not the same as testing a production model: mass production may change the quality of component parts and subtle variations from blueprints may occur at the production level. It had been learned too late that intensive testing within the company of prototype and production models had become absolutely essential with the mass production of increasingly sophisticated engines.

The company lost large orders for engines from manufacturers who decided sooner than they otherwise would have done to establish their own production facilities. Out of necessity Perkins had always attempted to have large volume customers, but it was a precarious business arrangement – one in which the customers were not really heavily dependent on Perkins, but Perkins was heavily dependent on its customers. Life for a company in such a position is never easy. It will be remembered that Massey-Ferguson had given hope to Perkins for orders for their P3 engine in 1953 and 1954, when the #35 tractor was being planned, and that in the end the company chose to stay with the engine produced by Standard. Yet if the company had decided to pursue a policy of catering only to the "conversion" market, and to supply engines to smaller manufacturers, it is questionable whether it could have reduced its costs sufficiently.

Quality problems of the R6 not only lost large orders but also were directly expensive through the need to honour warranties. And, as if all these developments were not enough, there came the Suez crisis which closed the Canal and blocked shipment of engines to Near East and Far East markets, particularly India.

It was this combination of forces – structural changes in industry, costly errors of judgment and planning within the company, and uncontrollable political developments – that lay behind the company's deteriorating position. From 1951 to 1957 inclusive the profits as a percentage of turnover declined persistently. The figures, beginning in 1951, were, 9.9, 6.2, 5.9, 6.9, 4.2, 3.6, and in 1957 there was a loss. The price of the company's shares on the exchange dropped to about a third of their highest level. In the language of men of finance, the company was "ripe for a take-over."

In an attempt to deal with the company's problems Frank Perkins asked M. I. Prichard (he had been Deputy Managing Director and became Joint Managing Director with Perkins) in August 1957 whether he would put himself in the position of Administrator appointed by a Shareholders' Committee to introduce changes necessary for shareholders to obtain a satisfactory return on their investment. Frank Perkins himself felt that a solution lay in reducing the size of the establishment.

That request led to a memorandum by Prichard headed "The Future of F. Perkins Limited." The memorandum pointed out that there had been a persistent decline in rate of profits even as sales were rising; that profits from the spare parts business had grown in importance because deteriorating quality had increased sales of spares; that the three biggest vehicle manufacturers had at one time been customers of Perkins but no longer were; and that competition in diesel manufacturing was still increasing. Worse still, it noted that to cut down expenses on developmental work would be suicidal because the company's engines were no longer adequate in existing competitive markets and no new engines were in sight. Finally, the company had a high overdraft, and with its shares depressed, its prospects for obtaining outside financing were not bright. The conclusion that Prichard felt could not be avoided was the following:

> To exist, the Company needs new products, technologically and commercially competitive. To introduce these it needs a substantial addition of capital during 1959 and 1960. This is unlikely to be available from normal sources. Amalgamation with a complementary organization with cash available would appear to present the only real solution.

It was not an easy conclusion to accept by a company that was still deeply imbued with the strong loyalties and commitments of a family enterprise. But there was nothing substantial to counter its logic. Prichard was permitted to explore the possibilities of some sort of association with another organization. Overtures were made to the Ford Motor Company and to Chrysler Corporation, but these were turned down.

For years Perkins had had commercial relations with Massey-Ferguson and the latter knew the company quite well. They were even supplying engines for the new MF-65 tractor. In mid-1958, while lunching with Eric Young, Managing Director of the Eastern Hemisphere Division of Massey-Ferguson, Prichard raised the question of an association with him. Thornbrough was informed, but although Prichard met both Thornbrough and Phillips that summer, the issue was not discussed with any sense of urgency on the part of Massey-Ferguson. At Toronto, however, management was beginning seriously to think about Perkins as a source for engines, particularly in view of the difficult negotiations with Standard. Meanwhile Perkins had begun to establish operations in Brazil and Prichard was intending to fly there on business in November. Prior to going there he wrote Thornbrough the following:

> From one or two remarks dropped to me by Eric Young, I am led to suggest I would very much like to see you personally on my next visit to North America . . . I return from [Brazil and Argentina] on the 1st of December to New York and could very easily get up to Toronto to see you. . . . There are one or two possibilities I would like to speak to you about in addition to the normal run of our business association.

Massey-Ferguson management by that time had become quite interested and began to consider seriously the sort of proposal that Prichard might have in mind. While Prichard was in Brazil, Thornbrough sent him a message asking him to call at Toronto on his way back. This he did, on December 6, and Prichard learned that Massey-Ferguson wished to buy the Perkins company. Massey-Ferguson executives were particularly interested in a swift agreement because of the rumours that competitive companies might be interested in Perkins. They were also most anxious to obtain full ownership of the company. After several meetings Prichard returned to Peterborough, discussed the company's proposal with Frank Perkins on December 8, and in a few days it really was all over.

The announcement was made on January 23, 1959. It said that Massey-Ferguson had offered to purchase all of the 5,200,000 issued shares of

10 shillings each for 17 shillings and 3 pence in cash, and that the Board of F. Perkins Limited as well as their advisors, Baring Brothers and Company Limited, recommended that shareholders accept the offer. This they did before long.

Just prior to the agreement, Perkins had made an arrangement with the Oliver Corporation of Chicago whereby the latter would sell its equipment for the manufacture of outboard motors in exchange for 650,000 ordinary shares in Perkins. This had resulted from a rather hastily considered attempt on the part of Perkins to diversify its production. Massey-Ferguson, however, was somewhat less than anxious to have a competitor or anyone else as a minority shareholder. Thornbrough telephoned the president of Oliver about it and soon arrangements were made to pay Oliver $1,183,000 in lieu of the 650,000 shares.

There was a brief reference in the announcement to the effect that Massey-Ferguson intended to continue to supply engines to the existing customers of F. Perkins Limited. Since this meant supplying engines to some of its own competitors, including Ford, it was a reference that had important implications for the way the Perkins operation would be fitted into the Massey-Ferguson operation. The size of the Perkins operation was such that for lowest-cost production it needed outside customers as well as the business emanating from Massey-Ferguson. Thornbrough recognized these points in his memorandum to Prichard on "Policy for Conduct – F. Perkins Limited." The policy objectives stated were:

1. To preserve the corporate and commercial integrity of the F. Perkins Limited. . . .
2. To take over as soon as possible the engine requirements of Massey-Ferguson, not only in the Eastern Hemisphere, but also in North America. . . .
3. To integrate with Massey-Ferguson such functions, services and facilities as may be accomplished without destroying or compromising the integrity of F. Perkins Limited.
4. To operate the F. Perkins Limited so as to return a reasonable profit.
5. To expand the sale of engines, to outside customers and new customers to the fullest extent so long as such provisioning can be accomplished profitably.

The memorandum ended with a word of welcome which reflected a genuine desire on the part of Thornbrough to minimize the impact on personnel of the kind he himself had experienced after the 1953 merger:

WELCOME TO MASSEY-FERGUSON LTD.

We should like to record the enthusiasm, sincerity, and appreciation of all Massey-Ferguson personnel in extending to you welcome into the Massey-Ferguson group. The contribution of F. Perkins Limited will be of immeasurable assistance in the forward programming of Massey-Ferguson. Should you or any of your staff or employees encounter any instance which would conflict with the intent and spirit of our new organization, I would like to be personally informed as soon as possible. It is of the utmost importance that there be no interruption or distraction from the forward progress of the F. Perkins Limited.

Integration of F. Perkins Limited was achieved with much less personnel disruption than had been the case with the Ferguson company. While this was partly because Massey-Ferguson had since then learned something about the repercussions of mergers, it was more because the character of the Perkins operations argued that it should remain a distinct and separate entity. Administrative changes for achieving greater efficiency were introduced much more gradually than in the rest of the company, in spite of visible impatience from time to time among executives of the corporate group. Engine development, however, was stressed.

Before long Massey-Ferguson began to realize how very useful Perkins was for its operations abroad. When Massey-Ferguson tractors began to be manufactured in India, it used locally produced Perkins engines. In France, Massey-Ferguson acquired full control of the local Perkins company, and made arrangements for Perkins to manufacture its engines there. In Brazil, too, the locally produced Massey-Ferguson tractor is powered with a locally produced Perkins engine, as is the Landini tractor in Italy – Landini being a wholly owned subsidiary of Massey-Ferguson. In some cases local content requirements made it necessary to have a locally manufactured engine for the tractor, and Perkins made it possible for Massey-Ferguson to depend on its own source for such engines.

Perkins output in total expanded greatly. In 1958, the Peterborough plant had manufactured 77,000 engines. Five years later the company manufactured 250,000 engines. Plant expansion became necessary. So in early 1965 a 200,000 square-foot self-contained plant was added to the Peterborough, Eastfield, factory and brought into production. In May of that year another plant for manufacturing engines (130,000 square feet) was purchased at Fletton, near Peterborough. Other expansion followed, so that a capacity of 400,000 engines at Peterborough was in sight of being achieved by about 1968.

What a difference all this made! Gone were the seemingly endless and exceedingly time-consuming discussions with suppliers over engine requirements; no longer was forward planning of tractors and combines compromised by inadequate control over forward planning of engines. Massey-Ferguson was to be repeatedly reminded just how astute its move had been in acquiring F. Perkins Limited – that is, in acquiring the foundations built by Frank Perkins and the people around him. Perkins, for its part, too, could not be dissatisfied with subsequent events, for in several years Massey-Ferguson was using a larger number of diesel engines than any other company in the world. Frank Perkins himself saw all this happen before his death on October 15, 1967, at the age of 78.

CHAPTER 15

Special operations

The 1959 manual on organization provided for a new office with a somewhat intriguing name – Director Special Operations. It was to be the senior executive position in a new division, the Special Operations Division. This Division was somewhat odd from the start for while it was part of the corporate management group – its director answering to the President of the parent company – and not a separate operations unit as were the operations in Australia, France, Germany, North America, United Kingdom, Perkins, and later Italy, it none the less was located far from Toronto. It was located at 35 Davies Street, London, England, an obviously acceptable address, being just around the corner from Claridge's and within sight of Berkeley Square.

The 1959 manual explains, in a general way, the purpose of the new office:

The Director Special Operations is directly responsible to the President. He is responsible for conducting negotiations and carrying out the initial development of special situations of operational character and outside the marketing function in areas not already included in existing operations units. Such special situations would include, for instance, the establishment of local facilities or functions, either by the Company or by arrangements with local interests. The Director Special Operations is responsible for directing the co-ordination of the functional, economic and political aspects inherent in the development of these projects. When the Director Special Operations has guided situations through their initial development stage, it is possible that continuing responsibility for day-to-day supervision of these new activities would be assigned to existing operations units or to new operations unit management.... The Director Special Operations will also be responsible for performing such other tasks as may be directly assigned to him by the President.

The formation of this division may be regarded as an historically significant step in the development of Massey-Ferguson as an international corporation, a step that can best be placed in perspective by recalling briefly the long-term trends of the company's international activities. As an exporting organization the company had for decades been an international company. Threats to its exports led it to embark on local manufacturing operations beginning in the 1920s. But this, even by 1953, constituted only a relatively small part of the company's total manufacturing activity and was confined essentially to countries with advanced industrial structures – United Kingdom, France, Germany. The share interest in the H. V. McKay Company of Australia and SAFIM of South Africa for many years did not really involve management responsibilities.

With the Ferguson merger the company's markets outside North America immediately became much more important than they had been, exceeding in value its North American sales; but the merger did not change significantly the geographical distribution of its own manufacturing facilities – these remaining heavily concentrated in North America. However, the earlier experience of the Ferguson company, and its own experience after the merger, made the company aware of opportunities for reducing costs by international sourcing of major component parts. This was a prime economic incentive behind the acquisition of tractor and diesel engine manufacturing facilities, specifically in the United Kingdom. It was a significant new stage in the "internationalization" of the company. Acquisition of those facilities, it needs to be emphasized, was not prompted by "defensive" motives arising from any threats of local trade restrictions, as had previously been the case, but rather was prompted by the opportunity it provided to take advantage of relative cost differences between countries, and to do this not merely in the manufacture of complete machines but also in the manufacture of component parts for machines to be assembled elsewhere.

In planning for economically efficient allocation of its capital, the company began seriously to regard the whole of the world, rather than merely North America, as the potential area for capital investment in manufacturing facilities. While expansion in any particular country was still, of course, heavily influenced by the nature of that country's home market, plans for expansion began to be influenced much more by relative cost differences between countries than they had been before. Plans for expansion in the United Kingdom, France, Germany, United States, Canada, Italy, Australia, and South Africa began to be related to the objective of supplying the international, as well as the domestic, markets with machinery at minimum cost. Each operation can be an exporter if relative costs permit it, and the high degree of responsibility enjoyed by the management of those operations ensures that opportunities for competing in international markets will not long remain hidden from corporate management.

If all domestic and international markets could be supplied with machinery manufactured by operations units located in countries with relatively well-developed industrial economies, there would be no place for a Special Operations Division. But that is no longer possible. Other countries are increasingly becoming committed to developing their own industries and invariably the approach they are taking is to provide incentives to industry through import restrictions and direct assistance. If a company wishes either to retain markets already established in the lesser developed countries, or to participate in the growth of markets that government policies of those countries are designed to promote, it must contemplate participating in manufacturing projects in those countries. Projects of that kind, because they are not surrounded by well-developed industrial complexes, involve problems that are not encountered elsewhere, and so require managerial and technical skills of a type that must consciously be developed.

Participation in such projects may involve arranging local licensing agreements, acquiring a minority or majority interest in local companies, or establishing a wholly owned subsidiary company. Local participation in one form or another may be desirable for obtaining local capital, local "good will," and in some instances it may be necessary for meeting the requirements of local legislation. Negotiations with governments are necessarily very important. The success of such projects frequently depends on import licenses granted for certain components not available locally, on the fulfilment of government plans for providing direct aid to the project, and on government tariff, credit, and agricultural development policies designed to create a demand for the machinery to be manufactured.

The investigation of opportunities for such projects, the development of project plans for governments, the execution of plans, and the management and control of such operations seemed to involve knowledge, experience, skills, and abilities of a special sort and this essentially necessitated the formation of the Special Operations Division. London seemed to be an appropriate location because such projects already under way were geographically accessible to it, and because London is the natural commercial centre for many of the developing countries. Also the small group of personnel in the Export Division which had previously managed such operations was located in the United Kingdom, and relations with the Export Division at Coventry would necessarily have to be close. What is more, since new projects would almost certainly involve tractor manufacturing, easy access to the Coventry manufacturing operation was an advantage.

Actually the most immediate purpose for establishing the new division was to take several rather troublesome problems with common characteristics out of the hands of people who, because they had other major responsibilities, were dealing with them in a rather *ad hoc* fashion. J. W. Beith, for example, who was with the French company at the time, was "borrowed" in 1958 to investigate the possibilities for assembly/manufacturing operations in Spain. The Export Division had earlier concerned itself with arranging licensing agreements as a sort of side line, and had pioneered such agreements with Yugoslavia as early as 1955. But in 1958 and 1959 the company was drawn into considering the establishment of production facilities not in one but three countries – Brazil, India, and Spain. It appeared that these potential projects, as well as similar ones in other countries, would require continuing attention and that ideally they should receive the attention of management personnel with training and experience different from the kind likely to be found among the salesmen of the Export Division.

The Division was formed in 1960 with Beith as its first director and L. J. Boon as general planning manager, both long-time company employees. Boon became director in May 1963, at which time Beith became head of Export Operations. Very quickly the Division became the recognized authority within the company for exploring new opportunities for the assembly and manufacture of the company's products. The Division attained with remarkable speed a position where it seriously would consider new products wherever in the world opportunities for them seemed to exist. Single-minded attention to that type of activity, together with the willingness to consider production projects anywhere in the world, was a distinctly new development of the company as an international organization; and it was the modern parallel to a much earlier development which had led it to pay similar attention to, and adopt a similar geographical perspective for, the development of export markets for "home" produced machinery.

As already implied, special operations projects had arisen before the formation of a division to deal with them – some of them even in the early Ferguson and Perkins organizations. To trace the development of these special operations – mainly in India, Yugoslavia, Brazil, Italy, Turkey, Spain, and Argentina – it is necessary in some cases to examine developments that occurred long before 1959. Our discussion of these special operations activities is probably more detailed than would be justified by their immediate importance to the company as a whole, but they do reveal the character of the "frontier" activities of the international corporation and as such are rather interesting.

In India initial moves were made in 1953. Following the agreement between Harry Ferguson Limited and the Standard Motor Company Limited, giving the latter first refusal to begin assembling and manufacturing Ferguson tractors anywhere in the Eastern Hemisphere, a company called Standard Motor Products of India Limited began to import completely knocked down (C.K.D.) packs of TE-20 tractors and to assemble them in India in 1953. The assembled tractors were sold to Harry Ferguson (India) Limited (later Massey-Ferguson (India) Limited), which distributed them as well as Ferguson implements through its own dealers in one part of India and through an independent distributor – Escorts Limited – in the rest of India.

The government of India became increasingly anxious to begin local manufacturing of tractors and in 1956 Standard Motor Products of India Limited received a manufacturing licence. But that company did not proceed with the project, and when in 1959 Massey-Ferguson acquired the tractor facilities of Standard Motor, Standard ceased to be involved in plans to begin manufacturing Massey-Ferguson tractors in India or elsewhere. The Indian government continued to press for local manufacturing of tractors, and the responsibility for doing so, as far as Massey-Ferguson tractors were concerned, now lay indisputably in the hands of Massey-Ferguson. Its ability to proceed had been enhanced by having acquired a local source for engines through the purchase of the Perkins organization. Since 1954, Perkins had granted a licence for local assembly and manufacture of its P series engines to Simpson and Company Limited of Madras – one of a number of subsidiaries of Amalgamations Private Limited.

Massey-Ferguson decided that for operational and financial reasons it would be best to obtain local participation for embarking on a tractor manufacturing project. Because of the long association of Simpson and Company Limited of Madras with Perkins, Amalgamations Private Limited was a desirable partner to have for the project. This was particularly so because of the presence of S. Anantharamakrishnan, a well-known and highly respected industrialist at the head of Amalgamations Limited. It was a position he retained until his death on April 18, 1964.

Amalgamations agreed to participate but insisted on pursuing their group policy of retaining majority control of the venture. A new company, Tractors and Farm Equipment Limited (TAFE to everyone in the company), was formed, Simpson and Company Limited acquiring 51 per cent of the stock and Massey-Ferguson 49 per cent. Head Office was located in Madras, and the company started operations on January 1, 1961. It was decided that Massey-Ferguson (India) Limited would terminate its marketing activity, and that its assets and dealers would be transferred to TAFE. In this way the new company received the very considerable asset of a system of dealers. The distributor agreement with Escorts Limited was cancelled, and so TAFE became not only a tractor manufacturing company – with engines supplied by Simpson and Company Limited – but also the exclusive agent for marketing Massey-Ferguson Limited products in India.

At first TAFE sold only spare parts acquired from Massey-Ferguson (India) Limited and some trailers. But in August 1961 the company began to assemble tractors with 49 per cent local content – the 49 per cent made up of the engine, tires, battery, lubricants, assembly, and painting. It had been anticipated that the remainder of the components would be imported from England until local content could further be increased, and it was with English components that assembly began. But in late 1961 and early 1962, the shortage of sterling required the Indian government to reduce imports from the United Kingdom. This posed a serious threat to the TAFE tractor project.

TAFE attempted to circumvent it by negotiating with I.T.M. of Yugoslavia – a state organization with a Massey-Ferguson tractor licensing agreement – for supplying the components no longer available from the United Kingdom. Yugoslavia and India had a barter trade agreement which gave the former an incentive to sell to India. Actually the position of TAFE vis à vis I.T.M. was somewhat precarious. I.T.M. had already tried to market their MT-533 tractor – modelled on the MF-35 – in India, through Escorts Limited, the former Ferguson distributor, and while Massey-Ferguson had succeeded in halting that export activity because it was in contravention of the licensing agreement, there was always the possibility that if TAFE could not produce tractors the pressure on the government to permit the importation of I.T.M. tractors would become insurmountable.

However, agreement between TAFE and I.T.M. was reached, largely because of Massey-Ferguson insistence and pressure. Component parts began to flow and TAFE was kept in business. With that arrangement the name Massey-Ferguson was kept in the market in India. And, by the way, the advantages of interchangeability of component parts and of alternative sources of supplies had, in a small way, been demonstrated.

This importing arrangement was not without its difficulties. Quality problems with gears were encountered. Components from Yugoslavia were based on the metric system of measurement, those from Coventry were not. Also I.T.M. engineering development had not kept pace with similar development at Coventry. Documentation of components supplied frequently was inadequate, making it difficult to identify parts. Communication with the supplier by letter was not an easy matter when letters went unanswered. Many of the difficulties were overcome only through the supplier stationing a representative in Madras. But the arrangement made it possible for TAFE to expand its production. In 1961, 395 tractors were produced; in 1962, 1,308; and by 1966, the total had increased to 3,400. Doubling of that number in several years was anticipated because of the increased interest of the Indian government in agricultural development. TAFE has also begun to manufacture implements. It has developed a paddy disc harrow which has become quite popular, and it also manufactures a tiller and offset disc harrow. Other implements will be introduced in future.

Although the procurement of import licences for component parts and capital equipment was a major problem in establishing the project, the shortage of skilled technicians, and particularly of management personnel with technical experience, was a continuing difficulty. To help overcome it, Massey-Ferguson supplied TAFE with several senior management personnel and the latter began an internal training programme. Local employees who had experience in the United Kingdom or North

America appear to have been rather more sympathetic to the introduction of advanced managerial techniques than those without such experience.

Difficulties in increasing the local content of machinery have arisen for the same reasons in India as in most of the other lesser developed countries: the absence of a sufficiently wide range of industrial suppliers and the quality problems encountered in using locally manufactured components. Training an adequate labour force was not a particularly difficult problem. But the pace of work has been slower than in the North American or European operations units, and loss of time through frequent holidays and relatively high absenteeism, has been a troublesome problem.

All these influences, including the relatively small production runs, combine to make the cost of production of tractors substantially higher in India than the cost of tractors imported from Coventry – at existing exchange rates. But it is significant that while Massey-Ferguson tractors have received no protection against imports of tractors from "iron curtain" countries, TAFE has encountered no difficulty in selling all the tractors it could produce, and at a price substantially higher than those of the much less sophisticated imported machines.

There is little doubt that Massey-Ferguson has made greater progress in overcoming the problem of local manufacture of tractors in India than has any other company. It already enjoys a strong position in the market, partly because of the remarkably good record of the Ferguson tractor in India over the years, partly because of the high resale value of that tractor, and also because of the service that is readily available to farmers through the network of dealers built up over the years. Fundamentally, the market for tractors exists because farmers have found that, even at existing wages, farm mechanization reduces costs. Some farmers also believe that it increases crop yields.

Mechanization involves not merely the purchase of machines but also frequently necessitates the adoption of new tillage techniques. For this reason the marketing organization cannot remain as aloof from initiating changes in agricultural technology as can the marketing organization in more developed countries. This is a condition that the company is likely to encounter in many of the countries where it sells its products. In India TAFE has established a training school and farm just outside of Madras where it provides courses in product and service training to its dealers. It is encouraging to observe the speed with which advanced farm machinery technology is disseminated through that school to dealers in many parts of India.

Reference has already been made to a licensing agreement arranged in 1955 between Massey-Ferguson and a governmental organization in Yugoslavia. It was the company's first important experience with licensing agreements and negotiations leading to it probably uncovered all of the problems inherent in such an arrangement. Yugoslavia had vested the provision of tractors, implements and combines in two main enterprises – Industrija Traktors & Masina (I.T.M.) of Belgrade and Fabrika Poljoprivrednik Masina (Z.M.A.J.) of Zemun. Initial approaches were made to Harry Ferguson Limited by the Yugoslav authorities in 1953, and, after the merger, to Massey-Ferguson. Discussions were held in Belgrade in February 1954, a Yugoslav delegation visited Coventry in April, and a Massey-Ferguson and Standard Motor delegation visited Belgrade in June and July. A Yugoslav delegation concerned with combines visited the Massey-Ferguson plants at Marquette, Westhoven, Kilmarnock, and Coventry in August. Technical delegations from Yugoslavia went to Coventry in September and October. During all these 1954 negotiations, agreement in principle was reached on several of the main aspects of a licensing arrangement – royalties, technical assistance and aid, and conditions relating to export activity. On February 16, 1955, negotiations were begun for the purpose of concluding the required agreements.

For Massey-Ferguson the appeal of the arrangement was that for the life of the agreement it would enjoy a privileged position in the Yugoslav market; a market that, it then appeared, might absorb up to 45,000 tractors during the life of the agreement. The existing tractor population was small, so Massey-Ferguson would be entering on the "ground floor."

For the Yugoslav authorities the advantages lay in the rapid acquisition of machines of advanced design, and of a source of technical assistance in establishing a local manufacturing project at a time when progress in the development of the agricultural sector seemed urgently to be needed. Since F. Perkins Limited had already licensed a Yugoslav organization (I.M.R.) in 1954 for the manufacture of one of its P series engines, one of the major obstacles to producing tractors appeared to have been overcome.

Negotiations were protracted but agreement was reached in October 1955. I.T.M. was given a ten-year license for manufacturing certain tractors, and a seven-year license to manufacture certain implements; while Z.M.A.J. was granted a seven-year license for manufacturing the #630 and #780 self-propelled combines. In addition to royalty or technical assistance payments, Massey-Ferguson received orders for components – a most important part of many licensing agreements. Machines produced were not to be exported.

Upon expiration of the agreement the Yugoslavian licensees received the right to manufacture the relevant machines without payment of any royalty – a disadvantage for Massey-Ferguson that is softened somewhat by the fact that the models involved would probably be regarded by it as obsolete by the time the license expired. Two of the agreements expired in 1962 and were not renewed, but the other was extended to include a model of the new line of tractors.

One of the most interesting of the company's various special operations is its tractor and diesel engine manufacturing facilities in Brazil. Massey-Harris had had commercial relations with Brazil for many decades, but the most important of its South American markets had been in the Argentine. It had established a branch at Buenos Aires in 1917, replacing its distributor Moore & Tudor, and that branch soon became responsible for distribution of Massey-Harris products throughout South America. In 1937 and 1938 sales of the branch made a major contribution to maintaining the Massey-Harris company as a whole, and the vital role played by that branch in the development of the self-propelled combine has already been described. Nothing would have appeared more inevitable than for that branch to evolve into a manufacturing as well as a distribution centre.

With the outbreak of the Second World War, supplies of machines were suddenly greatly reduced, and in the Argentine military rule appeared. Juan Peron became President in January 1946. Policy guiding the allocation of import licenses was heavily influenced by politics, as were negotiations relating to the establishment of manufacturing operations. Eventually scarcity of foreign exchange also limited importation. The once proud and prosperous branch at Buenos Aires declined in size, and finally did little more than sell spare parts; more and more of the space in its large downtown Buenos Aires building became vacant. As a centre for expansion into manufacturing, its prospects seemed dim.

Yet the South American market was potentially so attractive that it could not be ignored, and Brazil, because of its agricultural and industrial potential, was an obvious possibility for a manufacturing project. Because it was an obvious possibility, it was discussed within the company over a long period. A selling company, Maquinas Massey-Harris Limitada, was incorporated as a limited liability partnership on June 29, 1940, with its head office at Porto Alegre. In addition an independent distributor, Veiculose e Maquinas Agricolas S.A. (VEMAG) of Sao Paulo also sold the company's products.

In the middle 1950s the Brazilian government introduced a programme, including necessary protective tariffs, for encouraging the local manufacture of trucks and cars. While a similar programme for tractors was not introduced at that time, it did not require much foresight to see that this would not be long in coming.

In 1955 Massey-Ferguson sold the assets of its Porto Alegre branch to a subsidiary of its distributor VEMAG S.A., since the distributor's facilities and prospects appeared to be better than its own in that city. It was intended that the funds generated in that way could be used to establish a new company, together with VEMAG, for the partial manufacture and assembly of tractors. VEMAG also began to negotiate with F. Perkins Limited for a joint engine-manufacturing project.

More detailed investigation into a manufacturing project was undertaken. A report on it in July 1956 referred to the dilemma facing the company in Brazil. It read:

14 / J. E. Mitchell, Group Vice President Industrial and Construction Machinery, and E. P. Taylor, Chairman of the Executive Committee, at the 1968 Annual Meeting of Shareholders.

15 / Dedication of a road in Peterborough, England, on May 27, 1968, honouring Frank Perkins, founder of the Perkins Engines Group. Left to right: Mayor of the City, Alderman A. W. L. Adams; Monty I. Prichard, Group Vice President, Engines; T. H. R. Perkins, General Manager Northern Europe, Engines Group, and son of the founder; and H. R. W. Laxton, Deputy Mayor.

16 / J. H. Shiner, former Vice President Marketing.

17 / R. A. Diez, General Manager Germany and the late W. Lattman, Chairman, Massey-Ferguson International AG.

18 / J. J. Jaeger, Vice President Research and Development.

19 / J. A. Belford, Vice President Personnel & Industrial Relations.

20 / J. W. Beith, Vice President and General Manager United Kingdom.

21 / E. W. Young, Chairman, Massey-Ferguson Holdings Limited

22 / L. J. Boon, Director Special Operations.

23 / H. A. R. Powell, Managing Director, Massey-Ferguson Holdings Limited.

24 / E. L. Barger, Director Agricultural Relations; S. R. Wilson, Director Management Structure and Processes and Director Logistics; and H. A. Wallace, Vice President Manufacturing, atop a new loader.

25 / Left, L. E. Elfes, Director Engineering, Farm Machinery, North America. Standing, G. K. Blair, Special Assistant to the President, and J. A. Evans, Director Legal Services, at launch of new construction machinery line, Akron, 1968.

Undoubtedly we must make a selection between only two alternatives, either –

(a) We establish an operation in Brazil which can be, within a nebulous time limit that will be imposed by the Government in keeping with conditions as they develop, in the position that it can produce a tractor and related implements from mainly Brazilian content which will be popularly accepted in the Brazilian market, or

(b) we reconcile ourselves to the fact that our plans to merchandise our equipment in Brazil can be made only for the limited time of that same time limit, which will likely be not more than 3 to 5 years.

The report recommended that the company in association with VEMAG should choose the first alternative. The Executive Committee of the Board of Directors of Massey-Ferguson Limited approved this move in principle and on October 15, 1956, Thornbrough wrote a letter to VEMAG for the purpose of recording their combined intention of establishing an industrial programme for the manufacture of certain Massey-Ferguson products in Brazil. Massey-Ferguson planned that VEMAG would supply the facilities to manufacture the FE-35. A Perkins engine to be manufactured locally would be used and a new local company, a wholly owned subsidiary of Massey-Ferguson Limited, would have technical responsibility for the products manufactured in Brazil and responsibility for the manufacture and assembly operations. VEMAG felt that a small participation by it in the proposed subsidiary would be desirable for many reasons, and Massey-Ferguson agreed.

A new company, Massey-Harris-Ferguson do Brasil S.A. Industria e Comercio was formed on April 24, 1957 ("Massey-Ferguson . . ." after March 24, 1958), with a substantial majority of its stock soon owned by Massey-Ferguson but with some local participation as well. That company was later not only to manufacture the company's products in Brazil but also to distribute them. VEMAG transferred its distribution system, spare parts inventory, and some equipment to the new company, which gave the latter a nucleus for its organization.

However, the tractor project did not advance swiftly. Perkins encountered difficulties in arranging for local manufacture of engines, which created a major obstacle for the project. But even more important, the company had to await the formulation of government policy including regulations relating to importation of necessary capital goods and components, the rate at which local content would have to be raised, price and credit conditions that would apply to sales of tractors to farmers, and protection available for locally produced machines against competition from imported machines. As long as such a policy was not declared, Massey-Ferguson continued to enjoy an export market in Brazil.

But in November 1957 the government of Brazil announced the formation of a Tractor Committee to study and regulate tractor production in Brazil. One of the first measures of the committee was to announce that companies interested in such a programme would have to present submissions by December 20, 1957, and would have to supply complete details by April 1958. Presentations would have to indicate planned annual production, capital expenditures anticipated, character of local participation, and the source of the engine. Local content of 30 per cent by weight was stipulated. Furthermore, it was decreed that beginning July 1958 only C.K.D. tractors, less deletion of "local content" components, would be permitted to be imported, and that after July 1959, companies not engaged in local manufacturing would not be permitted to import at all. Local manufacturers would receive protection and assistance through tariffs, quotas, and preferential exchange rates. While details of the government's proposed tractor programme changed and were not made final for several years, the principles enunciated did not change.

The time limit and "local content" requirements the committee tentatively had in mind appeared difficult to attain. To assist the government in developing detailed local content provisions and to impress on it the greater difficulties involved in attaining high local content for the more modern tractors compared with the older type, Massey-Ferguson do Brasil in 1958 dis-assembled an old Massey-Harris tractor and a newer Ferguson tractor for governmental inspection. Apparently other companies, too, were encountering difficulties in developing plans that would meet the time limit, for the government soon delayed, to June 1958, the deadline for initial presentations. Domestic difficulties later delayed even further the official decree outlining the details of the government's conditions relating to local tractor manufacturing.

In a sense these delays were fortunate for Massey-Ferguson, for preoccupation with negotiations with Standard Motors led it to pay less attention to the Brazilian project than would under any other circumstances have been justified. Indeed at times it seemed that management at Toronto was reluctant to proceed at all. The local management of Massey-Ferguson do Brasil, however, remained enthusiastic, and presented several tentative plans for tractor manufacturing to the Brazilian authorities to give evidence of the continued interest of Massey-Ferguson in Brazil. With the acquisition of Perkins and the Standard tractor facilities, Toronto management suddenly found itself preoccupied with operations in the United Kingdom; so it told management in Brazil that it would be six months, that is early 1960, before attention could be paid to the Brazilian project. With the Brazilian government anxious to adopt manufacturing plans of several manufacturers, and thus to begin its tractor programme, it was a period when Massey-Ferguson might well have lost its opportunity in Brazil.

That it did not do so was the result mainly of the efforts of the Managing Director of Massey-Ferguson do Brasil, J. E. Williams. Williams, a Welshman, who had joined Massey-Harris after having been a merchant seaman, had not only become completely fluent in Portuguese but had also become intimately acquainted with the economics and politics of Brazil. He kept the company's name to the fore among governmental authorities and, equally important, he seldom relented in reminding management at Toronto of the importance of the project and of significant new developments relating to it. Williams' enthusiasm was based essentially on his confidence in Brazil and its people and on his conviction that the project could be a commercial success.

One problem, that of obtaining a locally manufactured engine, appeared to be solved with the establishment in Brazil in 1959 of Motores Perkins S.A. That company, it was thought, might be able to manufacture an engine for the Massey-Ferguson tractor as well as engines for the growing truck and automobile industry. Commitments for it to be established, in conjunction with local interests, had been made before Massey-Ferguson acquired Perkins, but it had been viewed essentially as a company for producing a vehicle engine. Just as the revised engine project was about to commence, it had to be delayed because local capital contributions that had been expected were not forthcoming. Unless those difficulties could be overcome the whole tractor project would be in jeopardy.

A new note of urgency arose when on March 30, 1960, the Brazilian authorities finally approved tentative Massey-Ferguson plans for local manufacturing. Massey-Ferguson had to make final financial estimates so that a firm decision could be taken. When the Chairman of the Board of the parent company saw the tractor manufacturing plans and financial estimates, he found them unattractive and decided that he could not support them. Among the objections was that the project would require too much company capital. It was decided that the project should be dropped and that negotiations for only the Perkins engine project should be pursued.

Personnel in the Special Operations Division and in Massey-Ferguson do Brasil, who favoured proceeding with the project, were deeply disappointed. The latter had to inform both the official authorities in Brazil and their own dealers, of the company's decision; and they knew that the result could be a loss of any future opportunity for local manufacturing and also loss of their distribution system. There was a feeling among them that difficulties had arisen from inadequate liaison with the Financial Department of the parent company and from a failure on the part of parent company management to realize that alternative and satisfactory arrangements might be made if a position of flexibility was maintained and one of rigidity avoided.

Williams favoured flexible policy as did personnel of the Special Operations Division. They proposed alternative approaches which would help overcome parent company objections, suggesting that Brazilian capital might be available for the project. The parent company reversed its negative decision and permitted further exploration, although without visible enthusiasm. While successful attempts were being made to procure local capital for the Perkins engine project, Massey-Ferguson was also assisted by the general delay that its competitors were experiencing in embarking on a manufacturing project.

The change of government at the end of 1960 reduced the urgency of the project, since government tractor policy had again to be confirmed. This meant that tractors could be imported for a slightly longer period than had been anticipated. Massey-Ferguson was thereby given more time to persuade suppliers to tool up for some components deleted from imported machines. When the approach began to be successful VEMAG was encouraged to supply tooling for sheet metal against stock in the company. Imports as such were financed by the United Kingdom company, and earnings of Massey-Ferguson do Brasil from the import programme covered a large part of the subsequent operating expenses involved in commencing a manufacturing programme. Massey-Ferguson agreed to invest more funds in Brazil than it had earlier contemplated, and "swaps" provided an additional source of financing from abroad.

So in mid-1961 the Board of Directors of the parent company was able to approve the Brazilian project. A few tractors came off the assembly line before the end of the year. Volume production began in May 1962. In 1963 Massey-Ferguson do Brasil produced 2,982 of the MF-50 tractor, in 1965 a second model, the MF-65, had been added, and in 1966 a total of 4,051 tractors were produced.

The Brazilian company has been fortunate in its distribution system, for its dealers have invested substantial amounts in facilities. It would probably be surprising to most outside observers that some of the dealers are excellent even by North American standards. By 1965 the company had increased its share of the tractor market in Brazil to about 36 per cent. Profits were not yet adequate and repatriation of dividends had not been permitted. Great patience is a necessary characteristic of corporations embarking on such projects.

In 1965, with its lease on the building in which it manufactured tractors running out, the company had to decide what it would do if it wished to continue operations in Brazil. It decided to stay in Brazil and in 1966 purchased Terral S.A., a company with a suitable factory near the Perkins engine plant in Sao Paulo.

At times, the fledgling company has had to operate under conditions of extreme inflation. It has in consequence seen its cost of local funds rise as high as 6 per cent per month. Periodically management has had to spend an inordinately large proportion of its time in obtaining short-term funds, and even the parent company has had to become experienced in the techniques of financing under conditions of rampant inflation. Wage adjustments of about 20 per cent several times a year, and equivalent price adjustments, have had to be taken in stride. Political instability has not been absent. These instabilities have not made for very profitable operations to date.

Was it then a mistake for Massey-Ferguson to go into Brazil? Opinions may differ, but our view is that had Massey-Ferguson decided that it could not face the difficulties and uncertainties involved in manufacturing in Brazil and that it would leave that market to its competitors, it would have been difficult to take seriously its claim to being a truly international organization. For if it doubted its ability to adjust to the difficulties found in the Brazilian economy, it would undoubtedly have experienced similar fears in contemplating operations in other lesser developed countries. It seems certain that but for the efforts of the management of Massey-Ferguson do Brasil and of Special Operations Division, the parent company would not have proceeded with the project. At the same time financial results have not yet proven that the reluctance on the part of head office management was unjustified.

Massey-Ferguson tractors are being assembled in Turkey under license, with most of the components being imported by the licensee, British Overseas Engineering and Credit Company Limited (BOECC). That company began its associations with Massey-Ferguson through Harry Ferguson Limited, for in 1948 it began to distribute the TE-20 Ferguson tractor in Turkey. Massey-Harris, too, had enjoyed a strong market in Turkey through the efforts of its distributor Turkiye Zirai Donatim Kurumu (T.Z.D.K.), a government organization. The former distributed Massey-Ferguson tractors and the latter Massey-Ferguson harvesting machinery until the early 1960s, when the whole franchise was transferred to BOECC.

The government of Turkey, as so many other governments, has favoured developing a local tractor manufacturing industry. It began to restrict the importation of assembled tractors but permitted the importation of a certain volume of component parts – with a view to progressive reduction in future of the proportion of "foreign" content in locally assembled tractors.

So BOECC obtained a license from Massey-Ferguson to assemble the MF-35 and MF-65 tractors and it in turn arranged for them to be assembled by Uzel Limited, also of Istanbul. That company was a manufacturer of vehicle springs and, having excess plant and the required technical staff, was in a position to undertake the project whereas BOECC was not. In this way Massey-Ferguson tractors were kept in the market after import restrictions precluded their being imported in assembled form.

BOECC markets Massey-Ferguson products through five main dealers, fifty-eight subdealers, and sixteen sugar-beet co-operatives. It has succeeded in obtaining up to 25 per cent of the tractor market and has made more progress in local assembly of tractors than any of its competitors. The company supports its marketing programme with a product and service education programme for dealers. Frequently its travelling van will appear at village gatherings. A training centre is located next to the factory. This is a typical example of the train of communication by which the technology of advanced farm machinery, and the technique for operating it, are transmitted from Toronto, Detroit, and Coventry to the villages of lesser developed countries.

From time to time, since at least 1948, there had been discussion within the Massey-Harris and then Massey-Ferguson organization, about the necessity of establishing assembly and perhaps manufacturing operations in Spain. And, as elsewhere, the impetus for this discussion was the apparent threat to imports from governmental policies designed to promote domestic industrialization, and the persistent rumours that competitors were about to establish local assembly or manufacturing operations.

J. W. Beith, who was fluent in Spanish and in charge of the French Massey-Ferguson company, was to be the company's "chief negotiator" until a project emerged. From 1954 to 1956 protracted discussion took place between Massey-Ferguson and a local Spanish company over establishing such an operation. The time was wasted, for the two could not agree on terms. In early 1956 Massey-Ferguson began to think of the possibility of seeking the co-operation of another Spanish manufacturing company, Garteiz Hermanos y Compania of Bilbao, as well as of the Massey-Harris distributor, Pares Hermanos S.A. of Barcelona. This proved to be a useful approach. A plan involving formation of a new company for producing a small self-propelled combine, to be followed later by a Ferguson-type tractor and implements, with minimum Massey-Ferguson capital contribution but effective control, began to emerge.

A draft letter of intent was ready by December 1956 and it was signed soon thereafter. It reflected all the features of the above plan, and outlined them in detail. It envisaged first the assembly of a small self-propelled combine and gradual substitution of local for imported components, with stages of deletion determined by the Spanish Ministry of Industry. A similar process was envisaged for other combines and for tractors and implements, accessories, and spare parts. It was recognized that the company would have to be capable of manufacturing complete machines within the period designated by the General Directorate of Industry.

Massey-Ferguson was to render technical advice and assistance regarding choice of area and site, acquisition of land, construction and layout of factory buildings, purchase of machine tools and factory organization. It was also to provide information on manufacturing processes and technical advice on production problems, especially regarding organization of the technical departments of the factory. Staff assistance was to be provided. Massey-Ferguson was to grant the Spanish company necessary manufacturing and selling licences, including the right to certain trademarks.

In payment Massey-Ferguson would receive royalties or technical assistance fees, and also a payment as a fee for the licence. It would of course also participate in the profits of the company and would benefit from the sale of component parts to the company. A not inconsiderable advantage to Massey-Ferguson of this combine project was, that it would enable it to form the nucleus (i.e., "keep a foot in the door") for future tractor and implement manufacturing operations without a significant capital outlay.

While these negotiations were proceeding, F. Perkins Limited, still independent of Massey-Ferguson, was negotiating with Spanish interests for the formation of a local company that would produce Perkins engines under licence in Spain. Perkins had had the unfortunate experience of having its engine copied in detail by the Barreiros firm in Spain, but in spite of that it thought that it could profitably sell engines there. The company, Perkins Hispania S.A., with head office at Madrid, was formed with a majority of its stock initially owned by Spanish investors, but later by Perkins of Peterborough.

Returning to the Massey-Ferguson project, it had to be approved by the National Metal Syndicate, then by the local delegate of the Ministry of Industry in Pamplona, and finally by the Ministry itself in Madrid. This approval was granted in 1959. The new company, incorporated on September 5, 1958, was Motorizacion Agricola S.A. (M.A.S.A.). Spanish shareholders took up the majority of its stock. Land for offices and factory was acquired just outside of Pamplona at Noain (Navarra). That location was chosen because of its proximity to component part suppliers, its convenient rail facilities for shipping components from the Massey-Ferguson combine factory at Eschwege, Germany, and also because Navarra, historically an independent territory, enjoys certain tax advantages. There had also been disagreement among the Spanish shareholders about location of the factory, and agreement was reached by choosing Pamplona.

Little progress in production was made in 1959. Five #630 combines were imported and local SEAT engines fitted to them. By January 1960 a new building was completed at Noain, and in that year forty-seven of those combines in "rump" form were imported from Eschwege, to which the company fitted local engines, tables, reels, and bagger attachments. This number increased to 150 in 1962, and in the following year the company began to use the locally assembled Perkins engine and introduced the #685 self-propelled combine in addition to the #630. Disc plows were also manufactured, as were spinner broadcasters.

Initial production difficulties were great because the labour force was untrained. Within three years many of the workers had become relatively skilled. But the small volume of production and the cost of components purchased from other suppliers made continued tariff protection necessary for the project. While the project was small, together with the Perkins engine project, it constituted an important beginning that led to a major step forward in 1965.

This last development really began in 1964 when Massey-Ferguson became aware that an association with Motor Iberica S.A., a Spanish company of Barcelona manufacturing and selling tractors and trucks, might be possible. It was an unexpected development, for that company had been closely associated with the Ford Motor Company for over forty years. Its Ebro tractor in fact was a Ford tractor, its truck a Ford truck, and the engine powering both a Ford diesel engine. In 1962 the Ebro tractor accounted for over 40 per cent of tractors sold in Spain and, in its class, the truck had over half of the market. In 1964 over 6,000 locally produced Ebro tractors were sold, against less than 700 imported Massey-Ferguson tractors, and about 7,000 Ebro trucks were sold. The relatively low Massey-Ferguson sales reflected the increasing restriction on the importation of tractors into Spain, for even two years earlier Massey-Ferguson had sold almost 3,000 tractors in that country. This import restriction, as well as the strong distribution and large manufacturing facilities of Motor Iberica, made an association with that company seem very attractive.

On May 31, 1965, a joint press release by Motor Iberica and Ford announced that their association of long-standing had come to an end in complete harmony. The next day, June 1, Massey-Ferguson announced the merger of Massey-Ferguson's Spanish interests with those of Motor Iberica S.A., although agreements were not finalized until December 31, 1966. The merger involved bringing together the assets of Perkins Hispania, S.A., in which Massey-Ferguson had a majority interest, of Motorizacion Agricola S.A. in which it had a minority interest, and of Motor Iberica S.A., and the subsidiaries owned by Spanish shareholders. This necessitated increasing the capital of Motor Iberica S.A. since the new organization was to operate under that name. Massey-Ferguson acquired 36.6 per cent of the stock of that company, paying for it with its share of the assets of Perkins Hispania and Motorizacion Agricola, with a cash investment and the future income from technical assistance fees.

The new company has factories at Barcelona (tractors and trucks), Madrid (Perkins engines), Noain (combines), and Zaragoza (implements). It will produce Massey-Ferguson tractors as soon as it has phased out of its Ebro production and will greatly enhance the position of Massey-Ferguson in Spain. For Massey-Ferguson perhaps the most interesting aspect of the arrangement is that it will, for the first time, involve the company in truck manufacturing.

In 1966 the company also became associated as a minority shareholder with a company assembling Massey-Ferguson tractors and implements under licence in Morocco. In the same year the company's Mexican subsidiary was transformed from a purely selling company into a manufacturing operation and it began constructing a new plant at Queretaro; and a minority interest was acquired in a Mexican company that now produces Perkins engines under licence.

Massey-Ferguson, from time to time, has negotiated licensing agreements in various countries which have not involved any investment of capital on its part, and these have also been supervised by the Special Operations Division. They include agreements in Finland, Denmark, New Zealand, and Pakistan as well as those already discussed. For a period the wholly owned Italian operation and the partially owned South African operation, discussed in succeeding chapters, and also the Brazilian operation, were the responsibility of the Special Operations Division until they became full-fledged operations units. The division still oversees the company's interest in the Indian, Spanish, and Moroccan companies in which Massey-Ferguson has a minority interest.

As a result of the varied experiences of the Special Operations Division that have arisen from all this activity, it is possible to make certain general observations about the nature of its work and of the character of its rather interesting field of operation. A significant portion of the time of the Division has been devoted to investigating potential opportunities for introducing new assembly or manufacturing projects into countries planning to introduce measures favouring local production. From time to time investigation has been undertaken to consider such projects for Pakistan, Morocco, Japan, Algeria, Egypt, Ceylon, Thailand, Mexico, Argentina, Tunisia, Iraq, and Syria. Much experience has been gained in devising methods of appraising the attractiveness of such opportunities, in negotiating with government authorities, and in developing business arrangements with local corporations.

Typically a recommendation to the Special Operations Division to undertake an investigation emanates from the Export Division, which is moved to do so by events that appear to it to threaten the company's export sales in particular markets. Enquiries, of course, can also come directly from government authorities, and preliminary discussions almost invariably take place with them. This makes it particularly useful to have offices of the Division located in London, close to the various embassies, and in a city that is visited frequently by foreign trade missions.

As a result of establishing tractor manufacturing facilities in Brazil and India, assisting licensees to establish similar projects in Yugoslavia, Turkey, Morocco, and Pakistan, and particularly as a result of investigating the feasibility of such projects in many countries, the company has accumulated considerable knowledge and experience concerning manufacturing activities in lesser developed countries. The character of its discussions and negotiations with those countries permits one to make some generalizations both about the objectives of the government authorities in seeking to establish manufacturing facilities, and about the economics of such projects.

Most governments are interested in establishing a tractor manufacturing operation, rather than an operation involving any of the other more important agricultural machines. Such a project seems to have a kind of glamour to it that makes it politically attractive. It is closely associated with the desire to increase supplies of food; it is useful in lifting the labour force to semi-skilled levels; and it is regarded (usually mistakenly) as attractive because of the opportunities it provides for employing surplus labour.

Typically such projects involve an annual production of several thousand tractors in one or two sizes. The authorities involved usually want to maximize "local content" in the tractor at the beginning of the project, and the rate at which deletions of imported components are planned to take place. They also usually aim at minimizing the price of the final product to the farmer, and hope, in vain, that it will not be higher than the imported tractor; they strive to reduce the amount of protection to be given to the local project; they hope to develop an export market for the tractors produced, as well as a home market; and frequently they favour some local financial participation in the project.

Proposals designed to meet those aspirations immediately face the problem of relatively high local costs of production. It seems that the cost of producing tractors will be about twice as high in countries without a well-developed automobile industry as in the United Kingdom, even assuming that local content will not exceed 50 per cent compared with local content in the United Kingdom of virtually 100 per cent. Depending on degree of local content and annual output, the figure could range from double the cost upward. In countries with a relatively well-developed automobile manufacturing industry, costs will tend to be lower, but the crucial matter of volume still impedes economically efficient production. This includes the problem of volume both in the tractor project, and in the automobile industry from which the tractor project might obtain components.

How much employment do such projects generate? In a general way, the value of a medium-sized tractor may be regarded as being composed of the following: assembly and paint, 5 per cent; proprietary items, 17 per cent; easily made parts, 7 per cent; parts requiring complicated machining, 35 per cent; certain specialized items, 10 per cent; and the engine, 26 per cent. It is obvious from this that a purely *assembly* operation which relates to the first cost item listed will generate very little employment. At the same time to increase local content beyond about 40 per cent requires large expenditures in tooling, either on the part of the company itself, or on the part of its suppliers, and it also requires a local engine. If the engine is assembled entirely from imported parts, then again the employment content is exceedingly small, while if its content is increased, heavy tooling expenditures are required. The high unit overhead costs arising from such capital expenditures in low-volume projects, as well as the surprisingly high cost of unskilled labour, make it inevitable that the production of tractors, outside of a few highly industrialized nations, requires protection from foreign competition.

This also means, of course, that the export opportunities of such projects are typically non-existent. The crucial decision for individual countries to make is whether the degree of protection required – which typically has to be accompanied by special credit facilities for farmers to permit them to buy the higher-priced tractors – is more than offset by longer-run advantages. It does not seem likely that the "infant industry" argument would justify many of the projects that have been considered, since optimum output levels achieved in the United Kingdom and the United States would never be attained by them. Advantages have to be sought in the benefits of training a local labour force and, through experimentation, utilizing that labour force in local industries that can be competitive, including companies supplying some components to the tractor project. The experience of the company in Brazil, India, and Spain has been that this kind of effort may be an important aspect of economic development. Some countries with a substantial potential local market may benefit from any move towards larger free trade groupings. For example, a Latin American free trade area could conceivably raise the output of a tractor project in Brazil to optimum production levels.

What should the reaction of a company like Massey-Ferguson be when it is faced with official proposals which it regards as economically disadvantageous to the country concerned? Essentially the approach evolved by the Special Operations Division has been to outline clearly the cost involved in establishing a project requiring heavy protection from foreign competition, and also to outline clearly the limited contribution that such projects are likely to make to employment. In many cases the result has been that the governmental authorities have temporarily, or permanently, postponed plans that they had earlier favoured with impatient enthusiasm. Massey-Ferguson, on the other hand, frequently has benefited from the candour of its approach, through increased export sales. Since government authorities typically negotiate with other companies as well as Massey-Ferguson, it is imperative from a commercial point of view, if from no other, that Massey-Ferguson's proposals be carefully constructed and based on accurate estimates of costs and volume.

If it is decided that a project should go forward Massey-Ferguson favours formation of a wholly owned local company, although as we have seen, for particular reasons, including sometimes its desire to minimize its local capital expenditure, it will proceed with either a majority controlled company or a company in which it has a minority interest, or even with merely a licensing agreement. Even in cases where the company is wholly owned by Massey-Ferguson, it favours minimizing its actual capital investment. Profits to the parent company will arise from technical assistance fees, sales of components and whole goods to the new company, royalties, and finally, dividends paid by the company. Contributions by the parent company, in the form of "know-how" in technical and managerial areas, are almost invariably much more important for the success of the projects than actual capital supplied.

Generally speaking, Massey-Ferguson strongly favours close co-operation with local interests, and it also strongly favours drawing on local personnel for staffing the managerial positions of the new company. It will, however, "lend" experienced personnel for varying periods to hasten the launching of a project, and also to rescue a project from specific difficulties. Because of this, the parent company and its individual operations units will, in future, have among their senior management personnel an increasing number of people experienced in the problems of establishing such operations. For example, the many difficulties encountered in launching the Brazilian project have provided the company with experience that will make it substantially easier for it to establish similar projects in future. The need consciously to develop such personnel is, however, still great and much remains to be done by the company.

From preceding discussion it will be obvious that the type of person most suitable for the work that the Special Operations Division must undertake will be quite different from the type of person found in most other areas of the company. Frequent and complex negotiations with government authorities, detailed appraisal of industrial opportunities in relatively underdeveloped economies, continuing surveillance of the efforts of competing companies in their own negotiations, shrewd judgment of the political risks involved in becoming associated with particular governmental powers – powers that may not necessarily be permanent – close co-operation with local capitalists, careful planning to ensure that projects will not unnecessarily undermine the company's export activities, and readiness to "put out fires" in operations already under way: these are the characteristics required of, and the tasks to be undertaken by, personnel of the Special Operations Division. The Division is, as it were, at the frontier of the growth of the company as an international or multinational organization.

CHAPTER 16

Landini of Italy

Neither Massey-Harris nor Harry Ferguson had ever been able to develop a strong market for their products in Italy. Protection of local manufacturing companies and the existence of well-established manufacturers had constituted major obstacles to penetrating that market. One of the Italian tractor manufacturers that had been supplying the market for many years was the firm of the Landini family in northern Italy known after 1955 as Officine Meccaniche Giovanni Landini e Figli S.P.A. It was by acquiring that company in 1960, that Massey-Ferguson established itself in the Italian market.

The founder of the firm was Giovanni Landini, born in Scandiano in the province of Reggio Emilia in 1859. He moved to Fabbrico in 1878, became an apprentice in a blacksmith's shop and opened his own shop there in 1884. That year marked the beginning of the Landini company. There was some experimentation in Italy even then with primitive forms of mechanical power for farming and Landini became interested in it. The large United States Mogul steam engine particularly influenced him, and in 1911 he made his first steam engine suitable for farming. At about this time the farm tractor with an internal combustion engine was coming to the fore in the United States and in Germany. However in Germany it was the semi-diesel type of engine of the kind for which the Lanz organization became well known that was most frequently employed. Landini turned his attention toward it rather than the North American gasoline tractors. In 1917 he began to manufacture a semi-diesel internal combustion engine for industrial purposes and to experiment with applying it to tractors. His death in October 1924 came just over a year before the first Landini semi-diesel tractor was manufactured. (See illustration p. 406.) His sons carried on the family business.

The original tractor was a 30-horsepower tractor. It was followed by a much improved 40-horsepower tractor in the early 1930s, then by the Super Landini and the Velite, and in 1950 by the L25 and L45. Other L series models followed, the L35 in 1953, the L55 in 1954, an improved L55/B in 1955, and the L44/M a successor to the L35 in 1957. All these models were of the semi-diesel type, even though the tractor industry generally had shifted heavily towards diesel and, earlier, gasoline engines. Failure to follow that trend was one of the reasons why the company began to experience difficulties.

Until after the Second World War, Landini had concentrated its production in the Fabbrico factory in northern Italy. However, the immediate post-war period was one of political unrest, and Communist strength was particularly great in northern Italy, as it still is. One of the Landinis, it is said, at that time received a Communist delegation just before a general election which generously informed him that they would not shoot him after the election. They liked him and so would permit him to work as a labourer in the factory. It was in that kind of environment that the decision was taken to open another plant even further to the North, at Como, probably with a view to transferring all the facilities there if conditions deteriorated further. They did not deteriorate further, the Italian Communist party did not win the election and, with production rising (see Table 14), both plants were kept in operation.

As all the output of the factories could easily be sold, the company stayed with its semi-diesel tractors. Competitors did not. They introduced the diesel tractor and, slowly, also began to be influenced by the revolution in tractor design that Harry Ferguson had unleashed. Landini made profits, but its share of the market declined sharply. With the first indication of a decline in sales after 1955, the company's financial position began to weaken.

At the same time, the company suffered from problems frequently encountered by family firms in which a number of members of the family are involved, with varying degrees of interest and authority. In 1957 Dr. Flavio Fadda was asked to join the organization and assume control. He emphasized the need to change to diesel engines and to tractors of new design, changes that were being advocated by the company's dealers.

A good diesel engine was required for such a programme. Fadda went to F. Perkins Limited and negotiated a licencing agreement with them for the local production of the P3 and P4 engines as well as for a two-cylinder engine designed but not manufactured by Perkins. It was planned to manufacture these at the Como plant. Then a completely new line of tractors was designed including the R4500, the L7000, the F300, all wheeled tractors, and the C4000 crawler tractor. The latter was introduced because Italy's hilly terrain required a crawler tractor. Fortunately for the company, the wheeled tractors enjoyed substantial success, and the transition from the semi-diesel tractor was effected without great

TABLE 14

Landini Tractor Production (1945–1966) Units

	Semi-Diesel	Diesel	Crawler Diesel	Total
1945	171	—	—	171
1946	444	—	—	444
1947	586	—	—	586
1948	520	—	—	520
1949	662	—	—	662
1950	749	—	—	749
1951	1174	—	—	1174
1952	1630	—	—	1630
1953	2308	—	—	2308
1954	2500	—	—	2500
1955	2516	—	—	2516
1956	2112	50	—	2162
1957	1062	950	—	2012
1958	1209	947	—	2156
1959	851	1435	—	2286
1960	177	2789	—	2966
1961	98	3605	674	4377
1962	—	4307	812	5119
1963	—	3750	1664	5414
1964	—	4261	1741	6002
1965	—	3975	2252	6227
1966	—	3192	2812	6004

difficulty. But the crawler tractor posed design problems that were new and therefore difficult for the company.

With the success of the new wheeled tractors in Italy, the company was encouraged to export them. It reached an agreement with a Yugoslav organization to ship most of the components for the tractor to Yugoslavia where a locally produced Perkins engine manufactured by the Yugoslavian organization under licence would be fitted to it. The resulting tractor was then marketed in Yugoslavia in competition with the Massey-Ferguson tractor built under licence by another Yugoslav organization. A further export market was developed in South Africa where the firm of Malcomess, which had just lost its Massey-Ferguson distributor franchise, began to market it. The Landini tractors were also sold in France, Belgium, and Holland.

In spite of the company's rejuvenation several major problems remained. It had not been possible to overcome all the difficulties involved in family control. But more than that, there were the worrying implications for the Landini tractor of the developing European Common Market. Tariffs had provided Italian-produced tractors with strong protection –

26 per cent – from the competition of other Common Market countries, but this would inevitably decline, and in 1960 was already down to 22 per cent. In addition, there was the apparent trend throughout the world towards increased concentration of tractor production in the international companies. So while Landini would certainly be able to operate profitably for several more years, it was doubtful in 1960 whether it could do so permanently.

In early 1960 J. H. Shiner, Vice-President Marketing of Massey-Ferguson, went to Italy to decide what to do about the sagging performance of the company's distributor there. He also enquired into the licencing arrangement that Perkins had with Landini – since Perkins by that time was owned by Massey-Ferguson – and in the course of doing so met Fadda.

Fadda raised the matter of a closer association between their two companies. He was able to point out that his company had about 10 per cent of the Italian tractor market, that Landini was a well-known and respected name in the market, and that it had an old and well-established sales organization selling its own tractors, and also implements and combines (mainly German) purchased from other suppliers. He also mentioned the potentialities of the crawler tractor that they had been developing. The factory at Fabbrico, fifteen miles from Reggio Emilia where the offices of the company were located, manufactured most of the tractors except castings, forgings, sheet metal and hydraulic gear pumps, while the Como factory made the necessary Perkins engines.

Shiner was quite interested and arranged for Thornbrough, Mawhinney (head of Export Operations), himself, and Fadda to inspect the Landini Company in detail. The Massey-Ferguson executives did not feel that the trip had been wasted. Indeed, for Massey-Ferguson, an association with Landini – provided it meant control of the company – had immediate appeal. The Italian market was fourth in importance in Europe, and yet Massey-Ferguson had not participated satisfactorily in it. The heavy import duties, the Fanfani finance plan which restricted preferred farmer finance to machinery produced in Italy, and the weak dealer organization that had resulted from limited sales, might all be overcome by such association. To be associated in Italy with the Landini name could not help but be beneficial, because of the reputation for quality if not technology that Landini enjoyed.

At the same time, if Massey-Ferguson did not act, its position in the Italian tractor market might deteriorate even further should the United Kingdom, where it manufactured most of its tractors, stay out of the Common Market. It is true that Massey-Ferguson was expanding its French tractor production, but costs were still too high there to sell those tractors in the Italian market, and would probably not be low enough until tariffs had virtually disappeared in the Common Market. Equally important, the Landini company had a crawler tractor, which Massey-Ferguson did not have. Since it could be sold not only in Italy but also in export territories

Modern agriculture's increasing demand for power is creating an expanding market for tractors in the 100 horsepower range such as this MF-1100.

it seemed a good reason for acquiring the company and one that would remain even after trade barriers inside the Common Market had disappeared. Never in all its long history had Massey-Ferguson been in a position to sell a crawler tractor and since it was beginning to move more heavily into industrial and construction machinery this gap in its line could become a serious matter.

The company decided it would attempt to purchase all the issued share capital of Landini. Agreement to that effect was reached. Landini tractors would continue to be marketed since they enjoyed a valuable name in Italy and were different in design and appearance from Massey-Ferguson tractors. The effect of the acquisition of Landini on the Massey-Ferguson organization, as for example the possibility of increased concentration on manufacturing crawler tractors, and the nature of its future role, were discussed, but no firm policies were defined. The name of the company was changed, but only slightly, for in June 1961 it became Landini s.p.a., and it continued to operate under the direction of Dr. Fadda. A new company, Massey-Ferguson Italiana s.p.a. was formed to sell products other than those of Landini and Perkins, but it too was under Fadda's direction.

Incorporating a small but proud organization, with its own traditions and methods, into the massive Massey-Ferguson organization was a matter of great sensitivity and importance. A memorandum of the President dated October 21, 1960, deals with the subject, and with the approach that the company decided to take:

> It is of the greatest importance that the change of control is accompanied by the minimum of disturbance to the day-to-day activities of the Landini staff under its General Manager. Too rapid an exposure of the Landini people to Massey-Ferguson procedures and too many visits by Massey-Ferguson executives will have adverse effects until the philosophies and policies of Massey-Ferguson are explained and understood, and in particular until the Landini staff realize that their organization and therefore their jobs have an expanding future under Massey-Ferguson control. For this reason, the organizational relationships to be established immediately between Massey-Ferguson and Landini will be restricted to those which are essential to establish effective financial control, to formulate strategic plans and carry through certain urgent marketing plans.

The Director Special Operations had co-ordinated all matters concerning Landini during the period leading up to the purchase agreement, and he was asked to establish liaison facilities between Landini management and Massey-Ferguson. This arrangement continued until late 1963, when Landini became an operations unit in the organizational structure of the company, alongside the United Kingdom, Perkins, France, Germany, and Australia. As such, its Managing Director was made directly responsible to the President of the parent company.

The operation in Italy did improve Massey-Ferguson's position in that market, and development of the crawler tractor has advanced. Massey-Ferguson has gained experience in operating a manufacturing company in an environment quite different from its other plants – including a labour force the majority of whose members belong to the Italian Communist party. It has also prompted Massey-Ferguson management to consider whether some kind of "two-line" policy in tractors – different from the disastrous one of the 1953 to 1956 period – might not be incorporated into its operations. In the meantime, operations in Italy are expanding. Table 14 has shown that the company manufactured 6,004 tractors in 1966, compared with 2,286 in 1959, and the operation has been profitable.

As some of the crawler tractors manufactured were for industrial purposes, in 1965 the company decided that Italy should be its source for such tractors and for some of its other industrial equipment as well. For this purpose the company formed a new subsidiary in Italy, Massey-Ferguson (I.C.M.) S.p.A., located its head office at Aprilia, just south of Rome, and there on a 62-acre site it began in 1966 to build a new plant with a floor space of 350,000 square feet. The decision to locate the plant in the south of Italy rather than in the north was influenced by the lower labour costs and the more stable labour conditions that exist there. This concentration on industrial and construction machinery gave permanency to the Italian manufacturing operation, something which had always remained in doubt as long as the operation depended solely on the Landini line of tractors.

CHAPTER 17

Reorganization in South Africa and Australia

In 1957, parent company management decided to close all its branches in South Africa and to franchise independent distributors, despite its majority interest in a South African manufacturing company. On the face of it, this move was quite contrary to the company's policy in other countries, and also surprising in view of the company's long association with South Africa. In 1960 the company reversed its decision. By 1964 it was firmly established in the South African market and was planning further expansion. This was strange behaviour. But let us begin at the beginning.

Massey-Ferguson's association with South Africa, as with a number of other countries, began in the 1880s through the sudden increase in the export activity of the Massey company. Massey machines, it is reported, were at work in South Africa by 1887. The then famous commercial house of R. M. Ross and Company at Cape Town became distributors for the Massey machines, particularly for reapers and binders. By 1892 a local newspaper reported that "200 Massey self binders and reaping machines are now in use in South Africa."

Invariably, Massey-Harris machines were displayed at agricultural shows and entered in contests, thereby receiving a remarkable degree of publicity, even before the turn of the century. Describing their participation in the Western Provincial Agricultural Show one paper wrote:

> Messrs. R. M. Ross & Company made a brave show. They had, among other implements, a Massey-Harris self binder, which machine is now a favourite in South Africa, being used in every grain district in the colony also in the Transvaal and Orange Free State. Its construction is simple, being built of steel and malleable iron, its lasting power is great.

Even the short-lived Massey-Harris bicycles were marketed in South

Africa, as was the Massey-Harris cream separator. Other distributors in addition to the Ross company were appointed in the Union, although Ross and Company remained as chief agents for South Africa.

This division of the market between different distributors reflected dissatisfaction with the sales distribution results of the Ross company, and a major step was taken in 1925 to increase further the effectiveness of the company's sales organization. In that year Massey-Harris decided that it would establish a branch office, rather than a subsidiary company, at Durban. The first manager, W. E. Gypson, came out from the company's Calgary, Alberta, branch, as did several other employees. This reflected an approach introduced by the Masseys, and prominent in the Massey-Harris company until very recent years – the appointment of Canadian employees of the company to senior posts abroad.

In December, 1926, Massey-Harris formed its first new local company, Massey-Harris (South Africa) Proprietary Limited, with its head office in Durban. Not quite a year later a depot was opened at Johannesburg, and in 1929 the East London branch was opened. By 1929 the Massey-Harris staff numbered over forty, and the turnover at Durban had grown to such an extent that it became necessary to build a large Massey-Harris warehouse at a cost of £25,000. Possession of the new building was taken in May 1930.

For a few years Massey-Harris and R. M. Ross both sold the company's machines, but Ross went out of existence during the crisis of 1931. Massey-Harris succeeded him at Cape Town, and opened a branch there, as well as one at Maitland. Even as early as 1927, Massey-Harris had begun to replace several of the other distributors and had established a branch outside the Union at Salisbury, Southern Rhodesia. The East African branch was opened first at Nairobi in 1930, then was closed because of economic adversity in 1934, and finally reopened at Nakuru, Kenya, in 1940. The sphere of responsibility of the South African company and its branches continued to expand, and by 1939 it embraced all African countries south of the equator.

During the 1930s the area of cultivated land increased in the Union, so Massey-Harris sales enjoyed an expanding market. All this time the South African market was served by implements imported from the company's factories in Canada and the United States, and it faced no competition from local producers because there was no local implement industry.

However, during the Second World War the supply of imports dwindled ominously, and since demand was expanding rapidly, basic conditions suddenly favoured the establishment of a local implement manufacturing industry. The move in this direction came from a local business man. The venture that was subsequently undertaken was soon to become the most important local manufacturer of implements, and eventually a subsidiary of Massey-Ferguson. It was South African Farm Implement Manufacturers Limited (SAFIM).

In the early 1930s Lieutenant-Colonel K. Rood began to consider the possibilities of manufacturing farm implements in South Africa. He was then Chairman of the Union Steel Corporation, the pioneer steel company of South Africa established at Vereeniging near Johannesburg, a suitable location because of abundant local supplies of water, power, and coal. Rood came to the conclusion that successful local manufacturing would require tariff protection against foreign competition, and he approached the government to grant him this protection. The Minister of Economic Affairs took an encouraging attitude, without making any definite moves to provide protection. Rood embarked on a rudimentary manufacturing operation, hoping for the best. But protection was not forthcoming and the operation collapsed. He tried again in 1939 and now the government, because of war, was much more interested in the project than it had been before. On this occasion Rood received an indication in writing that protection would be forthcoming, and the company, soon called SAFIM by everyone, was incorporated at the end of 1939 with Rood as its chairman.

Financial backing for the venture came essentially from two groups, the Federale Volksbeleggings, an investment corporation, and Champions Limited, an old firm of merchants and retailers at Bloemfontein. A Czechoslovakian engineer, Dr. S. Kux, became the manager and technical expert of the company, and played an important part in bringing the factory into production.

Facilities of 26,000 square feet were built at Vereeniging, near the source of steel; second-hand machine tools were obtained from the United States after the first order for such machinery had been intercepted on Belgian docks by the German army; and production itself started in 1941. The lack of skilled and experienced men gave the new venture disconcerting and prolonged teething troubles, and its reputation for a time was somewhat tarnished. SAFIM never built up its own system of retail distributors or dealers, and the Federale and Champion groups together formed Farm Implement Distributors Limited (F.I.D.) to act as one of the many distributors of SAFIM goods, and also to control the company. F.I.D. held a majority of the stock, later transferred to a new company formed by Federale and Champions, SAFIM Holdings Limited. Only the most simple ox-drawn and mule-drawn implements – plows, harrows, cultivating implements and single row planters – were manufactured, and while there was some expansion, the company did not grow at a particularly impressive rate. By 1947 the facilities amounted to 117,000 square feet, sales had grown to £597,212 involving about 138,-000 small implements, and paid up capital and reserve was £395,177.

The company's total dependence on animal-drawn implements did not provide a base for successful future operations. In view of competition it would have to move rapidly into tractor-drawn implements. However, SAFIM's difficulty in this respect was that it did not possess the technical knowledge and skills required to make that transition quickly, and it was this that encouraged Rood to consider an association with a more mature foreign company. Massey-Harris was an obvious possibility, and its management was contacted.

Massey-Harris showed immediate interest. M. F. Verity and H. H. Bloom went to South Africa in April 1947 to examine at first hand the possibilities of forming an association. Negotiations were entered into and there was the usual amount of bargaining over fees that Massey-Harris would charge for rendering technical assistance, and over the price that Massey-Harris would pay for SAFIM shares. Agreement was eventually reached and Massey-Harris acquired a 20 per cent interest in SAFIM by purchasing 50,000 of its common shares valued at $225,884. By the agreement, Massey-Harris also became a distributor of SAFIM products in South Africa (there were several other distributors in South Africa) and it became exclusive distributor for SAFIM products in other markets. As a result of the association, SAFIM immediately gained access to the experience and skills of Massey-Harris engineers and technicians, while Massey-Harris acquired a manufacturing facility in its major market for animal-drawn equipment. Requirements for this type of equipment had practically vanished from the home market, and so could not profitably be produced in North American factories.

SAFIM now was available to Massey-Harris for manufacturing Massey-Harris brand implements, as well as SAFIM brand implements. At one time it was suggested that the name "Massey-Harris" should be incorporated in the name of SAFIM, but the South African shareholders did not want this, and thereby saved the company in South Africa some of the difficulties that emerged under such an agreement in Australia. Massey-Harris negotiators felt that their position in South Africa would be improved because of the prestige of being associated with the local organization, and also because of the protection that it would give the company in the event of stronger import controls being imposed. The company assured itself of a voice in the SAFIM operation by an agreement that gave it two directors in the company.

In 1948 SAFIM acquired a controlling interest in a small implement manufacturing company in Southern Rhodesia – Rhodesian Plough and Machinery Company (1948) Limited. This company had been started in the 1930s by Dr. Kux's former employer, and when the owner was about to retire, SAFIM accepted a request to take an interest in it. It is still in operation. SAFIM made profits but the business showed little growth. Sales were £597,212 in 1947 and £577,000 in 1950. Net value per share was only slightly larger in 1950 than it had been in 1947. The capital of the company was increased in 1948, 1949, and 1950, and on each occasion Massey-Harris took up its share of the increase, acquiring additional shares. Consequently by 1950 it held a 27 per cent interest in the company. In 1952 this was abruptly changed when the company purchased the Federale shares in SAFIM Holdings Limited. Actually, what happened was that Federale had held 51 per cent of the shares of SAFIM Holdings which Massey-Harris acquired, but to satisfy the wishes of Champions

– the other major shareholder in SAFIM Holdings – Massey-Harris sold them 1 per cent of the Holdings's stock. By an agreement between Champions and Massey-Harris, the latter received the right to nominate the chairman and the majority of the Board of SAFIM, to appoint the general manager and executive officers, and to determine technical and management policy, but not general policy. All this came about because of some uneasiness between the two major South African partners, and some discontent among Massey-Harris management over the management of the SAFIM company. As a result of these arrangements, Massey-Harris now held directly and through SAFIM Holdings, 52.6 per cent of SAFIM's common shares, and had effective control of the manufacturing operation in South Africa. Federale again acquired a major interest in the company in 1960 by acquiring control of Champions.

As everywhere, the merger in 1953 with Ferguson created new complications in South Africa. In 1948 Ferguson had formed Harry Ferguson of South Africa Limited with headquarters in Durban, and the duties of this company were chiefly to represent the parent Ferguson company in its relation with local distributors, and to supervise imports. A new company, Tractors and Farm Tools Limited (TAFT), was granted the Ferguson distribution franchise for the Union. The Ferguson tractor was immediately popular, but currency restrictions severely limited sales. Still, by 1953 10,000 Ferguson tractors had been sold.

After the merger, TAFT continued to sell the Ferguson line because of the Massey-Ferguson policy of marketing two separate lines of machinery, and Harry Ferguson of South Africa Limited continued its supervisory role. From a distribution standpoint, however, Massey-Ferguson's position in South Africa had become a jungle: it was now being represented there by its own two companies (Massey- Harris and Harry Ferguson), by the Massey-Harris subsidiary (F.I.D.), by several independent distributors and co-operatives selling SAFIM products, and by TAFT selling Ferguson products. This confusion was compounded following the purchase by Massey-Ferguson of H. V. McKay of Australia, for that company had been selling its Sunshine machinery through Malcomess (Proprietary) Limited, another independent distributor. Nor was the situation made less complicated by the fact that SAFIM manufactured some implements for the Ford organization. The growth of sales of these various organizations was not encouraging over the period 1953 to 1956, and Massey-Ferguson sales showed no growth at all. Production in the SAFIM factory was increased by placing orders for Ferguson machinery with that company, rather than with other manufacturers.

SAFIM, however, had other problems. It had difficulties obtaining adequate management, it faced a serious cash problem, and encountered difficulties in production, quality, and product development. Under the circumstances it was not surprising that in 1954 a Massey-Ferguson employee, also from the Calgary branch, was sent out to assume the position of general manager of SAFIM. As a step towards simplifying the organization in South Africa and reducing costs, the head office of Massey-Harris (South Africa) Proprietary Limited was moved from Durban to Vereeniging in 1955, and in September of that year the new General Manager of SAFIM was also appointed its Managing Director. For the first time unified direction of the operations of the two companies was established – but it did not last long.

The parent company at Toronto began thinking about closing its branches in the Union. Through its sale of the SAFIM, Massey-Harris, Ferguson, and Sunshine lines it enjoyed a substantial portion of the market, and it had made some management changes; but still there were losses. This was hardly surprising considering the complex, even chaotic, state of the organization's distribution system there – exemplifying in an extreme form the disarray that grew up in the distribution system of the world-wide organization after 1953.

Management at Toronto decided that something had to be done. W. W. Mawhinney and M. F. Verity of Massey-Ferguson went to South Africa to investigate. Not long after a plan was devised. Since Toronto management was beginning to pursue its policy of moving towards one product line and one distribution system, it might have been expected that the plan would involve closing out the distributors and modifying and building up the company's branches and dealers. Actually, it involved quite the opposite approach, that is, of closing out the branches.

Management took the view that the Massey-Harris line had a limited future in South Africa without a tractor, and since TAFT was already distributing the Ferguson tractors, there would be little sense in offering those through Massey-Harris branches as well. It would be better to give all the tractor business to TAFT. Another distributor might be found to sell the company's harvesting machines. Such an approach appeared attractive to the parent company. Losses of the South African subsidiary would be halted and, through a reduction of its capital, a pay-out of dividends, and sale of its inventories to distributors, a large sum of money could be repatriated to the parent company. Such funds the parent company could use in rebuilding its North American operations. Some of the members of top management were also pessimistic over the political environment in South Africa.

After lengthy negotiations Massey-Ferguson granted a franchise (as of November 1, 1957) for its harvesting equipment and trailed implements to Malcomess (Proprietary) Limited, and the franchise for its tractors and mounted implements to TAFT. Massey-Harris (South Africa) Proprietary Limited purchased the assets of Harry Ferguson of South Africa Limited and changed its own name to Massey-Ferguson (S.A.) (Proprietary) Limited. This company became a shadow of its former self, existing merely to co-ordinate relations between Massey-Ferguson and TAFT, Malcomess, and African distributors south of the equator; to supervise the operation of Farm Implements Distributors Limited, its wholly owned subsidiary; and to subcontract some implements with local manufacturers. Distribution in Southern Rhodesia, Northern Rhodesia, and East Africa was taken over by the Ferguson distributors in those areas, and the company's branches at Salisbury, Southern Rhodesia and Nakuru, Kenya, were sold. Single-line distribution was thereby effected.

This reorganization plan was thought to be in harmony with new marketing policy, in that it provided for the sale of Ferguson-system tractors through only one group of dealers. But it violated another principle of that policy, for it did not give any one dealer a full line of equipment to sell. It also made the company's local manufacturing organization (SAFIM) depend even in its home market on independent distributors. Unfortunately for the South African operation, it was not till later that the principle was explicitly adopted that the company would control its distribution system in countries where it controlled manufacturing facilities.

The new arrangement did not generate the sales that had been expected. In 1959, sales fell drastically, and this could not entirely be attributed to depressed economic conditions. To base the future of the SAFIM manufacturing operation on the sales performance of an independent distributor organization seemed not to be sustainable policy.

A major reorganization was begun after Verity, Shiner, Mawhinney, and Beith visited South Africa in 1960. All the elaborate and time-consuming arrangements that only a few years earlier had resulted in the company divesting itself of its own distribution system, had now apparently, somehow, to be reversed. First of all, TAFT was purchased and that company ceased to exist as a distributing outlet. The distribution agreement with Malcomess (Proprietary) Limited was simply, and rather hastily, terminated. A new company, "Massey-Ferguson (South Africa) Limited," was formed to control all the interests of Massey-Ferguson in South Africa, although not in the rest of Africa. Also, Farm Implement Distributors Limited, a wholly owned subsidiary of Massey-Ferguson, became a dormant company. There had, of course, been South African shareholders of SAFIM, and when in 1961 SAFIM became a wholly owned subsidiary of Massey-Ferguson (South Africa) Limited, the South African shareholders received shares in the latter, equivalent to their holdings in SAFIM. Massey-Ferguson remained the majority shareholder, but Federale Volksbeleggings, the local shareholders, remained as active partners in the South African operations of the company.

The company thought that this reorganization would once and for all simplify its distribution system in South Africa, which, to a great extent, it did. Still, one complication soon emerged. When Malcomess lost its Massey-Ferguson and Sunshine franchise, it still had a Landini franchise, Landini being the Italian manufacturer of tractors. Shortly thereafter, Massey-Ferguson purchased the Landini company in Italy, so Malcomess again emerged, although indirectly, as a distributor for Massey-Ferguson, and the latter again found itself with two distinct lines of tractors in South Africa. In 1966 the President of Massey-Ferguson assured Malcomess of "... the continuation of our relationship within the framework of your franchise agreement and the availability of the Landini line of wheeled agricultural tractors to your organization."

The experience of the South African company illustrates one important advantage of the Massey-Ferguson system of internationally distributed, relatively independent, operations. Such an arrangement not only enables the local company to meet specific needs in specific markets, but also, in so doing, and in seeking to solve particular problems, it sometimes develops products and operational procedures that can with benefit be adopted more generally in the world-wide organization.

For example, over the years SAFIM endeavoured to design implements that were slightly more rugged, and somewhat more simple, than the ones obtained from the North American and United Kingdom factories, implements that could be operated by less educated operators. Some of these implements, such as the poly disc harrow, the planter, and the tool bar cultivator, also had unique characteristics to satisfy local demand. Massey-Ferguson found the South African designs useful in establishing manufacturing activities in various underdeveloped countries, where the appropriate machines for those new ventures were not always of North American or European design.

The South African operations unit has also sought to overcome a problem that is faced by most other operations units – that of meeting competition from local manufacturers of the simpler farm implements, and from manufacturers of spare parts that fit the machines of the major implement companies. When the operations of the company were consolidated in South Africa in 1961, the new Massey-Ferguson company none the less permitted its subsidiary, SAFIM, to continue to manufacture its line of SAFIM-brand implements in addition to the Massey-Ferguson brand implements. The SAFIM-brand implements can be introduced quickly to meet special needs and particular competitive situations. SAFIM also manufactures spare parts for various machines of competing companies, an activity not engaged in by any other of the Massey-Ferguson operations and one that it began a number of years ago when it was in desperate need of production volume.

In retrospect, it seems apparent that the 1961 reorganization was most successful. It created a greatly simplified and fully integrated organization, one that imported, exported, distributed and manufactured the company's products. The financial success of the new organization was greater than could have been forecast. Whereas Massey-Ferguson sales in 1956 had been about $11 million, in 1966 they were almost triple that amount, in spite of severe drought conditions in 1965 and 1966. Massey-Ferguson tractors in 1965 accounted for about 40 per cent of all tractors sold. Future plans include local production of the company's tractors.

The company's policy in Australia after 1956 was quite different from what it was in South Africa, for no thought was given to reducing its involvement there. Acquisition of the McKay company in late 1954 had given the company a large manufacturing plant at Sunshine, just out of Melbourne, but one that was very seriously run down. Average age of the staff was high; organization, procedures, and facilities were antiquated. Much needed to be done, and this at a time when sales had begun to subside from their abnormal post-war growth rates. It is interesting to trace the reorganization that occurred in the Australian operations unit because it typifies the kind of reorganization that was pursued in all the operations units, and exemplifies the way North American concepts of business management were transmitted to all the corporation's global operations.

By 1956, sales were definitely weak in Australia, in part because of severe floods. A cut-back in production was necessary, but because of the many years of buoyant employment in the factory, this was not an easy task. The public might well interpret it as resulting from acquisition of the McKay company. Yet this period of declining profits not only necessitated retrenchment, but also revealed inefficiencies in the labour force that could not now be ignored. A general "tightening up" of the organization was begun and by the end of August it seemed as if the programme of cutting costs and controlling inventories was relatively successful.

In September and October of 1956 management began to turn attention to the basic issue of two-line distribution. The word was going around that the parent company had definitely adopted a single-line policy. L. T. Ritchie, the Managing Director of the Australian company, made some revealing comments about it in a letter to Verity on October 26, 1956:

> I think we all recognize that the "distributor" system of distribution is expensive, and this system may be six or seven percent more expensive than the Branch Agent/Dealer system. Perhaps, because of the circumstance that the Ferguson goods only became a factor in the implement trade in the past ten or twelve years, and did not have an organised sales force, these goods attracted a distributor system made up largely of firms who were well grounded in the Motor trade! – It is natural that a product which seemed to have such obvious "price" and "quality" and "buy" appeal should attract major Motor distributing companies – even to the extent of their providing substantial finance: This was all to the good, but it is, nevertheless, a fact that they did create a problem in that they were a competitive distributing element once the merger was consummated.
>
> This situation has been a headache ever since, and I think several of us have realised that it will eventually be necessary to abandon the principle of competing with ourselves. It has created problems in the United States and Canada and we have so many "shall nots" or "must be's" in our policies that we never get around to benefiting from the "do's" that could be accomplished under single line representation. It must cost us terrific expense to provide what amounts to almost duplicate advertising – which means duplication of effort in that aspect of our effort as well as all others: When we introduce new lines, the problem is even magnified – You have seen my reply to representations by our

Ferguson people and by our H. V. McKay people that certain fences should be set up around certain products! I cannot subscribe to that view, and feel that we should all sell all our products to anyone who can use them, or we shall not benefit from the real aims of the merger.

It is not going to be easy to straighten out the situation in Australia, although I am personally convinced that the long-term problems will be minimised by working towards single line representation just as fast as we are able to do so. . . . The big problem, of course, will be the elimination of the Ferguson distributor, and this will not be easy. . . .

I have concluded that we must prepare, within the next four or five months, to make the fundamental moves towards single line representation in part of our Territory, so as to be able to give notices to those distributors whom we agree must be changed, by about April or May next year, so that the following twelve months' duration of their contracts would clear the decks for single line in those Territories by the time a new tractor becomes available.

Preliminary work on analyzing the implication of moving toward single line distribution was undertaken. In January 1957, Ritchie reported to Verity on the effect it might have:

The policy of establishing single-line representation through Branch dealer for each trade area, during 1957, is commendable in many ways: It gives effect to the merger. It places distributor profit where it rightfully belongs; and it will also reduce the many difficult administrative problems.

The latter point is already a real issue. It would be still greater if another new tractor were available to either Ferguson or S.M.H. dealers. It will be more intensified the moment another new tractor or two is introduced, which also would mean competing with ourselves.

The accomplishment of this policy will not be easy, but it could be handled. Many hours and weeks of consultation and analysis . . . have shown that there is no real short-cut or Houdini procedure that can replace a bold move to assert our distributive rights under the merger.

Ritchie pointed out that he saw no particular difficulties in terminating distribution agreements with distributors in Western Australia, South Australia, Queensland, and Tasmania. But he did see serious difficulties in terminating the distributor agreement with British Farm Equipment Proprietary Limited (an affiliate of Standard), the distributor in Victoria and New South Wales, for its contract was valid until January 21, 1960. This was an arrangement that had been entered into by letter at the time of Massey-Ferguson's acquisition of the McKay company to ensure continuity in the operations of that very important distributor. Now it was clear that the arrangement would interfere with new distribution policy.

In March 1957 a meeting was held with British Farm Equipment Proprietary Limited to discuss how the company's policy of having one strong dealer in each sales territory could be implemented through co-operation between that company and Massey-Ferguson. While it was agreed that it would be desirable for either the distributor or the company to merchandise the complete line in a given territory, discussion was of an unofficial nature and no specific recommendations were made. All the while Massey-Ferguson was very conscious of the fact that the sales territories covered by that distributor accounted for 60 per cent of Ferguson sales and 50 per cent of Sunshine-Massey-Harris sales in the Commonwealth. The more management thought of procedures to be taken, the more uncertain they seemed to become. On April 30, 1957, the Managing Director wrote to the parent company:

I have been thinking a great deal about the Distribution, and I am by no means certain that we have found a solution for our problem. I believe you feel as I do that we could agree to a procedure that would seem to be logical based on theory and we would find it extremely difficult of administration on a practical basis! . . . I hate to sound discouraging, but I think it is going to be extremely hard to find a compromise that is workable, let alone profitable or bring about the Company's policy of Single-line Distribution.

Continuing attemps were made to arrive at some agreement with British Farm Equipment Proprietary Limited over distributing one line in New South Wales and Victoria, but without success, even after a meeting in Toronto in August between Massey-Ferguson executives and Arthur Crosby of that distributor company. It began to appear as if the company would have to proceed with terminating the other distributor agreements and simply wait for the expiration in 1960 of the agreement with British Farm Equipment Proprietary Limited. By September 1957 Massey-Ferguson had available the financial information necessary to negotiate with the distributors.

A concrete plan for acquiring the facilities of the distributors was completed by December 1957, one that included the reorganization of dealers upon the dissolution of distribution agreements, product education, servicing, advertising, head office and branch personnel, as well as the approach to be taken to negotiate dissolution of the agreements themselves.

On January 30, 1958 letters were sent to the various Ferguson distributors inviting them to a meeting with Thornborough, Klemm, Verity and Alexander, all of the parent company, and Ritchie, to negotiate the points at issue. Thornbrough came with some experience for he had borne the brunt of the negotiations with the distributors in the United States. There followed a period of intensive negotiation, ending in the termination of all the old agreements through purchase of the assets of the distributors as well as additional compensation, except for British Farm Equipment Proprietary Limited. Negotiations with that company were not successful and the contract simply ran its course. Until its expiry date the company was in the curious position in Victoria and New South Wales of háving its own dealers competing with British Farm Equipment Proprietary Limited dealers, both retailing the same products. In some cases dealers were even cross-franchised between the two companies. However, eventually this problem took care of itself, and the move towards a single system of Massey-Ferguson dealers was completed.

Reorganization of the operations of the factory took just as long as the reorganization of the distribution system. The Australian operation had wanted the Vice President Manufacturing of the parent company, H. A. Wallace, to come out for a visit to review its operations, but problems in North America delayed his visit. This was unfortunate, for it was not until he came and made his report in September 1958, that the groundwork was laid for a more or less permanent organizational structure for the manufacturing function in Australia. The major points in Wallace's report reflected the approach that was being adopted throughout the company's operations units. It emphasized that the general factory manager should be given full responsibility for the operation of the factory and full authority commensurate with this responsibility. Only in this way, it was thought, would he be able to perform adequately, and if under these conditions he did not perform adequately he would have to be replaced. At the same time it was emphasized that only the general factory manager should be the managing director's source of information for all manufacturing matters. Other staff people should acquire necessary manufacturing information through him, not directly from the factory. The report emphasized that " . . . there is a basic principle of management which is very clearly defined in Colonel Phillips' Memorandum on organization, which was issued in December 1956, and that is – with responsibility must go the corresponding authority."

Having begun by emphasizing the need to define unmistakably where responsibility lies and to allocate authority commensurate with responsibility, the report went on to make a number of detailed recommendations. It recommended that since the purchase of material and outside services is an important part of the cost of manufacturing, it should be placed directly under the general factory manager. It would be the duty of the purchasing department, in a sense, to compete with the factory, in that it would have an incentive to buy from the outside when the factory was unable to manufacture specific items as cheaply as others could. The report also emphasized that factory cost accounting, because of its importance to factory cost control, was a very necessary tool of factory management. For this reason, it recommended that the chief factory accountant should be made a part of the factory organization, reporting to the general factory manager, even though the policies and procedures by which the factory accounting department was to be governed should emanate from the Comptroller's office.

The report then discussed the difficult issue of production engineering, the engineering function that bridges the gap between the design engineer and the factory. The report expressed the view that:

> Good design engineers are creative men, and I would not change them for the world, because if you did you would lose the value of their creativeness. The shop man on the other hand is a practical fellow, generally with less academic training and imagination, and it cannot realistically be expected that he have the same point of view as the creative design man. As a consequence there is

always a certain amount of conflict between these two decidedly different types of men, which results in production delays, increased costs, and many other situations detrimental to the welfare of the overall Company. The Australian situation is no different than that found elsewhere. . . .

To assist in the solution of this problem, the report recommended the practice that had been adopted in North America and was being advocated for the Eastern Hemisphere: the establishment of a Production Engineering Department. Its prime responsibility would be to serve as a liaison between factory and engineering, and to estimate production costs and tooling and capital expense so that management could make decisions on new products before production drawings were in existence. It was stressed that the Production Engineering Department should be physically located within the Engineering Department but directly responsible to the general factory manager.

As far as factory organization was concerned, the report recommended the same type of organization as was being introduced throughout the company's world-wide operations. Briefly, it involved separation of staff and line in the factory with the staff side headed by a manager of facilities and the line side by a manager of production. A detailed organization chart for those divisions was developed. The report emphasized that because of the sprawling nature of the factory, material handling was of great importance and special provision would have to be made to minimize costs in this area. To raise the morale and the prestige of the position of foreman, which seemed to have deteriorated in past years, it was recommended that a training programme for foremen be established, that supervisors be placed on salary, that supervisors need not punch time clocks, that foremen be paid more than the people they supervise, and that all effort be made to recognize the supervisory staff as important members of the management team.

The report then turned its attention to the matter of factory facilities. In general, it emphasized that capital should be spent on putting the physical plant in reasonable shape, and on carrying out the necessary rearrangement and regrouping of machines and departments, rather than on machine tools. The operations unit was told that the amount of money allocated to such a programme would depend upon the financial success of the Australian company. It was thought that the organization and installation of a proper cost control system, which had been exceedingly crude in the past, would greatly enhance the ability of management to make wise decisions in allocating capital. It was also thought that, while great strides had been made in introducing such an accounting system, it would still take up to a year before the value of that programme could really be appreciated.

The report had rather interesting advice to offer about the use of consultants, and it arose because the Australian company had used such services in improving its factory operation:

Good consulting firms can be of a great deal of service to industry, particularly where the "know-how" is lacking in a Company and the acquisition of this "know-how" would be a long and time consuming chore. They can bring to a Company immediately the proper amount of the right kind of "know-how" to establish necessary controls. However, it must be borne in mind that it is the responsibility of management to operate a Company, and it is not the responsibility of a consulting firm, so, therefore, at some point it must be decided that the management personnel of the Company are capable of taking over completely on their own. I put this out as a word of caution because if a firm position is not taken you will have consultants with you for ever.

The report also stated that it was of the utmost importance to establish proper standards for an incentive plan if a worthwhile cost control programme was to be achieved; in doing so it supported management in its attempt to establish a standard incentive plan.

From this time onward, the basic organizational structure of the manufacturing operation was understood, responsibility was fixed, and the direction that improvement had to take was known. When, therefore, it was later discovered that stock obsolescence and rectification losses were substantially more than had been provided for in the budget, and that this had resulted from inadequate control by factory management, the action that had to be taken was clear – management in manufacturing had to be changed, and it was.

At about the same time that Wallace investigated the manufacturing operation (1958), L. H. Pomeroy came from North America and was appointed Director Marketing of the Australian operations unit. When it seemed clear that it was only a matter of time before the company would have achieved its objectives of one-line distribution and of dealer rather than agency retail points (a change in retail distribution begun in 1957), serious attention could be turned to establishing a full-fledged marketing function for the Australian operation. A sharp decline in the company's share of the market in 1958 and 1959 made it an urgent matter.

Pomeroy's two basic objectives were to establish and activate a marketing organization and a marketing function, and to find and train an Australian replacement for himself. The latter objective was in harmony with the company's policy of staffing local operations units with nationals. There had been no Director Marketing before Pomeroy came, only a general sales manager with the traditional type of sales organization. Market research and product planning did not exist. But there were aspects to the marketing problem that were highly favourable to the company. Through having combined Ferguson, Sunshine, and Massey-Harris lines, the company found itself with a substantial share of the total market. Also, many of the dealers, in terms of volume, capitalization, and experience, were better than the dealers the company had in the United States. It appeared that, potentially at least, the company was on the threshold of highly profitable marketing performance if only the necessary reorganization could be effected.

We have already seen that part of this reorganization involved closing out the independent distributors. It also involved appointing the best Ferguson and McKay dealers as Massey-Ferguson dealers, a process that was not completed until well into 1960. One of the first marketing moves made was to initiate professionally planned sales campaigns, such as the "Power Line Drive" and "Early Bird Campaign," to increase pre-season purchases by dealers and to speed sales of the newly introduced MF-65 tractor. This was not the first time such campaigns had been used by the company in Australia. To bolster the Marketing Division and to speed its reorganization, another member of corporate management went to Australia in mid-1959 and made recommendations for changes. His report spoke of the need to strengthen the Marketing Division, so that it could support a more vigorous marketing function. It recommended that a general sales manager be appointed with responsibility and authority over line operations, thereby freeing the Director Marketing to implement a comprehensive marketing function. It recommended that a Marketing and Economic Research Department be established and adequately staffed. Dealer sales potentials were to be established for each dealer, and a product planning manager was to be designated. The distinction between product training and product planning was clearly drawn. The product planning manager should have several regional product research specialists. A dealer development manager should be appointed.

To develop a more effective retail distribution system, emphasis was placed on establishing a dealer development programme, a programme for merchandising used implements, a retail salesmen programme, a dealer parts and service programme, and a programme that would lead to adequate Massey-Ferguson dealer inventory policy. Experimentation with company retail stores was recommended. To broaden the sales and profit base of the Australian operation, consideration was given to reducing the dependency of the company on tractors and combines or headers by developing a comprehensive local manufacturing programme. These recommendations, of course, reflected the approach to marketing that had been evolved by the parent company and had gradually begun to be implemented in North America. While many of its detailed characteristics have changed, it remains the basis of the approach to marketing in Australia, as in the other operations units.

The Australian company had for many years been engaged in developing products unique to that country. It had a unique tradition of design engineering, and this tradition was not broken when the company became part of the world-wide Massey-Ferguson organization. At first, Massey-Ferguson management thought of appointing a North American engineer to head that department, but before long it was decided that this was not necessary. J. K. Gaunt, who had been Chief Engineer of the McKay Company, became Director Engineering and Chief Engineer.

In addition to pursuing normal product improvement research, the Australian operations unit had begun to develop a new machine. This was the #515 sugar cane harvester. The apparent demand for such a harvester was noticed by the first Managing Director of the Massey-Ferguson organization in Australia, L. T. Ritchie, during a visit to North Queensland in 1954. In 1955 J. K. Gaunt spent several months investigating methods of harvesting sugar cane, and formed a view of what kind of machine would be acceptable to the industry and how it should be constructed. It would have to be a machine within a price range that would suit the individual farmer, that would be capable of harvesting all types of sugar cane, and that would top, cut, and load the cane. It was also important that such a machine could be integrated with existing cane transport systems.

The company began to design the machine in February 1956, and a prototype, side-mounted on a #744D Massey-Harris tractor, was tested in North Queensland in October 1956. It embodied a short cross-conveyor behind the choppers to convey the chopped cane to the main bucket-type elevator. While its fundamental principles have not changed to this day, many of the details have undergone major modifications. The cross-conveyor, elevator, the twin-base cutting discs, the flail-type topping unit, the independent hydraulic pump, the reservoir on machine, and fixed shoes on front gatherers, were all discarded when the machine was returned to Sunshine and rebuilt in 1957. Modifications included the redesigning of a base cutting disc, straight-through rear elevator, moving side walls, and floating ground shoes.

In the autumn of 1957 the machine harvested successfully down and sprawled cane, and was demonstrated before dealers. A total of 150 tons of cane was harvested during 1957, before the machine was returned to Sunshine. It was further modified and streamlined and was now fitted to the new MF-65 tractor with power steering. In October 1958, it received its first major public demonstration.

Five new prototype cane harvesters had been built for the 1959 season trials, and the tests in that year were largely aimed at evaluating, demonstrating, and operating the machine in various areas and under a wide variety of conditions. In 1960, the 1959 machines were rebuilt to new specifications, and six additional machines were assembled from factory-made components. Ten of the machines were sold to farmers in different areas, and one machine was sent to the United States for evaluation in Louisiana and Florida. In the following year thirty-four Massey-Ferguson cane harvesters, including twenty-three new machines manufactured in the factory, were in operation in Queensland, and they harvested 235,000 tons of cane, or about 2.3 per cent of total Australian production. Their number increased to ninety-four in 1962, and 8.7 per cent of the total Australian sugar cane tonnage was harvested mechanically in that year. This expansion continued in 1963 and machines were exported on an experimental basis to Trinidad, Venezuela, the Philippines, Madagascar, Mozambique, Martinique and Brazil. Unfortunately after all this effort the cane harvester was found not to be suitable for export markets.

Development of the sugar cane harvester led to an innovation in marketing. The Massey-Ferguson world-wide organization essentially develops and manages its export markets through a separate export division centred in Coventry. But the Australian company had for many years extended its interest beyond Australia, because of the distance from the United Kingdom and North America, to some of the small nearby markets such as Papua/New Guinea.

That policy has persisted to the present and so the Australian operations unit is the only one, of the various operations units, that is responsible for an export territory. But the Australian operations unit was also given responsibility for attempting to develop export markets for its cane harvester. Australian management, being fully committed to the success of the cane harvester, believed that it would be more successful in marketing that machine in other countries than would the Export Division. At first, the parent company in Toronto over-ruled the wishes of local management, but relented when local management persisted in its view.

In 1961, upon the retirement of L. T. Ritchie, the parent company achieved its long term objective of finding an Australian to fill the position of Managing Director of the Australian operations unit. The man was H. P. Weber, a graduate in science of Melbourne University, who had been general manager of Felt and Textiles of Australia Limited and prior to that a senior executive of Monsanto Chemicals (Australia) Limited. At the time of his appointment there was only one person in senior management who was not an Australian. He had come from the parent company because the Australian company did not, immediately, have available an appropriate candidate.

It had taken a long time to reorganize operations in Australia, much longer than could have been imagined when the McKay company was acquired in 1955. But encouraging results finally began to emerge. Sales had fallen to a low of $26 million in 1958, and the company's share of the market to an estimated low of 33 per cent in 1959. In 1963 the company's sales were valued at $43 million or something like 41 per cent of industry sales. By 1966 sales had risen to $58 million in spite of two years of severe drought.

CHAPTER 18

Progress in perspective

Massey-Ferguson today is a large, integrated, multinational, industrial corporation managed by professionally trained executives who employ sophisticated concepts of organization and control to achieve planned profit objectives. All the important events that have made it such a corporation must be regarded as milestones in its history. It was necessary in past pages to abandon chronology and also to explore numerous by-ways, in the course of which those milestones may have been obscured. Yet with a moment's reflection they stand out clearly.

The practice of the Massey and Harris family companies, at first separately and then after 1891 together, to acquire advanced technology from the United States and to modify it for local conditions was of great historical importance to the company. So was their decision long before the turn of the century to develop export markets abroad and then, after 1910, in the United States. Obtaining the Wallis tractor in 1927 kept the company in a market that was soon to dominate the farm machinery business, while the ideas that Harry Ferguson developed in the 1920s were to place the company in the forefront of tractor technology many years later. The company in the 1920s reacted to threats to the export markets in France and Germany by embarking on local manufacturing, which added a new and significant dimension to its international activities.

Within the perspective of history the company made two of its most important decisions during the Depression, both emanating from the influence of the new, young General Manager (later President and Chairman of the Board), J. S. Duncan: to continue manufacturing operations in Europe and the United States in spite of an accumulated deficit of $21.3 million, and to support the design of a self-propelled combine. As things turned out, the entry of E. P. Taylor and W. E. Phillips into the affairs of the company in 1941 was also an important development.

The war period removed the company's financial burdens, but it was the decision of management during the war to plan for greatly expanded peacetime activity that ensured its post-war success, including particularly the decision to put the #21 self-propelled combine into mass production and to make a significant inroad into the highly competitive U.S. market with it. At about the same time, that is, in 1946, the company also decided to enter into manufacturing operations in the United Kingdom, while Harry Ferguson in that year arranged for the Standard Motor Company to begin manufacturing his famous TE-20 tractor at Coventry – both events that were greatly to influence the shape of the company's operations in future years. Then came the 1953 merger with Ferguson which for the first time made the Massey-Harris Company an organization to be reckoned with in the world tractor market.

Paradoxically, however, the company experienced its worst period since the Depression after the merger, which period ended with another historically significant event – the change in top management of 1956. As we have seen, the financial deterioration of the company after the merger arose from a number of influences of both a short-term and long-term nature. But from an historical point of view the important question raised in that difficult period was whether a group of executives that came to the fore prior to the emergence of professional training in management science as a dominating influence could cope with the competitive environment that was emerging and with the problems of size and complexity created by the merger of 1953. The Executive Committee of the Board, and particularly E. P. Tayor and W. E. Phillips, concluded that they were not coping with those difficulties and decided that the President and other executives would have to go. A. A. Thornbrough emerged immediately as the senior operating executive, and before long a large number of executive personnel were brought in from other companies. The long-term significance of this development was that it changed the internal attitude of suspicion towards professionally trained managers and towards formalized procedures in organization and control, including detailed planning, to one of complete acceptance. Phillips took a special interest in this area and in the years following 1956 new approaches to it were painstakingly introduced into every one of the company's operations units around the world.

But other developments after 1956, within the various functional divisions of the company, are of lasting importance to the company as well. The abandonment of the two-line distribution policy in the marketing division rid the company of a heavy burden inherited from the Ferguson merger, and this involved difficult and time-consuming negotiation, particularly in North America and Australia. Phillips had been particularly concerned over the company's lack of control over its manufacturing activity through its heavy dependence on outside suppliers – a development that had begun in the 1930s, accelerated during the war, and had reached extreme proportions with the 1953 merger with Ferguson. After long and frequently acrimonious negotiations with the Standard Motor Company, Massey-Ferguson finally achieved control over its source of tractors in the United Kingdom and France. Control over diesel engine production was achieved by purchasing Perkins, a most significant event, since Massey-Ferguson was soon to use more diesel engines than any other company in the world. And increased control over its North American manufacturing activity came from the acquisition of a transmission and axle plant and from using major tractor components manufactured in its overseas, particularly its Coventry, facilities. The decision to maintain Perkins as a separate operation and sell diesel engines to other companies was also of long-term significance, as was the one of developing the policy of international sourcing of machine components, and more recently of moving seriously into the industrial equipment business.

Several developments made the company even more of an international enterprise than it had been. Together with emphasis on international sourcing of machine components there emerged a policy of basing plant location decisions more explicitly on relative costs of manufacturing in various countries than had previously existed. The new emphasis on decentralization of decision-making and of hiring nationals to run individual operations units moved the company in the direction of being a multinational, and not just an international, corporation. Full commitment to the international environment, in the sense of examining seriously opportunities for the sale and the manufacture of its products wherever on earth they might arise, led naturally to the development of the Special Operations Division. It has assisted in establishing the company in Italy, India, South Africa, Brazil, Spain, and Mexico, and in arranging a number of licensing agreements.

Not everything went well after 1956. In the turmoil of multi-directional change, management at first failed to cope adequately with two important problems – the tractor expansion programme in France and the design of combines. Its share of the combine market in North America and Europe declined. The sad experience of Massey-Harris with tractors and the more recent experience of Massey-Ferguson with combines underlined the perils in a competitive market of slowing the pace of product development, and the company is not likely soon to forget it. Its initial success in the United States market up to 1959 was partially fortuitous, and it was soon recognized that that highly competitive market demanded substantially more improvement in the company's marketing organization. While establishing its centrally controlled spare parts operation it ran down its spare parts inventory too far and service deteriorated seriously. Nor was the production of the company's new line of combines at Brantford, and tractors at Coventry, begun without serious programme problems and certain management mistakes. Its industrial products programme was at first a confused one and took a long time to become effectively defined and implemented. In some parts of the company, particularly North American operations, frequent changes of senior executives created undesirable uncertainty.

But the company grew rapidly. From 1956 to 1965 its sales doubled, and future prospects made it certain that more capacity would be required. So Perkins moved to increase its capacity by purchasing additional factory premises. More facilities were obtained on long-term lease to expand tractor production at the Banner Lane, Coventry, plant. The tractor plant at Beauvais was greatly expanded, and the foundry capacity at Marquette was also increased. Facilities were purchased at Des Moines for an assembly plant and for the company's North American administrative headquarters, all part of a strategy to strengthen its position in the United States corn belt, and similar steps were taken to make Cuyahoga Falls, Ohio, the centre of its North American industrial and construction equipment operation. Facilities for producing spare parts were put into operation in Bendigo, Australia. In Brazil changes in diesel engine and tractor facilities permitted the introduction of the MF-65 tractor into the market and the company moved from a leased factory to one it purchased. In Spain the company entered into an association with Motor Iberica S.A. for the production and marketing of Massey-Ferguson machinery, Perkins diesel engines, and Ebro trucks in that country. And in association with local interests it began a tractor and engine operation in Mexico. By 1967 its manufacturing capacity was sufficient to accommodate anticipated needs for several years ahead.

The record of the company's performance can be seen in the trend of its costs, sales, and profits. Sales of Massey-Ferguson throughout the world totalled $355 million in 1956. In 1966, they were valued at $932 million. A glance at Chart 11 shows that these sales were derived from the markets of many countries in the world – 11 per cent coming from Canada in 1963, 31 per cent from the United States, 38 per cent from Europe. The chart also shows the advantages of geographical dispersion of markets, for seldom was there a period when the rate of growth of sales did not differ noticeably between the various major markets. Full commitment to supplying those markets, not only through exports but also through local manufacturing operations, gives the company the opportunity to perpetuate its past growth rate. The high cost of manufacturing had plagued the company and improvement after 1956 in its manufacturing divisions did not come overnight. But Chart 12, which shows manufacturing costs as a proportion of sales, indicates that steady improvement in manufacturing costs occurred until 1964. Although further improvement is not likely to come quickly, room for improvement has not been exhausted, as is correctly suggested by the slight deterioration in costs in 1965 and 1966.

That deterioration of costs arose from a number of external and internal influences. In some markets, particularly Germany and certain export areas, competitive forces put unusual pressure on the price struc-

CHART 11
Massey-Ferguson Limited
Sales by territories 1955-1966

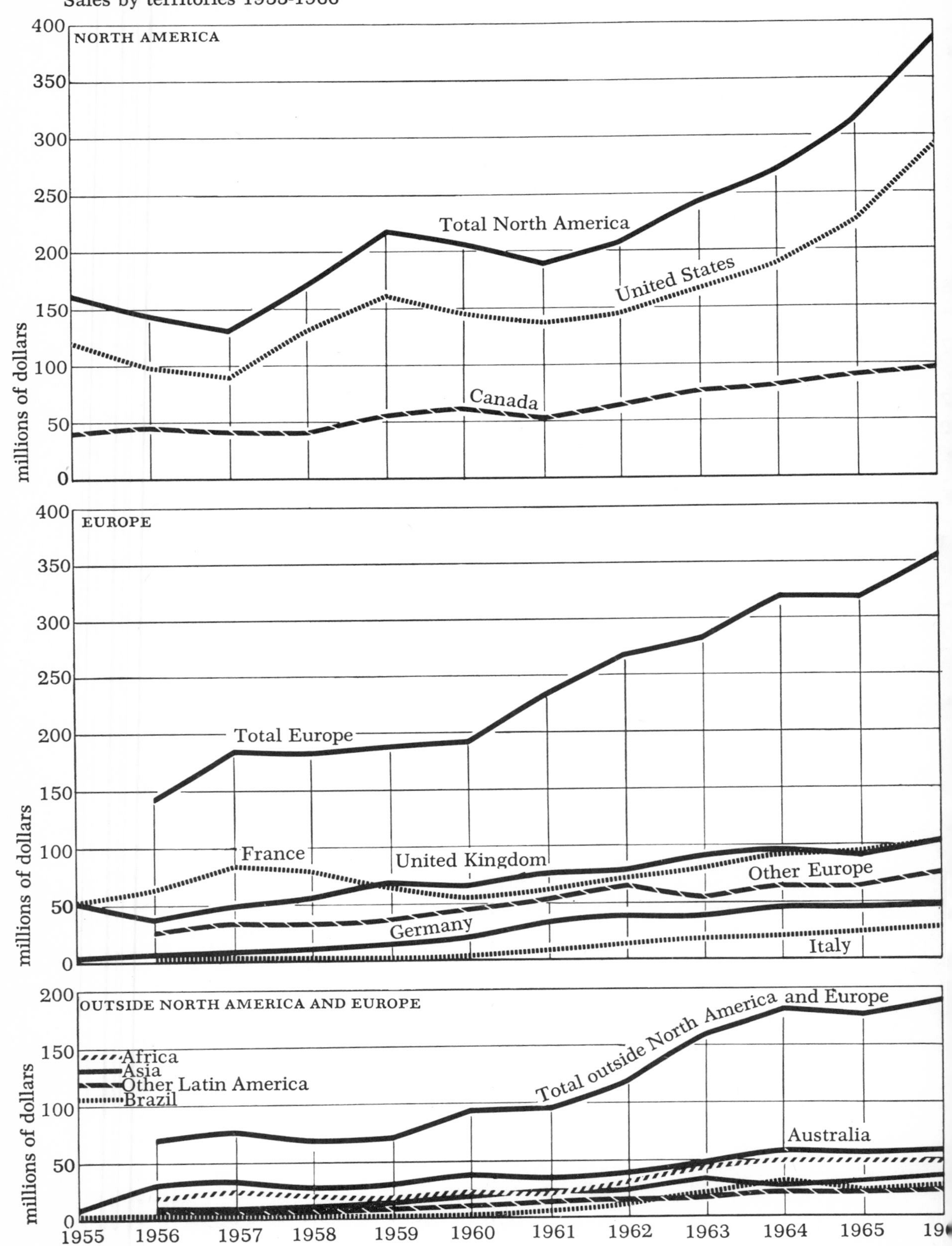

CHART 12
Massey-Ferguson Limited/World-wide operations
Costs 1955-1966

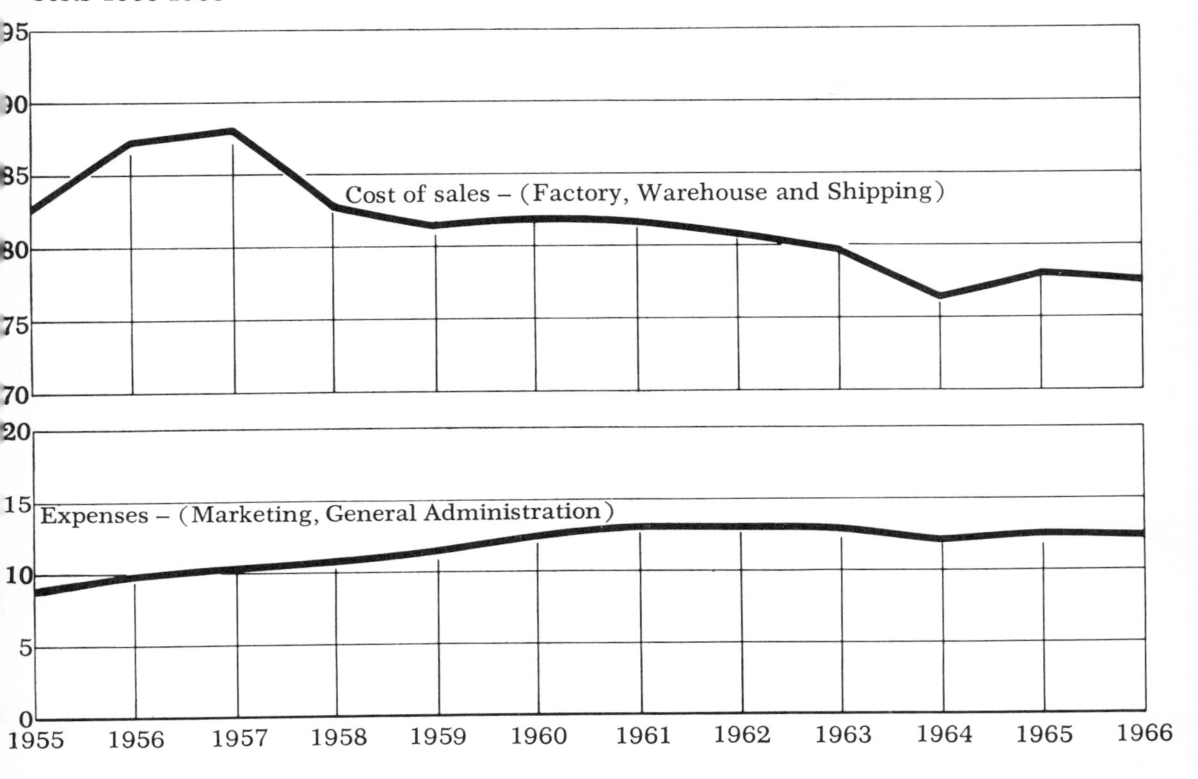

ture; in others – including North America – the company found itself in the later stage of an inflationary business cycle, that is, where cyclical costs frequently outpace practicable price increases; and in still others, particularly Australia and South Africa, severe drought conditions prevailed. Internally, there were cost problems in the Banner Lane, Coventry, tractor plant and in combine production generally. The Banner Lane costs on the new line of tractors exceeded estimates partly because some specification changes had not entered into original costs and could not, in the market environment in existence, be offset by price increases, and also because labour costs were higher than anticipated. The company has come to learn that management in the United Kingdom does not have the flexibility in reducing its labour force that it does elsewhere. The fear among labour of "redundancy" and the emphasis on "spreading the work" seem to have posed major difficulties in fully utilizing new procedures for increasing productivity. A programme designed to improve communication between management and labour in matters of productivity and job security has been introduced. As for combines, the problem was that the highly attractive design of the Massey-Ferguson machines was relatively costly to manufacture and the profit margin was consequently reduced. Finally the round of capital spending that began in about 1964 or 1965 gave the company capacity for sales of about $1¼ billion, if the mix of sales was such as to permit it to utilize all its available capacity; since sales were less than $1 billion, costs of carrying excess capacity entered into the cost of goods sold.

The year 1967 was really the first year in which the company's major product lines were not undergoing change and in which major changes in facilities were not underway. This stability in the company's manufacturing operations provided it with an opportunity to audit the relative efficiency of the various phases of its world-wide manufacturing operations. For that purpose it retained a firm of consultants.

Dangerously excessive inventories are usually an indication of inadequate management control. Persistently excessive inventories are also costly, for they have to be financed. Chart 13 gives an indication of the company's success in increasing its efficiency in this area. What is obvious is that until about 1960, new management had not really solved its inventory control problems. Up to 1957, the ratio of inventories to net sales declined simply because of a determined attempt to rid the company of excessive and obsolescent machines. In spite of an improved trend in sales, excessive inventories appeared in 1959. After that, new controls and procedures began to have a greater impact, and a steady

CHART 13
Massey-Ferguson Limited/World-wide operations
Inventories 1955-1966

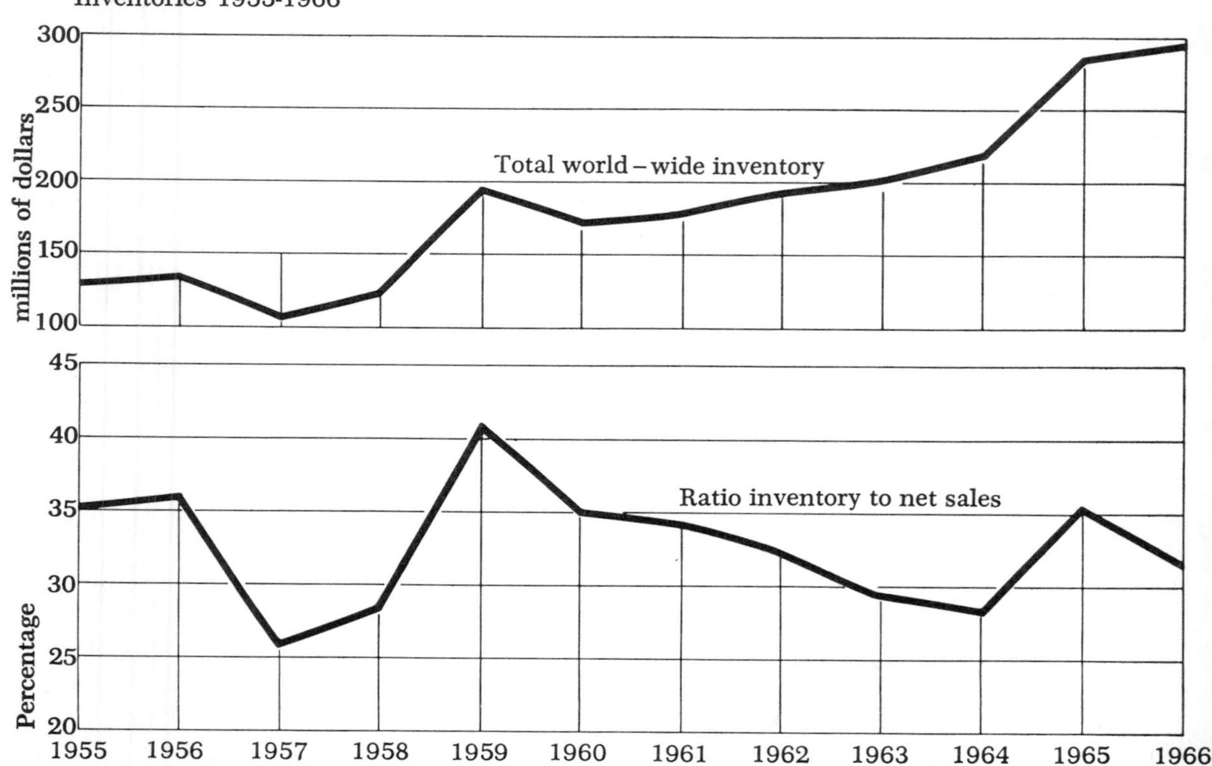

CHART 14
Massey-Ferguson Limited/World-wide operations
Assets 1955-1966

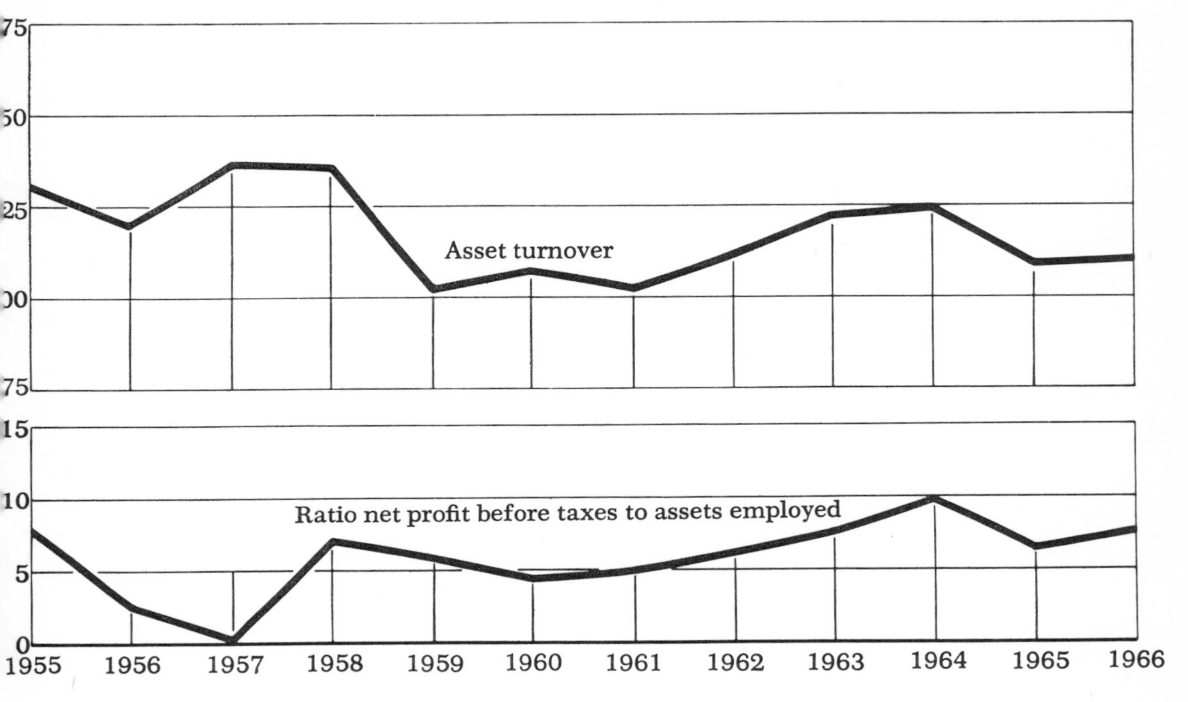

decline in the relative size of inventories occurred. The year 1965 was an exception, partly because of a smaller than planned increase in sales and partly because inventory control procedures still left room for improvement. The same general pattern is seen in Chart 14, which shows the extent to which the company has succeeded in turning over its assets (sales divided by assets) each year since 1955.

Net income per common share from 1956 to 1966 was $.22, $.61, $1.25, $1.65, $.97, $1.13, $1.13, $1.36, $1.68, $3.04, $2.66, $2.50. These figures, as is seen clearly on Chart 15, show that the company experienced an unusual improvement in 1959, but only modest performance for the next five years until 1964. There are other ways of looking at the company's profits. Chart 14 shows net income before taxes to assets employed; Chart 15 shows the rate of return on shareholders' equity and net income per common share; Chart 16 shows net income before taxes to net sales and it compares that ratio with net income before taxes to assets employed. All these trends reveal the same pattern – a high peak in 1959, then a period of deterioration followed by modest improvement up to 1964, substantial improvement in that year, and finally a period of somewhat lower profit rates.

There is no doubt that the 1959 improvement depended heavily on transitory factors involving the company's North American operations. Industry sales were rising rapidly, and the company, through market-

CHART 15
Massey-Ferguson Limited/World-wide operations
Return on investment 1955-1966

ing programmes of unusual but in some respects temporary effectiveness, increased the company's share of industry sales. Profits increased substantially as a result. But in the following year industry sales declined in North America, as did Massey-Ferguson's share of the market, and it became apparent that the company had a number of problems still to overcome.

The record of only modest profit performance from 1959 to 1963 may in retrospect be explained by a number of things. A durable organizational structure had only just appeared in 1959, and much remained to be done in developing concepts of planning and control. The I.P.C. procedures were only being worked out in 1960, and their application in the major operations units took several years. Moreover, the capital re-organization necessary to give each operations unit adequate independence in that area was not completed until 1964. Deterioration of the company's competitive position in combines was not halted immediately, and it had to depend on others for its supply of large tractors. The problems encountered in the French operations unit also reduced the company's total profits.

CHART 16

Massey-Ferguson Limited/World-wide operations

Net income before taxes 1955-1966

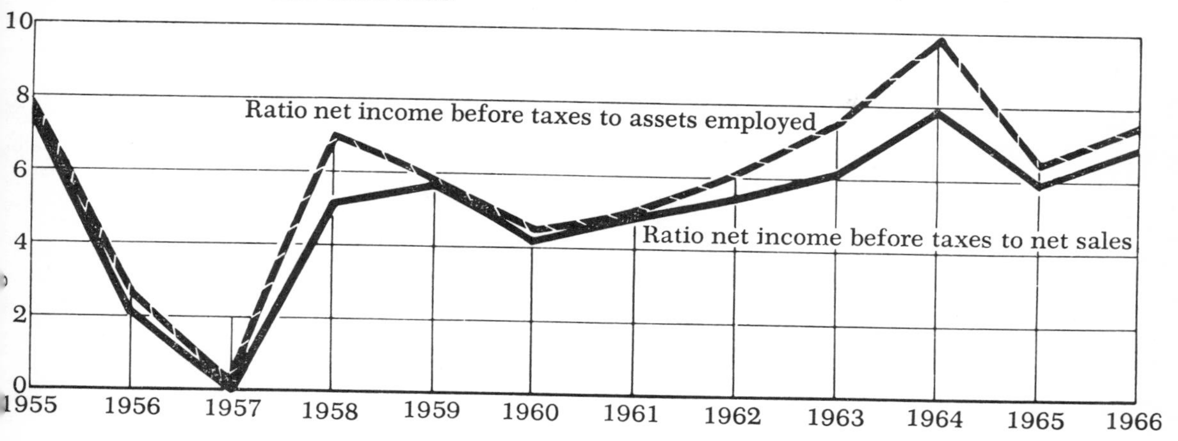

Then came the very good year of 1964, during which profits after taxes per share rose to $3.04 from $1.68 the preceding year. If these profits are viewed as a return on funds belonging to shareholders, they amounted to 15½ per cent of shareholders' equity – a rate that placed the company in about the top one-third of large North American corporations. There were a number of reasons for this performance, some fortuitous, some not. A strong demand for farm machinery existed; sales increased much more than expected; and prices in some markets could be raised. The obsolescence costs involved in selling out the old line of tractors to make way for the new proved to be much lower than had been estimated in advance. This took place because the old line was still selling exceedingly well, because the marketing strategy for selling out the old line was highly effective, and because of the generally buoyant economic conditions. It was also possible to keep costs down that year because amortization of new equipment for producing the new line of tractors did not begin until 1965, that is, until tractor production began, and cyclical cost increases were not yet in the upswing. Finally by 1964 the corporation was operating according to plans, with greatly improved supervision and control of all its operations units and of their functional divisions. The reduced rate of return on shareholders' equity after 1964 came as a consequence of the disappearance of the favourable influences present in 1964, the appearance of certain unfavourable cost increases noted earlier, and the general levelling off of business in some of the company's major markets.

Where do the challenges and the opportunities of the company lie in the years ahead? Will the not unreasonable target of 15 per cent return on shareholders' equity be achieved? Needless to say, this will depend partly on the company staying in the forefront of farm machinery technology and on continuing to emphasize change that will lead to further improvement in manufacturing efficiency. The company will also have to develop marketing strategies appropriate to each of its many markets and ensure that changes in organization and control will be evolutionary so that they need never be revolutionary. But in a more specific sense its progress in future may well depend on its success in penetrating the United States market more deeply than it has to date and in becoming increasingly well established in non-agricultural machinery areas – beginning with diesel engines and industrial and construction machinery.

While it is apparent that objectives still to be achieved can easily be identified, it has begun to appear that the concept of Massey-Ferguson as a global corporation is a financially viable one. Principles have been established and some success has been achieved in applying them. It is now for top management to ensure that the inevitable disappearance of internal momentum initially created by the challenge of a corporation in trouble will not be succeeded by complacency. A new source of momentum may be the company's approach to organization and management. This emphasizes the importance of defining executive responsibility and authority, expects management to achieve results by introducing change, and recognizes the need for measuring the performance of managers. The specific emphasis on "profit impact" of action taken by managers may just prove to be the way that the entrepreneurial spirit can be made to grow in the managerial ranks of this massive and complicated global corporation.

CHAPTER 19

The concept of the global corporation

The experiences of the company in the international environment, taken all together, and viewed in broad perspective, tell much about the fundamental character of international or multinational industrial companies in general and of the problems they encounter. This makes it difficult to resist the temptation to generalize on the concept of the global corporation from the special vantage point of the history of Massey-Ferguson.

It seems that the essence of a sophisticated international industrial corporation is that it, at all times is prepared and in a position to develop markets for its products wherever on earth such opportunities exist and seeks to deploy its manufacturing and engineering facilities internationally in a way that will minimize its global production and developmental costs and will assist in the development of particular markets. Arbitrary rules guiding asset deployment can only do harm to such a corporation. Rational judgments relating to prospective rates of return, based on detailed knowledge of local social, political, and economic conditions must lie at the heart of its international asset deployment decisions. Emotional attachments to the "home base" and misconceptions about conditions and opportunities abroad have no place.

What induces a corporation to move beyond the boundaries of its domestic market? Obviously opportunities for increased profit are of paramount importance. But this simple explanation is not very illuminating; the profit motive can express itself in various ways. Massey-Harris discovered early in its history that export business was not only profitable but frequently even more profitable than business in its home markets. It also recognized early that expansion of sales through the development of export markets reduced manufacturing costs through increased scale of operation.

Once a company has established valuable export markets, and has geared its total organization to fit the expanded volume of business that export markets have produced, it becomes very difficult to sit idly by when those markets are suddenly threatened by local trade restrictions. Management is therefore given a compelling incentive to contemplate establishing manufacturing facilities outside its domestic market. In the process of developing export markets and embarking on some manufacturing operations outside the home country, a corporation inevitably becomes increasingly familiar with, and therefore less frightened about, the commitment of its assets to international business operations in general. Personnel with experience in international business, and with necessary language qualifications are gradually acquired, and their views are likely to be biased in favour of sustaining and expanding international operations.

Also, a corporation that does not develop business abroad when its competitors do, or when foreign companies invade its own market, may find that its smaller scale of operation places it at a cost disadvantage even in its domestic market. In other words, if full economies of scale are attained only if a corporation's activity is extended into international markets, then a purely domestic company in the industry may encounter cost disadvantages. In the manufacture of technically complicated machines, or complicated components for machines, optimum-sized production may be very large. To a certain extent, a corporation may therefore be pushed into the international environment in its fight for survival, and this may take the form of export activity, licensing agreements, or assembly and manufacturing operations.

A convenient bridge for entering manufacturing operations abroad, for Massey-Ferguson at least, has been the company's local independent distributor. Such local independent companies can often supply capital themselves and interest other local investors. They also have an incentive to participate in the project, for failure to do so could mean loss of their distribution rights. For the international company such associations are frequently invaluable as a source of both local capital and of specialized knowledge concerning negotiations with governmental authorities, the acquisition of local manufacturing facilities, and the availability of management personnel. Some local participation may be required by law, or may be desirable because it may cause governmental authorities to view a project more favourably than they would without it. Where risks are great, the international company may even be pleased to have some local contribution of risk capital. Finally, acquiring the facilities of an existing local company – often the simplest way to become established – may be possible only with some participation of former owners.

While the reasons for some form of association with local interests may, under certain circumstances, be compelling, there are also major disadvantages. If profits of a subsidiary represent not merely a return on capital but also an indirect payment for managerial and technical

An MF-55 wheel loader, one of several products introduced in 1968 comprising MF's new construction machinery line.

contribution of the parent company (a point to which we shall return), then sharing profits on the basis of financial capital contributed might reduce the profitability of the project from the parent company's viewpoint. The need to heed the predilections of local shareholders may also complicate the planning procedures and operations of a world-wide corporation. For example, local Board members may have their own views as to how the local company should invest its short-term funds or plan its expansion, and these may conflict with the policy of the parent company. There are no such complications with a wholly owned subsidiary, for the parent company elects all the directors.

Massey-Ferguson has favoured wholly owned subsidiaries in the industrially more advanced countries but, unlike some corporations, it has pursued a flexible policy, elsewhere. As we have seen, there are many reasons why such a policy may be necessary or desirable. An international corporation that pursues a rigid policy of operating only through wholly owned subsidiaries is likely increasingly to deny itself access to business opportunities. In some countries Massey-Ferguson owns a majority of the stock of the local company, in others a minority of the stock, while in some cases its only association is through a licensing agreement. But it has been careful in recent years not to lose control over its corporate name through permitting it to be incorporated into the name of a company that it does not control. We shall see shortly that one reason why an international corporation may profitably compromise with a policy of operating only through wholly owned subsidiaries is that its income from such operations need not and frequently is not confined to its share of the profits.

What is the fundamental role of a global corporation? What is it that the international company provides that justifies the profits it receives on its foreign business? The obvious answer is that it provides capital. This is also an incomplete answer. Indeed, it is so incomplete as to be largely incorrect. Massey-Ferguson, as so many other international corporations, generally depends heavily on local sources for loan capital. Its equity contribution, even where it operates through a wholly owned subsidiary, is minimized, commensurate with giving the local company an adequate credit standing in the capital market.

Providing funds is not, therefore, the unique contribution of the international corporation. Its unique contribution is the provision of advanced technology and managerial skills. It is much more a seller of "know-how" than a provider of funds, even though the funds it provides are at times substantial in amount. To us the most impressive characteristic of the global corporation is the speed with which it channels advanced technology into the far corners of the world. It is a characteristic that distinguishes it clearly from mere international "lending" agencies; it is also one that seems not to be clearly understood by all emerging nations. The profits a global corporation receives from the operations of a subsidiary constitute payment not merely for capital contributed but also for "know-how" provided. The same is true for the gains it may make through appreciation in market value of the investments it has in operations abroad.

But the income that an international corporation receives from operations abroad comes from more directions than this discussion implies. Besides the total profits of subsidiary companies, or the share of profits from companies in which it has merely a majority or minority interest, an international company may at times receive additional income from technical assistance agreements or royalties, from licensing agreements, and from the sale of component parts to the local company. It also not infrequently happens that import licences for whole machines are more generously given to companies with local manufacturing projects than to those without such a stake in the country. For all these reasons the reduction in income implied in an international company not being sole owner of a local company may to a degree be offset by other forms of income from the operation. Indeed, one would expect in theory that it would be entirely offset in the long run, and in practice it probably is.

This possibility is of considerable importance to international corporations because of the strong feeling in many countries over non-resident ownership of industry. Ideally, it seems that international corporations would in most cases wish to see local investors invest in the stock of the parent company rather than in stock of the local company. However, habits of local investors and sentiment surrounding ownership of industry have not become sufficiently "internationalized" in most countries to permit this to happen. Consequently international corporations are at times forced to deviate from operating only through wholly owned subsidiaries or forgo opportunities for expansion.

At the same time the possibility that local investors will begin more and more to be interested in parent company stock should not be entirely discounted. As the relative size of Massey-Ferguson in the United States market has increased, so has the relative amount of its total stock held by United States investors. While the exemption of the United States withholding tax on new issues has had a significant effect, it is none the less interesting that after the company's 1966 rights offering of one share for every five shares held, United States investors acquired about two-thirds of the rights issued, buying many of them from Canadian investors. Consequently United States investors now own just under 40 per cent of the common stock of the company, compared with just over 20 per cent in 1964.

The total commitment to the international environment implied in the concept of the global corporation outlined earlier has significant implications for the organizational structure of a company and for the system that is devised to control its many operations. Indeed, some writers have even defined international corporations in terms of their organizational structure rather than in terms of the nature of their commitment to the international environment. It does seem apparent that an organizational structure suitable for a purely domestic corporation is not likely to be suitable for world-wide operations.

There are several reasons for this. First of all permanent differences between nations demand that international corporations learn to cope with diversity while yet achieving efficiency through careful planning, co-ordination, and control of world-wide operations. The problems posed by international diversity are reduced if nationals predominate in the operations of local subsidiary companies and if they are given maximum responsibility and authority for achieving the defined objectives of the parent company. Such executives are more likely to have the language requirements, the knowledge of local conditions, and the cultural background that is necessary for being effective in the local business environment. All this suggests that there is merit in moving towards maximum separation of corporate or parent company executives from line responsibilities in local operations units, and in avoiding the temptation of sending executives from the "home base" to manage operations abroad.

Not that this is typically an easy task. The experience of Massey-Ferguson has been that managerial skills, while scarce around the world, are less scarce in the United States than elsewhere. The temptation to send North Americans abroad is therefore always great, and only a conscious and determined effort to seek out and develop local managerial talent will prevent it. It may be recalled that when, in 1956, Massey-Ferguson, a Canadian company, suddenly required a number of senior executives quickly, it found most of them in the United States. A managerial talent "gap" seems to be a major problem in most countries, and it is one gap that the international corporation is in a unique position to fill – given a chance.

Maximum separation of corporate executives from operating responsibilities seems desirable for other reasons. The executives of local operations units must be able to feel satisfied that each operations unit will be treated fairly when world-wide decisions concerning, say, capital expansion and export activity are made. If corporate executives are more closely associated with some operations units than with others, this feeling of confidence is difficult to create. Such an association may also make it difficult for corporate executives to appraise objectively the operations of each subsidiary and it may leave them mentally unprepared for planning the global strategy of the corporation. Detailed attention to operational matters might also mean inadequate attention to long-term planning. Finally, it is difficult to hold local subsidiaries substantially responsible for operations if corporate intervention in local operations is detailed and continuous.

Massey-Ferguson has accepted the principle that operations units should enjoy maximum responsibility and authority and that exceptions to this principle should be defined and understood. Massey-Ferguson Limited, the parent company, has been transformed into a holding company and its executives constitute the corporate group. The North American operations unit has the same relationship with the corporate or parent company executives as do operations units in the rest of the world. True, corporate executives retain some detailed line responsibilities and, in emergency cases, temporarily assume others in individual operations units. But generally their role is to examine and eventually approve annual plans of operations units, to examine performance against plans, and to plan long-term strategy for the continuing development of the company's world-wide operations.

Long-term planning of a global corporation involves considerations that are not encountered by a domestic company. Efficient location of plant requires a knowledge of conditions in a number of countries. Massey-Ferguson produces various types of tractors at Detroit (United States), Coventry (England), Beauvais (France), Sao Paulo (Brazil) and (with associates) at Madras (India), Barcelona (Spain), Queretaro (Mexico). It has centred combine production at Brantford (Canada), Kilmarnock (Scotland), Marquette (France), Eschwege (Germany), and Sunshine (Australia). Decisions regarding the number of units and the models that should be produced at each factory, and where expansion should occur, involve strategy on an international plane. By pursuing a policy of maximum interchangeability of component parts – particularly for tractors and combines – Massey-Ferguson has increased its flexibility in international production strategy and has reduced its costs of production.

This global strategy in plant location gives the international corporation flexibility not enjoyed by a domestic company. Over the years, for example, Massey-Ferguson has come to locate its labour-intensive North American operation in Canada where wages are lower than in the United States. Some operations are located in the United States partly because of the relationship that is believed to exist between local plant location and local market penetration. As a consequence, and in spite of over two decades of free trade and substantially increased company penetration of the United States market, the proportion of Massey-Ferguson's total North American employees located in Canada was about the same in 1966 as it was in 1939. Also, when the company required more combine capacity in Europe, it had to choose between expansion at Kilmarnock or Marquette. Marquette was chosen partly because of the long history of labour difficulties at the former. The international corporation is also in a position to acquire technology and new products quickly by considering the purchase of existing companies outside its home base. Massey-Ferguson did this on a number of occasions. For an international corporation, therefore, plant location decisions can be heavily influenced by relative cost differences, by the political environment, and by the opportunity for acquiring technology and products quickly wherever in the world they are available.

An international corporation must make similar global decisions in its financial operation. In what currencies should it assume new liabilities, and where should it invest its liquid assets? Appropriate strategy involves making assumptions about future exchange rates, interest rates, political stability, and about what constitutes a tolerable risk. Global tax liabilities may be minimized by taking full advantage of national tax structures by being aware of changes in them and, to the extent legally possible, by shifting profits to subsidiary companies facing the lowest tax rates. Unfavourable tax treatments in particular countries can more easily be minimized by international corporations than by domestic ones, and the possibility that even parent companies might shift locale for that reason, or for other ones, is not entirely remote. In the case of Massey-Ferguson, since the parent company is essentially a holding company with global management responsibilities, its staff is relatively small – numbering just over 100 in 1967 out of a world-wide total of about 46,000. The physical obstacles to shifting Head Office staff are therefore very small. There can be no doubt that the bargaining position of an international corporation in its relations with individual governments is inherently stronger than that of purely domestic companies. Much parochial legislation will face its moment of truth when it first encounters the international corporation.

The deep involvement of the corporate group in planning, co-ordinating, and controlling world-wide operations has rather interesting implications for the potential development of Massey-Ferguson in future. By acquiring the skills and developing the organizational and control procedures that are specifically suitable for supervising the operations of relatively independent and far-flung operations, it should become increasingly easier to add to the number of such operations units. Nor is there any inherent reason why such operations units should in future be confined to agricultural machinery and the industrial and construction equipment business. Indeed, the development of the Perkins organization with its large "external" sales of engines, the establishment in North America and Italy of distinct industrial equipment operations, and the acquisition of an interest in truck manufacturing in Spain may confirm that the organizational structure does lend itself to industrial diversification as well as to geographical diversification of corporate activity. A corporation of this kind is more properly thought of as being a highly competent managerial group than a segment of a specific industry. Now that Massey-Ferguson Limited is a holding company without fixed assets, this point is illustrated by its legal structure.

The corporate group cannot exercise adequate supervision and control over world-wide plans and operations without a comprehensive system of reporting. In Massey-Ferguson the required information is generated through the gradual adoption by each operations unit of the company's integrated planning and control system, by the submission of annual plans, and by the monthly submission of control reports that compare performance against approved plans and that highlight variances. This approach makes it virtually mandatory for executives in all the operations units to become intimately acquainted with essentially North American concepts of business planning and control. The application of this approach to operations units abroad, operations units that may previously have been quite unfamiliar with it, is time-consuming, frequently frustrating, and almost never completely successful until some new managerial personnel have been introduced into them. But experience has shown that it is not an impossible task.

To achieve adequate communication and co-ordination is a difficult task in any large corporation. These difficulties are compounded in a global corporation and require detailed attention. In the case of Massey-Ferguson, its operations units are located in countries with marked political, social and economic differences and with a variety of languages. The development of uniform organizational structures and reporting systems is only one way in which internal communication is improved. It is also improved by the practice of hiring local senior executives who have a working knowledge of English. At present almost all of the company's senior executives in each of the operations units abroad speak English even though many of them are nationals of the countries in which they work.

Co-ordination is achieved in a number of ways. It is achieved by the requirement that each operations unit must submit an annual plan to the corporate group and by the further requirement that the plan must be approved by corporate executives. It is enhanced by the organizational structure of the corporate group which ensures that there will be corporate executives with specialized knowledge of each one of the major functional divisions of the operations units. Co-ordination is also improved by the corporate group retaining certain defined line responsibilities, particularly engineering or product development, part of finance, and export activity. The corporate group also arranges conferences attended by executives of the various operations units. For example, there is an annual world-wide product planning conference. Occasionally there are also world-wide conferences for engineering, manufacturing, marketing, finance, personnel and industrial relations, and public relations. Corporate executives make frequent trips to the operations units and, to a limited extent, are transferred from operations units to the parent company, and from the parent company to the operations units.

This approach to managing a world-wide organization has important implications for the kinds of corporate executives that are required. A senior staff executive of such a group is in some respects in an unnatural position. To a large extent he exists to give advice. He must be careful not to give orders to operations units if the company's concept of decentralized responsibility is to be preserved. Accolades for increased profits must go not to him but to local line executives who chose to accept, or reject, his advice. Yet because of his background and because of the necessity of his having qualifications that will permit him to assume line responsibilities temporarily if serious trouble arises in particular operations units, his instincts may frequently be "line" rather than "staff." The borderline between "advising" and "ordering" can be thin – as thin as the tone of the voice, perhaps. But that borderline must be identified and respected at all times if a drift – often hardly observable in the short term – towards centralization is to be avoided. An international corporation is therefore particularly likely to have to devise means that will cause it to pay constant attention to the organizational principles that it has chosen, just as it must pay constant attention to introducing evolutionary changes into its organizational structure as the size and complexity of its operations increase.

A corporate executive must take it for granted that he will spend much of his time travelling by air and that he will frequently be away from his home and family. Permitting wives to accompany their husbands on such trips from time to time is a corporate necessity, not an act of friendly corporate paternalism or charity. Not all tax authorities see the problem quite this way.

In time the executive of the international corporation will probably develop a much more international outlook than an executive of a purely domestic corporation. His national economic loyalties are likely to be more diffused, and his views on trade more liberal than those of executives in domestic companies. Nor is one likely to find rabid nationalists among parent company executives of international corporations, although they can be found among the executives of the local companies of such a corporation. A corporate executive, for example, often can make a decision on capital allocation only after an objective examination of opportunities in several countries. His responsibilities and decisions force him into becoming an international man just as they force him to move physically in the international environment. Increasingly the parent company group to which he belongs will be drawn from a diversity of operations units and a diversity of countries. At that point the process of "internationalizing" the corporate group will attain its logical maturity.

The international industrial corporation is in many respects a remarkable phenomenon. It reflects the effect on business of a world that is getting smaller and of a technology that is becoming very complicated and universally demanded. It mirrors the remarkable adaptability of business institutions and businessmen in a rapidly changing international environment. It widens the horizon of businessmen contemplating new business opportunities and transforms them until they are more international than national. And in the course of all this it becomes intimately involved in distributing the advanced technology and managerial skills that so many nations seem to require. It is not a phenomenon that should be dismissed with superficial generalizations in the world that lies ahead.

Illustrations of tractors and combines

Big Bull, 1917
Massey-Harris' first entry in the Canadian tractor market, this machine, which weighed 3,000 lbs., was manufactured by the Bull Tractor Company in the U.S.

MH-1, 1918
The first tractor manufactured by Massey-Harris. It was patterned on the Parrett tractor and assembled at the Massey-Harris factory in Weston, Ontario.

Wallis Certified, 1929
A three-plow tractor with steel U-frame and magneto ignition produced by the J. I. Case plow works of Racine, Wisconsin. Massey-Harris acquired the company in 1928.

MH General Purpose, 1930
A four-wheel-drive general-purpose machine made in four widths of tread to conform to different row spacings. The rear axle was free to oscillate about an horizontal axis, giving flexibility when travelling over uneven ground.

MH Pacemaker, 1936
In later models it featured "Twin Power", which enabled the operator to vary horsepower by altering engine r.p.m.'s from 1,200 to 1,400. This three-plow tractor burned high octane (68-70) gasoline and was available with rubber or steel wheels.

MH Challenger (Row Crop), 1936
A 2-3 plow tractor with rubber or steel wheels and adjustable rear axle, powered by a four-cylinder, valve-in-head, all-fuel engine. The machine was available in row-crop or standard models.

MH-101, 1939
This 2-3 plow tractor employed an automobile-type, L-head, six-cylinder engine built by Chrysler. In some models rear wheels were adjustable from 52″ to 90″ on rubber and 52″ to 80″ in steel.

MH-44, 1946
A three-plow tractor powered with a four-cylinder overhead valve engine built for gasoline or for distillate fuel.

MH-55, 1946
A 4-5 plow tractor made at Racine featuring a heavy cast-iron frame and four-cylinder overhead-valve engine.

MH Pony, 1947
A one-plow tractor, designed primarily for market gardening, with a high-compression (6.5 to 1) engine and adjustable wheel settings on both front and rear axles.

MH Colt, 1952
It featured hydraulic lift and a three-point hitch introduced to compete with tractors employing the Ferguson System.

FE-35, 1955
A Ferguson-System tractor produced at the Banner Lane plant in Coventry, England.

MH-50, 1956
A Ferguson-System tractor in the 35-horsepower range, produced in North America.

MF-65, 1958
A medium-horsepower tractor available with either gasoline or diesel engine and featuring in later models "Multi-Power" which permitted gear change "on the go".

MF-95, 1958
A 90-horsepower diesel tractor produced for MF by Minneapolis-Moline.

MF-85, 1959
A 60-horsepower tractor available in both row-crop and standard four-wheel models and powered by a gasoline or diesel engine.

MF-25, 1961
A 25-horsepower diesel tractor produced in France.

MF-135, 1965
A diesel or gasoline tractor in the 40-horsepower range – a close derivative of the Ferguson 20. The MF-135 and all the current line of MF tractors incorporate Pressure Control which extends the principle of the Ferguson System to trailed implements.

MF-150, 1965
A diesel or gasoline tractor in the same power class as the MF-135, but differing in wheel base and front axle configuration.

MF-165, 1965
A medium horsepower tractor available with either gasoline or diesel engine; produced in North America, United Kingdom, and France.

MF-175, 1965
A medium-large gasoline or diesel tractor produced in North America and the United Kingdom.

MF-180, 1965
A North American tractor produced in tricycle or standard four-wheel models with a high operator platform.

MF-1130, 1965
A large North American diesel tractor in the over-100 engine horsepower range available in row-crop, standard four-wheel, and western versions.

Ferguson Black, 1933
The first prototype Ferguson tractor. It was called "black" simply because it was painted black.

Ferguson-Brown (Huddersfield), 1936
Incorporated the Ferguson system with Draft Control.

Ford 9N, 1939
A two-plow machine built by the Ford company in Detroit. It featured the Ferguson System and a four-cylinder, gasoline engine.

Ferguson TE-20, 1946
The original "Fergie" built in the United Kingdom by the Standard Motor Company. The U.S. version, known as the TO-20, was produced at the Ferguson Park plant in Detroit. Both tractors had adjustable wheel settings and four forward speeds from 2.48 to 13.13 m.p.h.

Ferguson TO-30, 1952
An improved version of the TO-20 produced in North America.

Landini N621, 1925
One of the first Landini agricultural tractors.

Landini L25, 1949
An agricultural tractor powered by a semi-diesel oil-burning engine.

Landini C25, 1958
The first Landini diesel crawler tractor.

Landini CL4000, 1960
A 40-horsepower crawler powered by a Perkins three-cylinder diesel engine produced under licence by Landini in Italy.

Landini R7000, 1960
Available with both two-wheel and four-wheel drive, powered by a Perkins four-cylinder diesel engine.

McKay Stripper Harvester, 1885
Developed by H. V. McKay in Australia, this was the first practical machine which combined the harvesting and winnowing processes.

Sunshine Auto Harvester, 1924
The first self-propelled combine. It cut a 12-foot swath and was propelled by a single driving wheel.

McKay MH-4 PT [pull-type] Stripper Harvester, 1930.

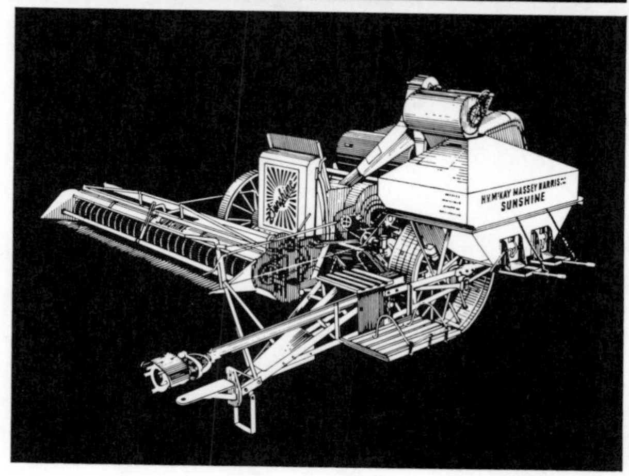

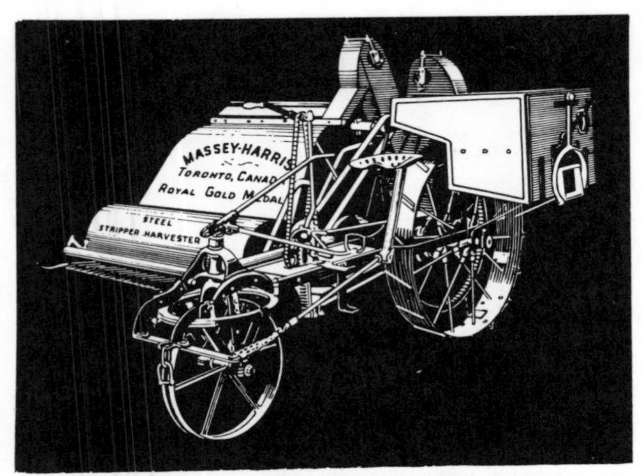

MH Stripper Harvester, 1901
Designed for the Australian market, this machine employed a stripper-type table and main wheel power.

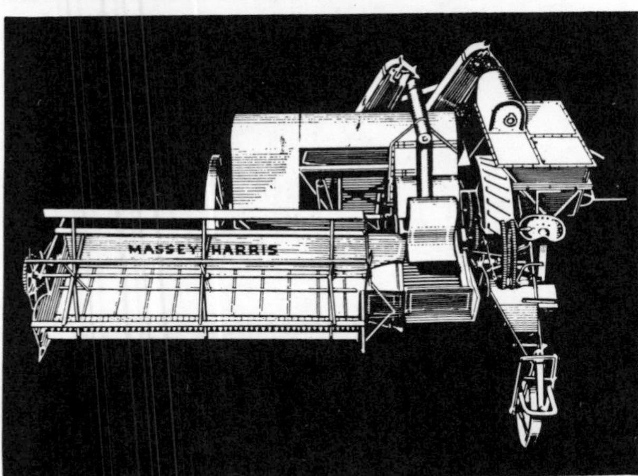

MF-1 Reaper Thresher, 1910
It combined the North American cutting principle and threshing machine and featured a reel-type table, cutter bar, and main ground wheel drive.

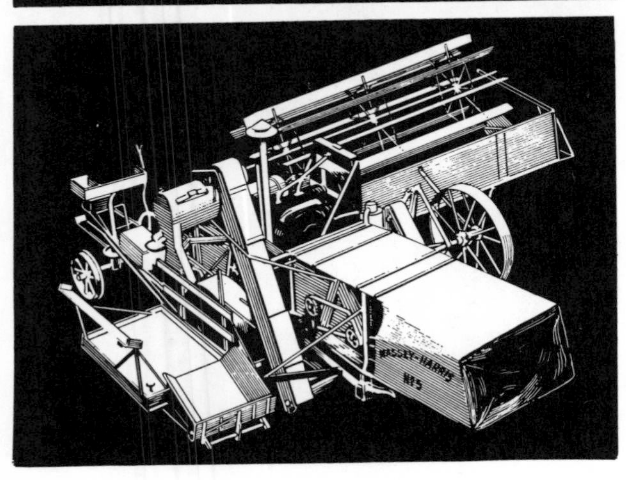

MH-5 Reaper Thresher, 1922
Employed a horse or tractor hitch, bagging platform, and auxiliary motor drive.

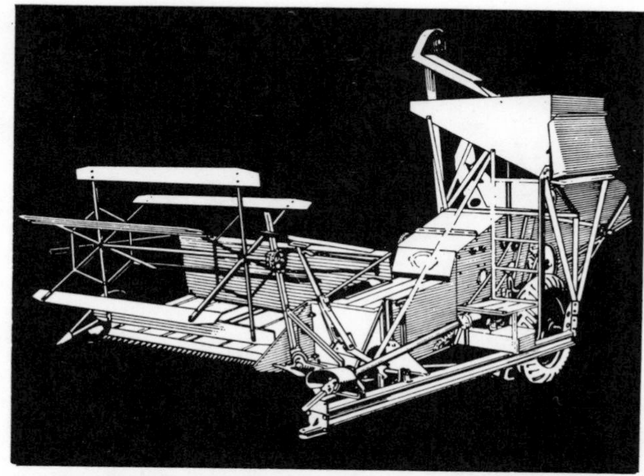

MH-15 Reaper Thresher, 1937
A tractor-drawn machine using power take-off from the tractor and equipped with rubber tires.

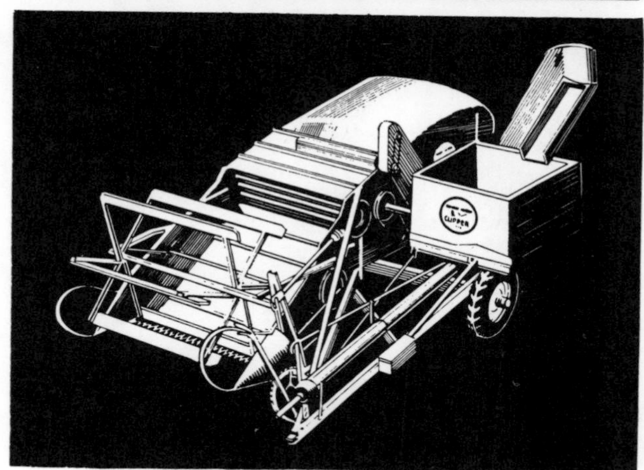

MH Clipper Combine PT, 1938
Tractor power take-off operated, scoop-type table, straight-through threshing and separation.

MH-20 Combine SP [self-propelled]
1939
The first practical North American self-propelled combine. It was developed in Toronto and the prototype was tested in Argentina. The machine cut a 16-foot swath.

MH-21 Combine (bagger version)
SP, 1940
First self-propelled combine produced in quantity; medium-small size; later built with auger table and grain tank.

MH-27 Combine SP, 1950
12- or 14-foot cut with large grain tank; featured balanced separation; medium size (32-inch cylinder width).

MF-780 Combine (bagger version) SP (U.K.), 1953
Produced at Kilmarnock, Scotland; a development from the MH-726, the first combine manufactured in the U.K.

MH-90 Combine SP, 1953
14- or 16-foot cut; 60-bushel tank; rapid unloading auger; medium-large size (37-inch cylinder width).

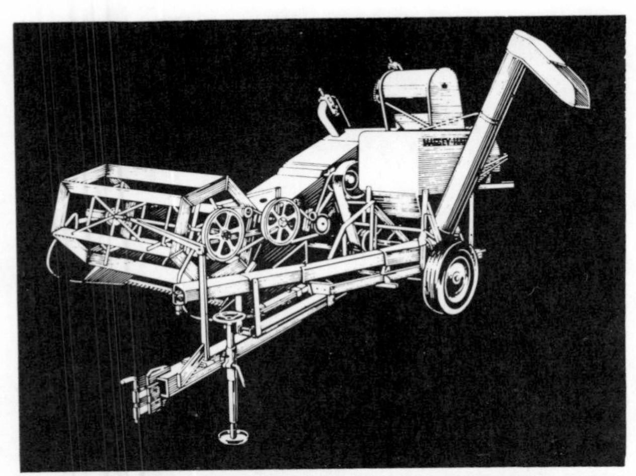

MH-60 Combine PT, 1953
A medium-sized pull-type combine, tractor power take-off operated.

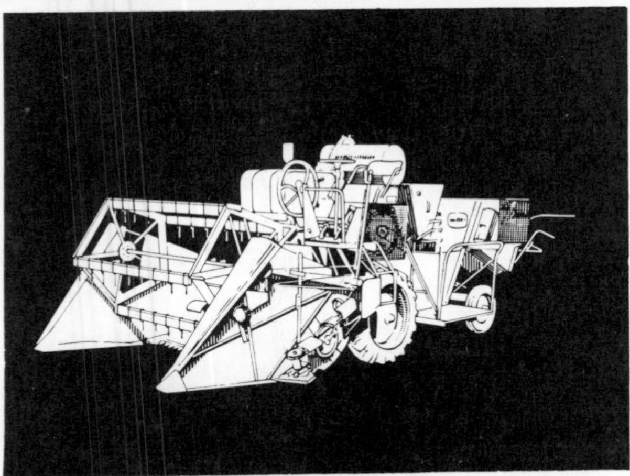

MF-630 Combine SP (Germany), 1954
A small combine (24-inch cylinder width) produced at the MF factory in Eschwege, Germany.

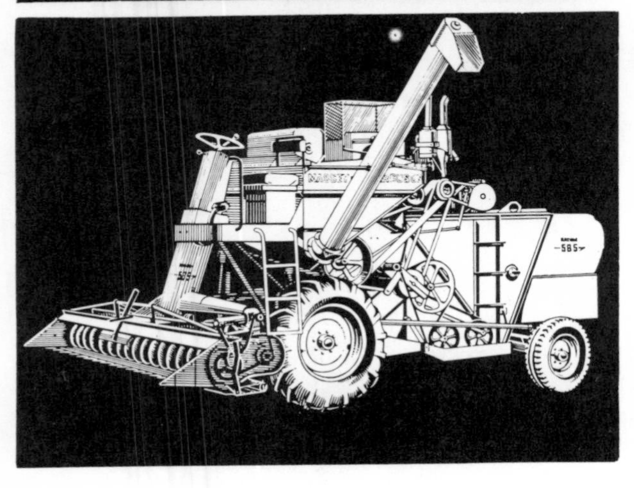

MF-585 Harvester, SP (Australia), 1957
Designed specifically to meet Australian farming conditions; employing both comb or stripper and open front.

MH-92 Combine SP, 1957
A medium-large self-propelled combine (37-inch cylinder width).

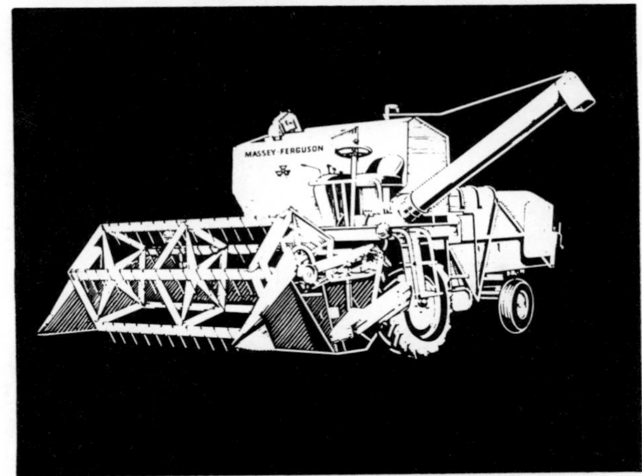

MF-685 Combine SP (Germany), 1957
A medium-sized combine produced in Germany.

MH-35 Combine SP, 1959
One of the smallest self-propelled combines (24-inch cylinder width) produced in North America.

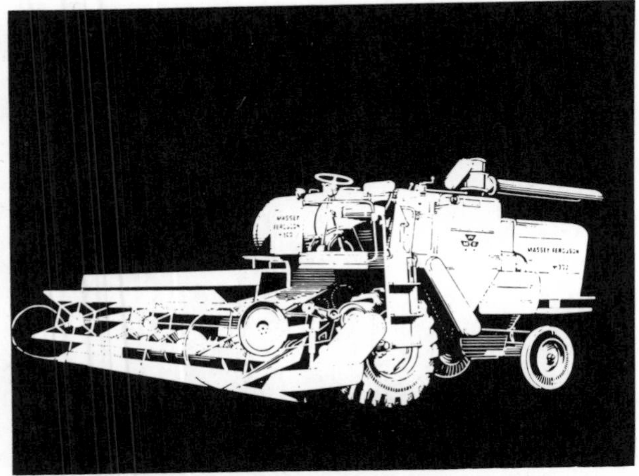

MF-300 Combine SP, 1962
A medium-sized combine (30-inch cylinder width) designed primarily for the U.S. midwest, featuring saddle-type grain tanks, formed steel welded body, and rotary air intake.

MF-510 Combine SP (NAO), 1963
A large combine (45-inch cylinder width), also featuring saddle-type grain tanks, formed steel welded body, and rotary air intake.

MF-410 Combine SP (NAO), 1963
A medium-large combine (37-inch cylinder width) similar to MF-510.

Index

Index

This book

was designed by

LESLIE SMART, FTDC

with the assistance of

LAURIE LEWIS

and was printed by

University of

Toronto

Press